The Photon

" The Elementary Quantum Particle of Light & Electromagnetic Radiation "

Edited by Paul F Kisak

Contents

1 Photon   **1**
    1.1 Nomenclature 1
    1.2 Physical properties 2
       1.2.1 Experimental checks on photon mass 3
    1.3 Historical development 3
    1.4 Einstein's light quantum 4
    1.5 Early objections 5
    1.6 Wave–particle duality and uncertainty principles 5
    1.7 Bose–Einstein model of a photon gas 7
    1.8 Stimulated and spontaneous emission 7
    1.9 Second quantization and high energy photon interactions 8
    1.10 The hadronic properties of the photon 9
    1.11 The photon as a gauge boson 10
    1.12 Contributions to the mass of a system 10
    1.13 Photons in matter 10
    1.14 Technological applications 11
    1.15 Recent research 11
    1.16 See also 11
    1.17 Notes 12
    1.18 References 12
    1.19 Additional references 16
    1.20 External links 17

2 Light   **18**
    2.1 Electromagnetic spectrum and visible light 18
    2.2 Speed of light 19
    2.3 Optics 20
       2.3.1 Refraction 20
    2.4 Light sources 20
    2.5 Units and measures 21

	2.6	Light pressure	22
	2.7	Historical theories about light, in chronological order	22
		2.7.1 Classical Greece and Hellenism	22
		2.7.2 Classical India	22
		2.7.3 Descartes	23
		2.7.4 Particle theory	23
		2.7.5 Wave theory	23
		2.7.6 Electromagnetic theory as explanation for all types of visible light and all EM radiation	24
		2.7.7 Quantum theory	25
	2.8	See also	25
	2.9	Notes	26
	2.10	References	26
	2.11	External links	27

3 Photon energy — 28

- 3.1 Formula — 28
- 3.2 Examples — 28
- 3.3 See also — 28
- 3.4 References — 29

4 Photon polarization — 30

- 4.1 Polarization of classical electromagnetic waves — 30
  - 4.1.1 Polarization states — 30
- 4.2 Energy, momentum, and angular momentum of a classical electromagnetic wave — 32
  - 4.2.1 Energy density of classical electromagnetic waves — 32
  - 4.2.2 Momentum density of classical electromagnetic waves — 32
  - 4.2.3 Angular momentum density of classical electromagnetic waves — 32
- 4.3 Optical filters and crystals — 32
  - 4.3.1 Passage of a classical wave through a polaroid filter — 32
  - 4.3.2 Example of energy conservation: Passage of a classical wave through a birefringent crystal — 33
- 4.4 Photons: The connection to quantum mechanics — 34
  - 4.4.1 Energy, momentum, and angular momentum of photons — 34
  - 4.4.2 The nature of probability in quantum mechanics — 36
  - 4.4.3 Uncertainty principle — 36
  - 4.4.4 States, probability amplitudes, unitary and Hermitian operators, and eigenvectors — 37
- 4.5 See also — 38
- 4.6 References — 38
- 4.7 Further reading — 38

5 Photon counting — 39
- 5.1 Measured quantities — 39
- 5.2 See also — 39
- 5.3 References — 39

6 Photonics — 41
- 6.1 History of photonics — 41
- 6.2 Relationship to other fields — 41
  - 6.2.1 Classical optics — 41
  - 6.2.2 Modern optics — 41
  - 6.2.3 Emerging fields — 42
- 6.3 Applications — 42
- 6.4 Overview of photonics research — 43
  - 6.4.1 Light sources — 43
  - 6.4.2 Transmission media — 43
  - 6.4.3 Amplifiers — 43
  - 6.4.4 Detection — 43
  - 6.4.5 Modulation — 43
  - 6.4.6 Photonic systems — 43
  - 6.4.7 Photonic integrated circuits — 43
- 6.5 See also — 44
- 6.6 References — 44

7 Electromagnetic radiation — 45
- 7.1 Physics — 45
  - 7.1.1 Theory — 45
  - 7.1.2 Properties — 47
  - 7.1.3 Wave model — 48
  - 7.1.4 Particle model and quantum theory — 49
  - 7.1.5 Wave–particle duality — 50
  - 7.1.6 Wave and particle effects of electromagnetic radiation — 50
  - 7.1.7 Propagation speed — 50
  - 7.1.8 Special theory of relativity — 50
- 7.2 History of discovery — 51
- 7.3 Electromagnetic spectrum — 51
  - 7.3.1 Interactions as a function of frequency — 52
- 7.4 Atmosphere and magnetosphere — 53
- 7.5 Types and sources, classed by spectral band — 54
  - 7.5.1 Radio waves — 54

	7.5.2	Microwaves	54
	7.5.3	Infrared	54
	7.5.4	Visible light	54
	7.5.5	Ultraviolet	54
	7.5.6	X-rays	54
	7.5.7	Gamma rays	54
	7.5.8	Thermal radiation and electromagnetic radiation as a form of heat	55
7.6	Biological effects		55
7.7	Derivation from electromagnetic theory		56
7.8	See also		57
7.9	References		58
7.10	Further reading		59
7.11	External links		59

8 Photoelectric effect — 60

- 8.1 Emission mechanism — 60
  - 8.1.1 Experimental observations of photoelectric emission — 61
  - 8.1.2 Mathematical description — 61
  - 8.1.3 Stopping potential — 61
  - 8.1.4 Three-step model — 62
- 8.2 History — 62
  - 8.2.1 19th century — 63
  - 8.2.2 20th century — 64
- 8.3 Uses and effects — 66
  - 8.3.1 Photomultipliers — 66
  - 8.3.2 Image sensors — 66
  - 8.3.3 Gold-leaf electroscope — 66
  - 8.3.4 Photoelectron spectroscopy — 66
  - 8.3.5 Spacecraft — 66
  - 8.3.6 Moon dust — 67
  - 8.3.7 Night vision devices — 67
- 8.4 Cross section — 67
- 8.5 See also — 67
- 8.6 References — 67
- 8.7 External links — 70

9 Wave–particle duality — 71

- 9.1 Brief history of wave and particle viewpoints — 71
- 9.2 Turn of the 20th century and the paradigm shift — 72

- 9.2.1 Particles of electricity . . . 72
- 9.2.2 Radiation quantization . . . 73
- 9.2.3 Photoelectric effect illuminated . . . 73
- 9.2.4 Einstein's explanation of the photoelectric effect . . . 74
- 9.2.5 De Broglie's wavelength . . . 75
- 9.2.6 Heisenberg's uncertainty principle . . . 75
- 9.2.7 de Broglie–Bohm theory . . . 76
- 9.3 Wave behavior of large objects . . . 76
- 9.4 Treatment in modern quantum mechanics . . . 77
  - 9.4.1 Visualization . . . 77
- 9.5 Alternative views . . . 77
  - 9.5.1 Both-particle-and-wave view . . . 78
  - 9.5.2 Wave-only view . . . 78
  - 9.5.3 Particle-only view . . . 79
  - 9.5.4 Neither-wave-nor-particle view . . . 79
  - 9.5.5 Relational approach to wave–particle duality . . . 79
- 9.6 Applications . . . 79
- 9.7 See also . . . 79
- 9.8 Notes and references . . . 79
- 9.9 External links . . . 81

10 Squeezed coherent state — 82
- 10.1 Mathematical definition . . . 82
- 10.2 Operator representation of squeezed coherent states . . . 83
- 10.3 Examples of squeezed coherent states . . . 83
- 10.4 Photon number distributions and phase distributions of squeezed states . . . 83
- 10.5 Classification of squeezed states . . . 84
  - 10.5.1 Based on the number of modes . . . 84
  - 10.5.2 Based on the presence of a mean field . . . 84
- 10.6 Experimental realizations of squeezed coherent states . . . 85
- 10.7 Applications . . . 85
- 10.8 See also . . . 85
- 10.9 References . . . 85
- 10.10 External links . . . 87

11 Uncertainty principle — 88
- 11.1 Introduction . . . 88
  - 11.1.1 Wave mechanics interpretation . . . 89
  - 11.1.2 Matrix mechanics interpretation . . . 90

- 11.2 Robertson–Schrödinger uncertainty relations 91
- 11.3 Examples 92
  - 11.3.1 Quantum harmonic oscillator stationary states 92
  - 11.3.2 Quantum harmonic oscillator with Gaussian initial condition 92
  - 11.3.3 Coherent states 93
  - 11.3.4 Particle in a box 93
  - 11.3.5 Constant momentum 94
- 11.4 Additional uncertainty relations 94
  - 11.4.1 Mixed states 94
  - 11.4.2 Phase space 94
  - 11.4.3 Systematic and statistical errors 95
  - 11.4.4 Quantum entropic uncertainty principle 96
- 11.5 Harmonic analysis 97
  - 11.5.1 Signal processing 97
  - 11.5.2 Benedicks's theorem 97
  - 11.5.3 Hardy's uncertainty principle 97
- 11.6 History 98
  - 11.6.1 Terminology and translation 99
  - 11.6.2 Heisenberg's microscope 99
- 11.7 Critical reactions 100
  - 11.7.1 Einstein's slit 100
  - 11.7.2 Einstein's box 100
  - 11.7.3 EPR paradox for entangled particles 100
  - 11.7.4 Popper's criticism 101
  - 11.7.5 Many-worlds uncertainty 101
  - 11.7.6 Free will 101
- 11.8 See also 101
- 11.9 Notes 102
- 11.10 External links 105

12 De Broglie–Bohm theory   106
- 12.1 Overview 106
  - 12.1.1 Double-slit experiment 107
- 12.2 The theory 107
  - 12.2.1 The ontology 108
  - 12.2.2 Guiding equation 108
  - 12.2.3 Schrödinger's equation 108
  - 12.2.4 Relation to the Born Rule 109
  - 12.2.5 The conditional wave function of a subsystem 109

- 12.3 Extensions 110
    - 12.3.1 Relativity 110
    - 12.3.2 Spin 111
    - 12.3.3 Quantum field theory 111
    - 12.3.4 Curved space 111
    - 12.3.5 Exploiting nonlocality 111
- 12.4 Results 111
    - 12.4.1 Measuring spin and polarization 112
    - 12.4.2 Measurements, the quantum formalism, and observer independence 112
    - 12.4.3 Heisenberg's uncertainty principle 113
    - 12.4.4 Quantum entanglement, Einstein-Podolsky-Rosen paradox, Bell's theorem, and nonlocality . . . 114
    - 12.4.5 Classical limit 114
    - 12.4.6 Quantum trajectory method 114
    - 12.4.7 Occam's razor criticism 115
    - 12.4.8 Non-equivalence 115
- 12.5 Derivations 115
- 12.6 History 116
    - 12.6.1 Pilot-wave theory 116
    - 12.6.2 Bohmian mechanics 118
    - 12.6.3 Causal interpretation and ontological interpretation 118
- 12.7 Experiments 118
- 12.8 See also 118
- 12.9 Notes 118
- 12.10 References 122
- 12.11 Further reading 123
- 12.12 External links 123

13 Bose gas — 124

- 13.1 Thomas–Fermi approximation 124
- 13.2 Inclusion of the ground state 125
- 13.3 Thermodynamics 126
- 13.4 See also 126
- 13.5 References 126

14 Stimulated emission — 127

- 14.1 Overview 127
- 14.2 Mathematical model 128
- 14.3 Stimulated emission cross section 129
- 14.4 Optical amplification 129

		14.4.1 Small signal gain equation	129

- 14.4.1 Small signal gain equation 129
- 14.4.2 Saturation intensity 130
- 14.4.3 General gain equation 130
- 14.4.4 Small signal approximation 130
- 14.4.5 Large signal asymptotic behavior 130

14.5 References 131

14.6 See also 131

15 Laser   132

15.1 Fundamentals 132

- 15.1.1 Terminology 133

15.2 Design 133

15.3 Laser physics 134

- 15.3.1 Stimulated emission 134
- 15.3.2 Gain medium and cavity 135
- 15.3.3 The light emitted 136
- 15.3.4 Quantum vs. classical emission processes 136

15.4 Continuous and pulsed modes of operation 136

- 15.4.1 Continuous wave operation 137
- 15.4.2 Pulsed operation 137

15.5 History 138

- 15.5.1 Foundations 138
- 15.5.2 Maser 138
- 15.5.3 Laser 139
- 15.5.4 Recent innovations 140

15.6 Types and operating principles 140

- 15.6.1 Gas lasers 141
- 15.6.2 Solid-state lasers 141
- 15.6.3 Fiber lasers 142
- 15.6.4 Photonic crystal lasers 143
- 15.6.5 Semiconductor lasers 143
- 15.6.6 Dye lasers 143
- 15.6.7 Free-electron lasers 144
- 15.6.8 Exotic media 144

15.7 Uses 144

- 15.7.1 Examples by power 145
- 15.7.2 Hobby uses 146
- 15.7.3 As weapons 146
- 15.7.4 Telecommunications in space 146

- 15.8 Safety ... 147
- 15.9 See also ... 147
- 15.10 References ... 148
- 15.11 Further reading ... 150
- 15.12 External links ... 151

16 Photon structure function — 152
- 16.1 Theoretical basis ... 152
- 16.2 Experimental analyses ... 153
- 16.3 Conclusion ... 154
- 16.4 References ... 154

17 Ballistic photon — 155
- 17.1 References ... 155

18 Photonic molecule — 156
- 18.1 Construction ... 156
- 18.2 Possible applications ... 156
- 18.3 Interacting microcavities ... 156
- 18.4 References ... 157
- 18.5 External links ... 158

19 Two-photon physics — 159
- 19.1 Astronomy ... 159
- 19.2 Experiments ... 159
- 19.3 Processes ... 160
- 19.4 See also ... 160
- 19.5 References ... 160
- 19.6 External links ... 161
- 19.7 Text and image sources, contributors, and licenses ... 162
  - 19.7.1 Text ... 162
  - 19.7.2 Images ... 169
  - 19.7.3 Content license ... 174

Chapter 1

Photon

A **photon** is an elementary particle, the quantum of all forms of electromagnetic radiation including light. It is the force carrier for electromagnetic force, even when static via virtual photons. The photon has zero rest mass and as a result, the interactions of this force with matter at long distance are observable at the microscopic and macroscopic levels. Like all elementary particles, photons are currently best explained by quantum mechanics but exhibit wave–particle duality, exhibiting properties of both waves and particles. For example, a single photon may be refracted by a lens and exhibit wave interference with itself, and it can behave as a particle with definite and finite measurable position and momentum. The photon's wave and quanta qualities are two observable aspects of a single phenomenon, and cannot be described by any mechanical model;[2] a representation of this dual property of light, which assumes certain points on the wavefront to be the seat of the energy, is not possible. The quanta in a light wave cannot be spatially localized. Some defined physical parameters of a photon are listed.

The modern concept of the photon was developed gradually by Albert Einstein in the early 20th century to explain experimental observations that did not fit the classical wave model of light. The benefit of the photon model was that it accounted for the frequency dependence of light's energy, and explained the ability of matter and electromagnetic radiation to be in thermal equilibrium. The photon model accounted for anomalous observations, including the properties of black-body radiation, that others (notably Max Planck) had tried to explain using *semiclassical models*. In that model, light was described by Maxwell's equations, but material objects emitted and absorbed light in *quantized* amounts (i.e., they change energy only by certain particular discrete amounts). Although these semiclassical models contributed to the development of quantum mechanics, many further experiments[3][4] beginning with the phenomenon of Compton scattering of single photons by electrons, validated Einstein's hypothesis that *light itself* is quantized. In 1926 the optical physicist Frithiof Wolfers and the chemist Gilbert N. Lewis coined the name photon for these particles. After Arthur H. Compton won the Nobel Prize in 1927 for his scattering studies, most scientists accepted that light quanta have an independent existence, and the term *photon* was accepted.

In the Standard Model of particle physics, photons and other elementary particles are described as a necessary consequence of physical laws having a certain symmetry at every point in spacetime. The intrinsic properties of particles, such as charge, mass and spin, are determined by this gauge symmetry. The photon concept has led to momentous advances in experimental and theoretical physics, including lasers, Bose–Einstein condensation, quantum field theory, and the probabilistic interpretation of quantum mechanics. It has been applied to photochemistry, high-resolution microscopy, and measurements of molecular distances. Recently, photons have been studied as elements of quantum computers, and for applications in optical imaging and optical communication such as quantum cryptography.

1.1 Nomenclature

In 1900, the German physicist Max Planck was studying black-body radiation and suggested that the energy carried by electromagnetic waves could only be released in "packets" of energy. In his 1901 article [5] in *Annalen der Physik* he called these packets "energy elements". The word *quanta* (singular *quantum*, Latin for *how much*) was used before 1900 to mean particles or amounts of different quantities, including electricity. In 1905, Albert Einstein suggested that electromagnetic waves could only exist as discrete wave-packets.[6] He called such a wave-packet *the light quantum* (German: *das Lichtquant*).[Note 1] The name *photon* derives from the Greek word for light, φῶς (transliterated *phôs*). Arthur Compton used *photon* in 1928, referring to Gilbert N. Lewis.[7] The same name was used earlier, by the American physicist and psychologist Leonard

T. Troland, who coined the word in 1916, in 1921 by the Irish physicist John Joly, in 1924 by the French physiologist René Wurmser (1890-1993) and in 1926 by the French physicist Frithiof Wolfers (1891-1971).[8] The name was suggested initially as a unit related to the illumination of the eye and the resulting sensation of light and was used later in a physiological context. Although Wolfers's and Lewis's theories were contradicted by many experiments and never accepted, the new name was adopted very soon by most physicists after Compton used it.[8][Note 2]

In physics, a photon is usually denoted by the symbol γ (the Greek letter gamma). This symbol for the photon probably derives from gamma rays, which were discovered in 1900 by Paul Villard,[9][10] named by Ernest Rutherford in 1903, and shown to be a form of electromagnetic radiation in 1914 by Rutherford and Edward Andrade.[11] In chemistry and optical engineering, photons are usually symbolized by $h\nu$, the energy of a photon, where h is Planck's constant and the Greek letter ν (nu) is the photon's frequency. Much less commonly, the photon can be symbolized by hf, where its frequency is denoted by f.

1.2 Physical properties

See also: Special relativity and Photonic molecule
A photon is massless,[Note 3] has no electric charge,[12] and is a stable particle. A photon has two possible polarization states. In the momentum representation of the photon, which is preferred in quantum field theory, a photon is described by its wave vector, which determines its wavelength λ and its direction of propagation. A photon's wave vector may not be zero and can be represented either as a spatial 3-vector or as a (relativistic) four-vector; in the latter case it belongs to the light cone (pictured). Different signs of the four-vector denote different circular polarizations, but in the 3-vector representation one should account for the polarization state separately; it actually is a spin quantum number. In both cases the space of possible wave vectors is three-dimensional.

The photon is the gauge boson for electromagnetism,[13]:29–30 and therefore all other quantum numbers of the photon (such as lepton number, baryon number, and flavour quantum numbers) are zero.[14] Also, the photon does not obey the Pauli exclusion principle.[15]:1221

Photons are emitted in many natural processes. For example, when a charge is accelerated it emits synchrotron radiation. During a molecular, atomic or nuclear transition to a lower energy level, photons of various energy will be emitted, ranging from radio waves to gamma rays. A photon can also be emitted when a particle and its corresponding

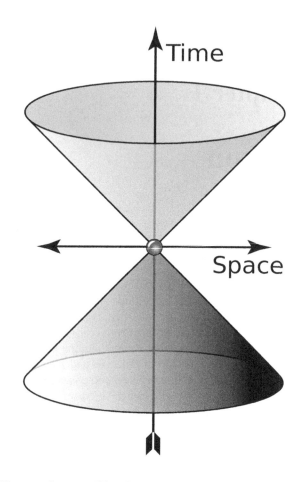

The cone shows possible values of wave 4-vector of a photon. The "time" axis gives the angular frequency (rad·s−1) and the "space" axes represent the angular wavenumber (rad·m−1). Green and indigo represent left and right polarization

antiparticle are annihilated (for example, electron–positron annihilation).[15]:572, 1114, 1172

In empty space, the photon moves at c (the speed of light) and its energy and momentum are related by $E = pc$, where p is the magnitude of the momentum vector **p**. This derives from the following relativistic relation, with $m = 0$:[16]

$$E^2 = p^2 c^2 + m^2 c^4.$$

The energy and momentum of a photon depend only on its frequency (ν) or inversely, its wavelength (λ):

$$E = \hbar\omega = h\nu = \frac{hc}{\lambda}$$

$$\boldsymbol{p} = \hbar\boldsymbol{k},$$

where $\boldsymbol{k}$ is the wave vector (where the wave number $k = |\boldsymbol{k}| = 2\pi/\lambda$), $\omega = 2\pi\nu$ is the angular frequency, and $\hbar = h/2\pi$ is the reduced Planck constant.[17]

Since **p** points in the direction of the photon's propagation, the magnitude of the momentum is

$$p = \hbar k = \frac{h\nu}{c} = \frac{h}{\lambda}.$$

The photon also carries a quantity called spin angular momentum that does not depend on its frequency.[18] The magnitude of its spin is $\sqrt{2}\hbar$ and the component measured along its direction of motion, its helicity, must be $\pm\hbar$. These two possible helicities, called right-handed and left-handed, correspond to the two possible circular polarization states of the photon.[19]

To illustrate the significance of these formulae, the annihilation of a particle with its antiparticle in free space must result in the creation of at least *two* photons for the following reason. In the center of momentum frame, the colliding antiparticles have no net momentum, whereas a single photon always has momentum (since, as we have seen, it is determined by the photon's frequency or wavelength, which cannot be zero). Hence, conservation of momentum (or equivalently, translational invariance) requires that at least two photons are created, with zero net momentum. (However, it is possible if the system interacts with another particle or field for the annihilation to produce one photon, as when a positron annihilates with a bound atomic electron, it is possible for only one photon to be emitted, as the nuclear Coulomb field breaks translational symmetry.)[20]:64–65 The energy of the two photons, or, equivalently, their frequency, may be determined from conservation of four-momentum. Seen another way, the photon can be considered as its own antiparticle. The reverse process, pair production, is the dominant mechanism by which high-energy photons such as gamma rays lose energy while passing through matter.[21] That process is the reverse of "annihilation to one photon" allowed in the electric field of an atomic nucleus.

The classical formulae for the energy and momentum of electromagnetic radiation can be re-expressed in terms of photon events. For example, the pressure of electromagnetic radiation on an object derives from the transfer of photon momentum per unit time and unit area to that object, since pressure is force per unit area and force is the change in momentum per unit time.[22]

1.2.1 Experimental checks on photon mass

Current commonly accepted physical theories imply or assume the photon to be strictly massless. If the photon is not a strictly massless particle, it would not move at the exact speed of light, c in vacuum. Its speed would be lower and depend on its frequency. Relativity would be unaffected by this; the so-called speed of light, c, would then not be the actual speed at which light moves, but a constant of nature which is the maximum speed that any object could theoretically attain in space-time.[23] Thus, it would still be the speed of space-time ripples (gravitational waves and gravitons), but it would not be the speed of photons.

If a photon did have non-zero mass, there would be other effects as well. Coulomb's law would be modified and the electromagnetic field would have an extra physical degree of freedom. These effects yield more sensitive experimental probes of the photon mass than the frequency dependence of the speed of light. If Coulomb's law is not exactly valid, then that would allow the presence of an electric field to exist within a hollow conductor when it is subjected to an external electric field. This thus allows one to test Coulomb's law to very high precision.[24] A null result of such an experiment has set a limit of $m \lesssim 10^{-14}$ eV/c^2.[25]

Sharper upper limits on the speed of light have been obtained in experiments designed to detect effects caused by the galactic vector potential. Although the galactic vector potential is very large because the galactic magnetic field exists on very great length scales, only the magnetic field would be observable if the photon is massless. In the case that the photon has mass, the mass term $\frac{1}{2}m^2 A_\mu A^\mu$ would affect the galactic plasma. The fact that no such effects are seen implies an upper bound on the photon mass of $m < 3\times10^{-27}$ eV/c^2.[26] The galactic vector potential can also be probed directly by measuring the torque exerted on a magnetized ring.[27] Such methods were used to obtain the sharper upper limit of 10^{-18} eV/c^2 (the equivalent of 1.07×10^{-27} atomic mass units) given by the Particle Data Group.[28]

These sharp limits from the non-observation of the effects caused by the galactic vector potential have been shown to be model dependent.[29] If the photon mass is generated via the Higgs mechanism then the upper limit of $m \lesssim 10^{-14}$ eV/c^2 from the test of Coulomb's law is valid.

Photons inside superconductors do develop a nonzero effective rest mass; as a result, electromagnetic forces become short-range inside superconductors.[30]

See also: Supernova/Acceleration Probe

1.3 Historical development

Main article: Light

In most theories up to the eighteenth century, light was pictured as being made up of particles. Since particle models cannot easily account for the refraction, diffraction and birefringence of light, wave theories of light were proposed by René Descartes (1637),[31] Robert Hooke (1665),[32] and

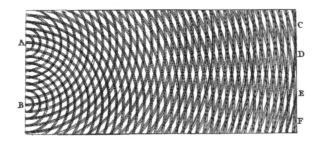

Thomas Young's double-slit experiment in 1801 showed that light can act as a wave, helping to invalidate early particle theories of light.[15]:964

Christiaan Huygens (1678);[33] however, particle models remained dominant, chiefly due to the influence of Isaac Newton.[34] In the early nineteenth century, Thomas Young and August Fresnel clearly demonstrated the interference and diffraction of light and by 1850 wave models were generally accepted.[35] In 1865, James Clerk Maxwell's prediction[36] that light was an electromagnetic wave—which was confirmed experimentally in 1888 by Heinrich Hertz's detection of radio waves[37]—seemed to be the final blow to particle models of light.

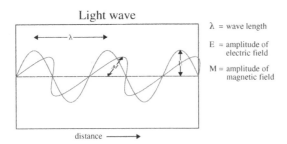

In 1900, Maxwell's theoretical model of light as oscillating electric and magnetic fields seemed complete. However, several observations could not be explained by any wave model of electromagnetic radiation, leading to the idea that light-energy was packaged into quanta described by E=hv. Later experiments showed that these light-quanta also carry momentum and, thus, can be considered particles: the photon concept was born, leading to a deeper understanding of the electric and magnetic fields themselves.

The Maxwell wave theory, however, does not account for *all* properties of light. The Maxwell theory predicts that the energy of a light wave depends only on its intensity, not on its frequency; nevertheless, several independent types of experiments show that the energy imparted by light to atoms depends only on the light's frequency, not on its intensity. For example, some chemical reactions are provoked only by light of frequency higher than a certain threshold; light of frequency lower than the threshold, no matter how intense, does not initiate the reaction. Similarly, electrons can be ejected from a metal plate by shining light of sufficiently high frequency on it (the photoelectric effect); the energy of the ejected electron is related only to the light's frequency, not to its intensity.[38][Note 4]

At the same time, investigations of blackbody radiation carried out over four decades (1860–1900) by various researchers[39] culminated in Max Planck's hypothesis[5][40] that the energy of *any* system that absorbs or emits electromagnetic radiation of frequency v is an integer multiple of an energy quantum $E = hv$. As shown by Albert Einstein,[6][41] some form of energy quantization *must* be assumed to account for the thermal equilibrium observed between matter and electromagnetic radiation; for this explanation of the photoelectric effect, Einstein received the 1921 Nobel Prize in physics.[42]

Since the Maxwell theory of light allows for all possible energies of electromagnetic radiation, most physicists assumed initially that the energy quantization resulted from some unknown constraint on the matter that absorbs or emits the radiation. In 1905, Einstein was the first to propose that energy quantization was a property of electromagnetic radiation itself.[6] Although he accepted the validity of Maxwell's theory, Einstein pointed out that many anomalous experiments could be explained if the *energy* of a Maxwellian light wave were localized into point-like quanta that move independently of one another, even if the wave itself is spread continuously over space.[6] In 1909[41] and 1916,[43] Einstein showed that, if Planck's law of black-body radiation is accepted, the energy quanta must also carry momentum $p = h/\lambda$, making them full-fledged particles. This photon momentum was observed experimentally[44] by Arthur Compton, for which he received the Nobel Prize in 1927. The pivotal question was then: how to unify Maxwell's wave theory of light with its experimentally observed particle nature? The answer to this question occupied Albert Einstein for the rest of his life,[45] and was solved in quantum electrodynamics and its successor, the Standard Model (see Second quantization and The photon as a gauge boson, below).

1.4 Einstein's light quantum

Unlike Planck, Einstein entertained the possibility that there might be actual physical quanta of light—what we now call photons. He noticed that a light quantum with energy proportional to its frequency would explain a number of troubling puzzles and paradoxes, including an unpublished law by Stokes, the ultraviolet catastrophe, and the photoelectric effect. Stokes's law said simply that the frequency of fluorescent light cannot be greater than the frequency of the light (usually ultraviolet) inducing it. Einstein eliminated the ultraviolet catastrophe by imagining a gas of photons behaving like a gas of electrons that he had previously considered. He was advised by a colleague to be careful how he wrote up this paper, in order to not challenge Planck, a powerful figure in physics, too directly, and indeed the warning was justified, as Planck never forgave him for writing it.[46]

1.5 Early objections

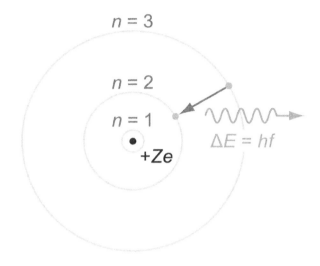

Up to 1923, most physicists were reluctant to accept that light itself was quantized. Instead, they tried to explain photon behavior by quantizing only matter, as in the Bohr model of the hydrogen atom (shown here). Even though these semiclassical models were only a first approximation, they were accurate for simple systems and they led to quantum mechanics.

Einstein's 1905 predictions were verified experimentally in several ways in the first two decades of the 20th century, as recounted in Robert Millikan's Nobel lecture.[47] However, before Compton's experiment[44] showed that photons carried momentum proportional to their wave number (1922), most physicists were reluctant to believe that electromagnetic radiation itself might be particulate. (See, for example, the Nobel lectures of Wien,[39] Planck[40] and Millikan.[47]) Instead, there was a widespread belief that energy quantization resulted from some unknown constraint on the matter that absorbed or emitted radiation. Attitudes changed over time. In part, the change can be traced to experiments such as Compton scattering, where it was much more difficult not to ascribe quantization to light itself to explain the observed results.[48]

Even after Compton's experiment, Niels Bohr, Hendrik Kramers and John Slater made one last attempt to preserve the Maxwellian continuous electromagnetic field model of light, the so-called BKS model.[49] To account for the data then available, two drastic hypotheses had to be made:

1. **Energy and momentum are conserved only on the average in interactions between matter and radiation, but not in elementary processes such as absorption and emission.** This allows one to reconcile the discontinuously changing energy of the atom (the jump between energy states) with the continuous release of energy as radiation.

2. **Causality is abandoned**. For example, spontaneous emissions are merely emissions stimulated by a "virtual" electromagnetic field.

However, refined Compton experiments showed that energy–momentum is conserved extraordinarily well in elementary processes; and also that the jolting of the electron and the generation of a new photon in Compton scattering obey causality to within 10 ps. Accordingly, Bohr and his co-workers gave their model "as honorable a funeral as possible".[45] Nevertheless, the failures of the BKS model inspired Werner Heisenberg in his development of matrix mechanics.[50]

A few physicists persisted[51] in developing semiclassical models in which electromagnetic radiation is not quantized, but matter appears to obey the laws of quantum mechanics. Although the evidence from chemical and physical experiments for the existence of photons was overwhelming by the 1970s, this evidence could not be considered as *absolutely* definitive; since it relied on the interaction of light with matter, and a sufficiently complete theory of matter could in principle account for the evidence. Nevertheless, *all* semiclassical theories were refuted definitively in the 1970s and 1980s by photon-correlation experiments.[Note 5] Hence, Einstein's hypothesis that quantization is a property of light itself is considered to be proven.

1.6 Wave–particle duality and uncertainty principles

See also: Wave–particle duality, Squeezed coherent state, Uncertainty principle, and De Broglie–Bohm theory

Photons, like all quantum objects, exhibit wave-like and particle-like properties. Their dual wave–particle nature can be difficult to visualize. The photon displays clearly wave-like phenomena such as diffraction and interference on the length scale of its wavelength. For example, a single photon passing through a double-slit experiment exhibits interference phenomena but only if no measure was made at the slit. A single photon passing through a double-slit experiment lands on the screen with a probability distribution given by its interference pattern determined by Maxwell's equations.[52] However, experiments confirm that the photon is *not* a short pulse of electromagnetic radiation; it does

Photons in a Mach–Zehnder interferometer exhibit wave-like interference and particle-like detection at single-photon detectors.

not spread out as it propagates, nor does it divide when it encounters a beam splitter.[53] Rather, the photon seems to be a point-like particle since it is absorbed or emitted *as a whole* by arbitrarily small systems, systems much smaller than its wavelength, such as an atomic nucleus ($\approx 10^{-15}$ m across) or even the point-like electron. Nevertheless, the photon is *not* a point-like particle whose trajectory is shaped probabilistically by the electromagnetic field, as conceived by Einstein and others; that hypothesis was also refuted by the photon-correlation experiments cited above. According to our present understanding, the electromagnetic field itself is produced by photons, which in turn result from a local gauge symmetry and the laws of quantum field theory (see the Second quantization and Gauge boson sections below).

A key element of quantum mechanics is Heisenberg's uncertainty principle, which forbids the simultaneous measurement of the position and momentum of a particle along the same direction. Remarkably, the uncertainty principle for charged, material particles *requires* the quantization of light into photons, and even the frequency dependence of the photon's energy and momentum.

An elegant illustration of the uncertainty principle is Heisenberg's thought experiment for locating an electron with an ideal microscope.[54] The position of the electron can be determined to within the resolving power of the microscope, which is given by a formula from classical optics

$$\Delta x \sim \frac{\lambda}{\sin \theta}$$

where θ is the aperture angle of the microscope and λ is the wavelength of the light used to observe the electron. Thus, the position uncertainty Δx can be made arbitrarily small by reducing the wavelength λ. Even if the momentum of the

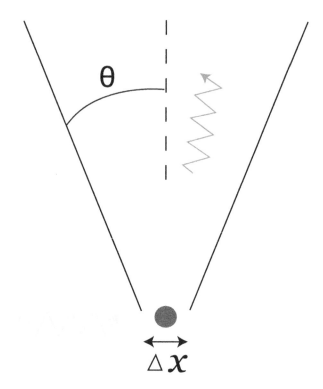

Heisenberg's thought experiment for locating an electron (shown in blue) with a high-resolution gamma-ray microscope. The incoming gamma ray (shown in green) is scattered by the electron up into the microscope's aperture angle θ. The scattered gamma ray is shown in red. Classical optics shows that the electron position can be resolved only up to an uncertainty Δx that depends on θ and the wavelength λ of the incoming light.

electron is initially known, the light impinging on the electron will give it a momentum "kick" Δp of some unknown amount, rendering the momentum of the electron uncertain. If light were *not* quantized into photons, the uncertainty Δp could be made arbitrarily small by reducing the light's intensity. In that case, since the wavelength and intensity of light can be varied independently, one could simultaneously determine the position and momentum to arbitrarily high accuracy, violating the uncertainty principle. By contrast, Einstein's formula for photon momentum preserves the uncertainty principle; since the photon is scattered anywhere within the aperture, the uncertainty of momentum transferred equals

$$\Delta p \sim p_{\text{photon}} \sin \theta = \frac{h}{\lambda} \sin \theta$$

giving the product $\Delta x \Delta p \sim h$, which is Heisenberg's uncertainty principle. Thus, the entire world is quantized; both matter and fields must obey a consistent set of quantum laws, if either one is to be quantized.[55]

The analogous uncertainty principle for photons forbids the simultaneous measurement of the number n of photons (see Fock state and the Second quantization section below) in an electromagnetic wave and the phase ϕ of that wave

$$\Delta n \Delta \phi > 1$$

See coherent state and squeezed coherent state for more details.

Both photons and electrons create analogous interference patterns when passed through a double-slit experiment. For photons, this corresponds to the interference of a Maxwell light wave whereas, for material particles (electron), this corresponds to the interference of the Schrödinger wave equation. Although this similarity might suggest that Maxwell's equations describing the photon's electromagnetic wave are simply Schrödinger's equation for photons, most physicists do not agree.[56][57] For one thing, they are mathematically different; most obviously, Schrödinger's one equation for the electron solves for a complex field, whereas Maxwell's four equations solve for real fields. More generally, the normal concept of a Schrödinger probability wave function cannot be applied to photons.[58] As photons are massless, they cannot be localized without being destroyed; technically, photons cannot have a position eigenstate $|\mathbf{r}\rangle$, and, thus, the normal Heisenberg uncertainty principle $\Delta x \Delta p > h/2$ does not pertain to photons. A few substitute wave functions have been suggested for the photon,[59][60][61][62] but they have not come into general use. Instead, physicists generally accept the second-quantized theory of photons described below, quantum electrodynamics, in which photons are quantized excitations of electromagnetic modes.

Another interpretation, that avoids duality, is the De Broglie–Bohm theory: known also as the *pilot-wave model*. In that theory, the photon is both, wave and particle.[63] "This idea seems to me so natural and simple, to resolve the wave-particle dilemma in such a clear and ordinary way, that it is a great mystery to me that it was so generally ignored",[64] J.S.Bell.

1.7 Bose–Einstein model of a photon gas

Main articles: Bose gas, Bose–Einstein statistics, Spin-statistics theorem, and Gas in a box

In 1924, Satyendra Nath Bose derived Planck's law of black-body radiation without using any electromagnetism, but rather by using a modification of coarse-grained counting of phase space.[65] Einstein showed that this modification is equivalent to assuming that photons are rigorously identical and that it implied a "mysterious non-local interaction",[66][67] now understood as the requirement for a symmetric quantum mechanical state. This work led to the concept of coherent states and the development of the laser. In the same papers, Einstein extended Bose's formalism to material particles (bosons) and predicted that they would condense into their lowest quantum state at low enough temperatures; this Bose–Einstein condensation was observed experimentally in 1995.[68] It was later used by Lene Hau to slow, and then completely stop, light in 1999[69] and 2001.[70]

The modern view on this is that photons are, by virtue of their integer spin, bosons (as opposed to fermions with half-integer spin). By the spin-statistics theorem, all bosons obey Bose–Einstein statistics (whereas all fermions obey Fermi–Dirac statistics).[71]

1.8 Stimulated and spontaneous emission

Main articles: Stimulated emission and Laser
In 1916, Einstein showed that Planck's radiation law could

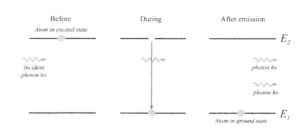

Stimulated emission (in which photons "clone" themselves) was predicted by Einstein in his kinetic analysis, and led to the development of the laser. Einstein's derivation inspired further developments in the quantum treatment of light, which led to the statistical interpretation of quantum mechanics.

be derived from a semi-classical, statistical treatment of photons and atoms, which implies a link between the rates at which atoms emit and absorb photons. The condition follows from the assumption that functions of the emission and absorption of radiation by the atoms are independent of each other, and that thermal equilibrium is made by way of the radiation's interaction with the atoms. Consider a cavity in thermal equilibrium with all parts of itself and filled with electromagnetic radiation and that the atoms can emit and absorb that radiation. Thermal equilibrium requires that the energy density $\rho(\nu)$ of photons with frequency ν (which is proportional to their number density) is, on average, constant in time; hence, the rate at which photons of any particular frequency are *emitted* must equal the rate at which they *absorb* them.[72]

Einstein began by postulating simple proportionality relations for the different reaction rates involved. In his model, the rate R_{ji} for a system to *absorb* a photon of frequency ν and transition from a lower energy E_j to a higher energy E_i is proportional to the number N_j of atoms with energy E_j and to the energy density $\rho(\nu)$ of ambient photons of that frequency,

$$R_{ji} = N_j B_{ji} \rho(\nu)$$

where B_{ji} is the rate constant for absorption. For the reverse process, there are two possibilities: spontaneous emission of a photon, or the emission of a photon initiated by the interaction of the atom with a passing photon and the return of the atom to the lower-energy state. Following Einstein's approach, the corresponding rate R_{ij} for the emission of photons of frequency ν and transition from a higher energy E_i to a lower energy E_j is

$$R_{ij} = N_i A_{ij} + N_i B_{ij} \rho(\nu)$$

where A_{ij} is the rate constant for emitting a photon spontaneously, and B_{ij} is the rate constant for emissions in response to ambient photons (induced or stimulated emission). In thermodynamic equilibrium, the number of atoms in state i and those in state j must, on average, be constant; hence, the rates R_{ji} and R_{ij} must be equal. Also, by arguments analogous to the derivation of Boltzmann statistics, the ratio of N_i and N_j is $g_i/g_j \exp{(E_j - E_i)/(kT)}$, where $g_{i,j}$ are the degeneracy of the state i and that of j, respectively, $E_{i,j}$ their energies, k the Boltzmann constant and T the system's temperature. From this, it is readily derived that $g_i B_{ij} = g_j B_{ji}$ and

$$A_{ij} = \frac{8\pi h \nu^3}{c^3} B_{ij}.$$

The A and Bs are collectively known as the *Einstein coefficients*.[73]

Einstein could not fully justify his rate equations, but claimed that it should be possible to calculate the coefficients A_{ij}, B_{ji} and B_{ij} once physicists had obtained "mechanics and electrodynamics modified to accommodate the quantum hypothesis".[74] In fact, in 1926, Paul Dirac derived the B_{ij} rate constants by using a semiclassical approach,[75] and, in 1927, succeeded in deriving *all* the rate constants from first principles within the framework of quantum theory.[76][77] Dirac's work was the foundation of quantum electrodynamics, i.e., the quantization of the electromagnetic field itself. Dirac's approach is also called *second quantization* or quantum field theory;[78][79][80] earlier quantum mechanical treatments only treat material particles as quantum mechanical, not the electromagnetic field.

Einstein was troubled by the fact that his theory seemed incomplete, since it did not determine the *direction* of a spontaneously emitted photon. A probabilistic nature of light-particle motion was first considered by Newton in his treatment of birefringence and, more generally, of the splitting of light beams at interfaces into a transmitted beam and a reflected beam. Newton hypothesized that hidden variables in the light particle determined which of the two paths a single photon would take.[34] Similarly, Einstein hoped for a more complete theory that would leave nothing to chance, beginning his separation[45] from quantum mechanics. Ironically, Max Born's probabilistic interpretation of the wave function[81][82] was inspired by Einstein's later work searching for a more complete theory.[83]

1.9 Second quantization and high energy photon interactions

Main article: Quantum field theory

In 1910, Peter Debye derived Planck's law of black-body

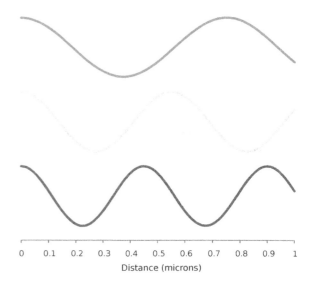

Different electromagnetic modes (*such as those depicted here*) *can be treated as independent simple harmonic oscillators. A photon corresponds to a unit of energy E=hν in its electromagnetic mode.*

radiation from a relatively simple assumption.[84] He correctly decomposed the electromagnetic field in a cavity into its Fourier modes, and assumed that the energy in any mode was an integer multiple of $h\nu$, where ν is the frequency of the electromagnetic mode. Planck's law of black-body radiation follows immediately as a geometric sum. However, Debye's approach failed to give the correct formula for the energy fluctuations of blackbody radiation, which were derived by Einstein in 1909.[41]

In 1925, Born, Heisenberg and Jordan reinterpreted Debye's concept in a key way.[85] As may be shown classically, the Fourier modes of the electromagnetic field—a complete set of electromagnetic plane waves indexed by their wave vector **k** and polarization state—are equivalent to a set of uncoupled simple harmonic oscillators. Treated quantum mechanically, the energy levels of such oscillators are known to be $E = nh\nu$, where ν is the oscillator frequency. The key new step was to identify an electromagnetic mode with energy $E = nh\nu$ as a state with n photons, each of energy $h\nu$. This approach gives the correct energy fluctuation formula.

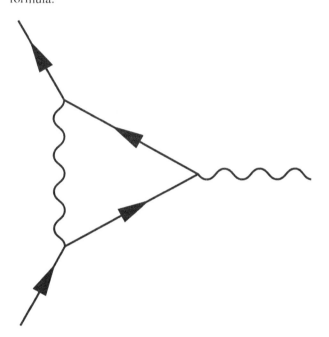

In quantum field theory, the probability of an event is computed by summing the probability amplitude (a complex number) for all possible ways in which the event can occur, as in the Feynman diagram shown here; the probability equals the square of the modulus of the total amplitude.

Dirac took this one step further.[76][77] He treated the interaction between a charge and an electromagnetic field as a small perturbation that induces transitions in the photon states, changing the numbers of photons in the modes, while conserving energy and momentum overall. Dirac was able to derive Einstein's A_{ij} and B_{ij} coefficients from first principles, and showed that the Bose–Einstein statistics of photons is a natural consequence of quantizing the electromagnetic field correctly (Bose's reasoning went in the opposite direction; he derived Planck's law of black-body radiation by *assuming* B–E statistics). In Dirac's time, it was not yet known that all bosons, including photons, must obey Bose–Einstein statistics.

Dirac's second-order perturbation theory can involve virtual photons, transient intermediate states of the electromagnetic field; the static electric and magnetic interactions are mediated by such virtual photons. In such quantum field theories, the probability amplitude of observable events is calculated by summing over *all* possible intermediate steps, even ones that are unphysical; hence, virtual photons are not constrained to satisfy $E = pc$, and may have extra polarization states; depending on the gauge used, virtual photons may have three or four polarization states, instead of the two states of real photons. Although these transient virtual photons can never be observed, they contribute measurably to the probabilities of observable events. Indeed, such second-order and higher-order perturbation calculations can give apparently infinite contributions to the sum. Such unphysical results are corrected for using the technique of renormalization.

Other virtual particles may contribute to the summation as well; for example, two photons may interact indirectly through virtual electron–positron pairs.[86] In fact, such photon-photon scattering (see two-photon physics), as well as electron-photon scattering, is meant to be one of the modes of operations of the planned particle accelerator, the International Linear Collider.[87]

In modern physics notation, the quantum state of the electromagnetic field is written as a Fock state, a tensor product of the states for each electromagnetic mode

$$|n_{k_0}\rangle \otimes |n_{k_1}\rangle \otimes \cdots \otimes |n_{k_n}\rangle \ldots$$

where $|n_{k_i}\rangle$ represents the state in which n_{k_i} photons are in the mode k_i. In this notation, the creation of a new photon in mode k_i (e.g., emitted from an atomic transition) is written as $|n_{k_i}\rangle \to |n_{k_i} + 1\rangle$. This notation merely expresses the concept of Born, Heisenberg and Jordan described above, and does not add any physics.

1.10 The hadronic properties of the photon

Measurements of the interaction between energetic photons and hadrons show that the interaction is much more intense than expected by the interaction of merely photons with the hadron's electric charge. Furthermore, the interaction of energetic photons with protons is similar to the interaction of photons with neutrons[88] in spite of the fact that the electric charge structures of protons and neutrons are substantially different. A theory called Vector Meson Dominance (VMD) was developed to explain this effect. According to VMD, the photon is a superposition of the pure electromagnetic photon which interacts only with electric charges and vector meson.[89] However, if experimentally probed at very short distances, the intrinsic structure of the photon is recognized as a flux of quark and gluon components, quasi-free according to asymptotic freedom in QCD and described by the photon structure function.[90][91] A comprehensive comparison of data with theoretical predictions is presented in a recent review.[92]

1.11 The photon as a gauge boson

Main article: Gauge theory

The electromagnetic field can be understood as a gauge field, i.e., as a field that results from requiring that a gauge symmetry holds independently at every position in spacetime.[93] For the electromagnetic field, this gauge symmetry is the Abelian U(1) symmetry of complex numbers of absolute value 1, which reflects the ability to vary the phase of a complex field without affecting observables or real valued functions made from it, such as the energy or the Lagrangian.

The quanta of an Abelian gauge field must be massless, uncharged bosons, as long as the symmetry is not broken; hence, the photon is predicted to be massless, and to have zero electric charge and integer spin. The particular form of the electromagnetic interaction specifies that the photon must have spin ± 1; thus, its helicity must be $\pm \hbar$. These two spin components correspond to the classical concepts of right-handed and left-handed circularly polarized light. However, the transient virtual photons of quantum electrodynamics may also adopt unphysical polarization states.[93]

In the prevailing Standard Model of physics, the photon is one of four gauge bosons in the electroweak interaction; the other three are denoted W^+, W^- and Z^0 and are responsible for the weak interaction. Unlike the photon, these gauge bosons have mass, owing to a mechanism that breaks their SU(2) gauge symmetry. The unification of the photon with W and Z gauge bosons in the electroweak interaction was accomplished by Sheldon Glashow, Abdus Salam and Steven Weinberg, for which they were awarded the 1979 Nobel Prize in physics.[94][95][96] Physicists continue to hypothesize grand unified theories that connect these four gauge bosons with the eight gluon gauge bosons of quantum chromodynamics; however, key predictions of these theories, such as proton decay, have not been observed experimentally.[97]

1.12 Contributions to the mass of a system

See also: Mass in special relativity and General relativity

The energy of a system that emits a photon is *decreased* by the energy E of the photon as measured in the rest frame of the emitting system, which may result in a reduction in mass in the amount E/c^2. Similarly, the mass of a system that absorbs a photon is *increased* by a corresponding amount. As an application, the energy balance of nuclear reactions involving photons is commonly written in terms of the masses of the nuclei involved, and terms of the form E/c^2 for the gamma photons (and for other relevant energies, such as the recoil energy of nuclei).[98]

This concept is applied in key predictions of quantum electrodynamics (QED, see above). In that theory, the mass of electrons (or, more generally, leptons) is modified by including the mass contributions of virtual photons, in a technique known as renormalization. Such "radiative corrections" contribute to a number of predictions of QED, such as the magnetic dipole moment of leptons, the Lamb shift, and the hyperfine structure of bound lepton pairs, such as muonium and positronium.[99]

Since photons contribute to the stress–energy tensor, they exert a gravitational attraction on other objects, according to the theory of general relativity. Conversely, photons are themselves affected by gravity; their normally straight trajectories may be bent by warped spacetime, as in gravitational lensing, and their frequencies may be lowered by moving to a higher gravitational potential, as in the Pound–Rebka experiment. However, these effects are not specific to photons; exactly the same effects would be predicted for classical electromagnetic waves.[100]

1.13 Photons in matter

See also: Group velocity and Photochemistry

Light that travels through transparent matter does so at a lower speed than c, the speed of light in a vacuum. For example, photons engage in so many collisions on the way from the core of the sun that radiant energy can take about a million years to reach the surface;[101] however, once in open space, a photon takes only 8.3 minutes to reach Earth. The factor by which the speed is decreased is called the refractive index of the material. In a classical wave picture, the slowing can be explained by the light inducing electric polarization in the matter, the polarized matter radiating new light, and that new light interfering with the original light wave to form a delayed wave. In a particle picture, the slowing can instead be described as a blending of the photon with quantum excitations of the matter to produce quasi-particles known as polariton (other quasi-particles are phonons and excitons); this polariton has a nonzero effective mass, which means that it cannot travel at c. Light of different frequencies may travel through matter at different speeds; this is called dispersion (not to be confused with scattering). In some cases, it can result in extremely slow speeds of light in matter. The effects of photon interactions with other quasi-particles may be observed directly in Raman scattering and Brillouin scattering.[102]

Photons can also be absorbed by nuclei, atoms or molecules, provoking transitions between their energy levels. A classic example is the molecular transition of retinal ($C_{20}H_{28}O$), which is responsible for vision, as discovered in 1958 by Nobel laureate biochemist George Wald and co-workers. The absorption provokes a cis-trans isomerization that, in combination with other such transitions, is transduced into nerve impulses. The absorption of photons can even break chemical bonds, as in the photodissociation of chlorine; this is the subject of photochemistry.[103][104]

1.14 Technological applications

Photons have many applications in technology. These examples are chosen to illustrate applications of photons *per se*, rather than general optical devices such as lenses, etc. that could operate under a classical theory of light. The laser is an extremely important application and is discussed above under stimulated emission.

Individual photons can be detected by several methods. The classic photomultiplier tube exploits the photoelectric effect: a photon of sufficient energy strikes a metal plate and knocks free an electron, initiating an ever-amplifying avalanche of electrons. Semiconductor charge-coupled device chips use a similar effect: an incident photon generates a charge on a microscopic capacitor that can be detected. Other detectors such as Geiger counters use the ability of photons to ionize gas molecules contained in the device, causing a detectable change of conductivity of the gas.[105]

Planck's energy formula $E = h\nu$ is often used by engineers and chemists in design, both to compute the change in energy resulting from a photon absorption and to determine the frequency of the light emitted from a given photon emission. For example, the emission spectrum of a gas-discharge lamp can be altered by filling it with (mixtures of) gases with different electronic energy level configurations.

Under some conditions, an energy transition can be excited by "two" photons that individually would be insufficient. This allows for higher resolution microscopy, because the sample absorbs energy only in the spectrum where two beams of different colors overlap significantly, which can be made much smaller than the excitation volume of a single beam (see two-photon excitation microscopy). Moreover, these photons cause less damage to the sample, since they are of lower energy.[106]

In some cases, two energy transitions can be coupled so that, as one system absorbs a photon, another nearby system "steals" its energy and re-emits a photon of a different frequency. This is the basis of fluorescence resonance energy transfer, a technique that is used in molecular biology to study the interaction of suitable proteins.[107]

Several different kinds of hardware random number generators involve the detection of single photons. In one example, for each bit in the random sequence that is to be produced, a photon is sent to a beam-splitter. In such a situation, there are two possible outcomes of equal probability. The actual outcome is used to determine whether the next bit in the sequence is "0" or "1".[108][109]

1.15 Recent research

See also: Quantum optics

Much research has been devoted to applications of photons in the field of quantum optics. Photons seem well-suited to be elements of an extremely fast quantum computer, and the quantum entanglement of photons is a focus of research. Nonlinear optical processes are another active research area, with topics such as two-photon absorption, self-phase modulation, modulational instability and optical parametric oscillators. However, such processes generally do not require the assumption of photons *per se*; they may often be modeled by treating atoms as nonlinear oscillators. The nonlinear process of spontaneous parametric down conversion is often used to produce single-photon states. Finally, photons are essential in some aspects of optical communication, especially for quantum cryptography.[Note 6]

1.16 See also

- Advanced Photon Source at Argonne National Laboratory
- Ballistic photon
- Dirac equation
- Doppler effect
- Electromagnetic radiation

- EPR paradox
- High energy X-ray imaging technology
- Laser
- Light
- Luminiferous aether
- Medipix
- Phonon
- Photography
- Photon counting
- Photon energy
- Photon polarization
- Photonic molecule
- Photonics
- Quantum optics
- Single photon source
- Static forces and virtual-particle exchange
- Two-photon physics

1.17 Notes

[1] Although the 1967 Elsevier translation of Planck's Nobel Lecture interprets Planck's *Lichtquant* as "photon", the more literal 1922 translation by Hans Thacher Clarke and Ludwik Silberstein *The origin and development of the quantum theory*, The Clarendon Press, 1922 (here) uses "light-quantum". No evidence is known that Planck himself used the term "photon" by 1926 (see also this note).

[2] Isaac Asimov credits Arthur Compton with defining quanta of energy as photons in 1923. Asimov, I. (1966). *The Neutrino, Ghost Particle of the Atom*. Garden City (NY): Doubleday. ISBN 0-380-00483-6. LCCN 66017073. and Asimov, I. (1966). *The Universe From Flat Earth To Quasar*. New York (NY): Walker. ISBN 0-8027-0316-X. LCCN 66022515.

[3] The mass of the photon is believed to be exactly zero. Some sources also refer to the *relativistic mass*, which is just the energy scaled to units of mass. For a photon with wavelength λ or energy E, this is $h/\lambda c$ or E/c^2. This usage for the term "mass" is no longer common in scientific literature. Further info: What is the mass of a photon? http://math.ucr.edu/home/baez/physics/ParticleAndNuclear/photon_mass.html

[4] The phrase "no matter how intense" refers to intensities below approximately 10^{13} W/cm^2 at which point perturbation theory begins to break down. In contrast, in the intense regime, which for visible light is above approximately 10^{14} W/cm^2, the classical wave description correctly predicts the energy acquired by electrons, called ponderomotive energy. (See also: Boreham *et al.* (1996). "Photon density and the correspondence principle of electromagnetic interaction".) By comparison, sunlight is only about 0.1 W/cm^2.

[5] These experiments produce results that cannot be explained by any classical theory of light, since they involve anticorrelations that result from the quantum measurement process. In 1974, the first such experiment was carried out by Clauser, who reported a violation of a classical Cauchy–Schwarz inequality. In 1977, Kimble *et al.* demonstrated an analogous anti-bunching effect of photons interacting with a beam splitter; this approach was simplified and sources of error eliminated in the photon-anticorrelation experiment of Grangier *et al.* (1986). This work is reviewed and simplified further in Thorn *et al.* (2004). (These references are listed below under #Additional references.)

[6] Introductory-level material on the various sub-fields of quantum optics can be found in Fox, M. (2006). *Quantum Optics: An Introduction*. Oxford University Press. ISBN 0-19-856673-5.

1.18 References

[1] Amsler, C. (Particle Data Group); Amsler; Doser; Antonelli; Asner; Babu; Baer; Band; Barnett; Bergren; Beringer; Bernardi; Bertl; Bichsel; Biebel; Bloch; Blucher; Blusk; Cahn; Carena; Caso; Ceccucci; Chakraborty; Chen; Chivukula; Cowan; Dahl; d'Ambrosio; Damour; et al. (2008). "Review of Particle Physics: Gauge and Higgs bosons" (PDF). *Physics Letters B*. **667**: 1. Bibcode:2008PhLB..667....1P. doi:10.1016/j.physletb.2008.07.018.

[2] Joos, George (1951). *Theoretical Physics*. London and Glasgow: Blackie and Son Limited. p. 679.

[3] Kimble, H.J.; Dagenais, M.; Mandel, L.; Dagenais; Mandel (1977). "Photon Anti-bunching in Resonance Fluorescence". *Physical Review Letters*. **39** (11): 691–695. Bibcode:1977PhRvL..39..691K. doi:10.1103/PhysRevLett.39.691.

[4] Grangier, P.; Roger, G.; Aspect, A.; Roger; Aspect (1986). "Experimental Evidence for a Photon Anticorrelation Effect on a Beam Splitter: A New Light on Single-Photon Interferences". *Europhysics Letters*. **1** (4): 173–179. Bibcode:1986EL......1..173G. doi:10.1209/0295-5075/1/4/004.

[5] Planck, M. (1901). "Über das Gesetz der Energieverteilung im Normalspectrum". *Annalen der Physik* (in German). **4** (3): 553–563. Bibcode:1901AnP...309..553P. doi:10.1002/andp.19013090310. English translation

1.18. REFERENCES

[6] Einstein, A. (1905). "Über einen die Erzeugung und Verwandlung des Lichtes betreffenden heuristischen Gesichtspunkt" (PDF). *Annalen der Physik* (in German). **17** (6): 132–148. Bibcode:1905AnP...322..132E. doi:10.1002/andp.19053220607.. An English translation is available from Wikisource.

[7] "Discordances entre l'expérience et la théorie électromagnétique du rayonnement." In Électrons et Photons. Rapports et Discussions de Cinquième Conseil de Physique, edited by Institut International de Physique Solvay. Paris: Gauthier-Villars, pp. 55-85.

[8] Helge Kragh: *Photon: New light on an old name.* Arxiv, 2014-2-28

[9] Villard, P. (1900). "Sur la réflexion et la réfraction des rayons cathodiques et des rayons déviables du radium". *Comptes Rendus des Séances de l'Académie des Sciences* (in French). **130**: 1010–1012.

[10] Villard, P. (1900). "Sur le rayonnement du radium". *Comptes Rendus des Séances de l'Académie des Sciences* (in French). **130**: 1178–1179.

[11] Rutherford, E.; Andrade, E.N.C. (1914). "The Wavelength of the Soft Gamma Rays from Radium B". *Philosophical Magazine.* **27** (161): 854–868. doi:10.1080/14786440508635156.

[12] Kobychev, V.V.; Popov, S.B. (2005). "Constraints on the photon charge from observations of extragalactic sources". *Astronomy Letters.* **31** (3): 147–151. arXiv:hep-ph/0411398. Bibcode:2005AstL....31..147K. doi:10.1134/1.1883345.

[13] Role as gauge boson and polarization section 5.1 in Aitchison, I.J.R.; Hey, A.J.G. (1993). *Gauge Theories in Particle Physics.* IOP Publishing. ISBN 0-85274-328-9.

[14] See p.31 in Amsler, C.; et al. (2008). "Review of Particle Physics". *Physics Letters B.* **667**: 1–1340. Bibcode:2008PhLB..667....1P. doi:10.1016/j.physletb.2008.07.018.

[15] Halliday, David; Resnick, Robert; Walker, Jerl (2005), *Fundamental of Physics* (7th ed.), USA: John Wiley and Sons, Inc., ISBN 0-471-23231-9

[16] See section 1.6 in Alonso, M.; Finn, E.J. (1968). *Fundamental University Physics Volume III: Quantum and Statistical Physics.* Addison-Wesley. ISBN 0-201-00262-0.

[17] Davison E. Soper, Electromagnetic radiation is made of photons, Institute of Theoretical Science, University of Oregon

[18] This property was experimentally verified by Raman and Bhagavantam in 1931: Raman, C.V.; Bhagavantam, S. (1931). "Experimental proof of the spin of the photon" (PDF). *Indian Journal of Physics.* **6**: 353.

[19] Burgess, C.; Moore, G. (2007). "1.3.3.2". *The Standard Model. A Primer.* Cambridge University Press. ISBN 0-521-86036-9.

[20] Griffiths, David J. (2008), *Introduction to Elementary Particles* (2nd revised ed.), WILEY-VCH, ISBN 978-3-527-40601-2

[21] E.g., section 9.3 in Alonso, M.; Finn, E.J. (1968). *Fundamental University Physics Volume III: Quantum and Statistical Physics.* Addison-Wesley.

[22] E.g., Appendix XXXII in Born, M. (1962). *Atomic Physics.* Blackie & Son. ISBN 0-486-65984-4.

[23] Mermin, David (February 1984). "Relativity without light". *American Journal of Physics.* **52** (2): 119–124. Bibcode:1984AmJPh..52..119M. doi:10.1119/1.13917.

[24] Plimpton, S.; Lawton, W. (1936). "A Very Accurate Test of Coulomb's Law of Force Between Charges". *Physical Review.* **50** (11): 1066. Bibcode:1936PhRv...50.1066P. doi:10.1103/PhysRev.50.1066.

[25] Williams, E.; Faller, J.; Hill, H. (1971). "New Experimental Test of Coulomb's Law: A Laboratory Upper Limit on the Photon Rest Mass". *Physical Review Letters.* **26** (12): 721. Bibcode:1971PhRvL..26..721W. doi:10.1103/PhysRevLett.26.721.

[26] Chibisov, G V (1976). "Astrophysical upper limits on the photon rest mass". *Soviet Physics Uspekhi.* **19** (7): 624. Bibcode:1976SvPhU..19..624C. doi:10.1070/PU1976v019n07ABEH005277.

[27] Lakes, Roderic (1998). "Experimental Limits on the Photon Mass and Cosmic Magnetic Vector Potential". *Physical Review Letters.* **80** (9): 1826. Bibcode:1998PhRvL..80.1826L. doi:10.1103/PhysRevLett.80.1826.

[28] Amsler, C; Doser, M; Antonelli, M; Asner, D; Babu, K; Baer, H; Band, H; Barnett, R; et al. (2008). "Review of Particle Physics*". *Physics Letters B.* **667**: 1. Bibcode:2008PhLB..667....1P. doi:10.1016/j.physletb.2008.07.018. Summary Table

[29] Adelberger, Eric; Dvali, Gia; Gruzinov, Andrei (2007). "Photon-Mass Bound Destroyed by Vortices". *Physical Review Letters.* **98** (1): 010402. arXiv:hep-ph/0306245. Bibcode:2007PhRvL..98a0402A. doi:10.1103/PhysRevLett.98.010402. PMID 17358459. preprint

[30] Wilczek, Frank (2010). *The Lightness of Being: Mass, Ether, and the Unification of Forces.* Basic Books. p. 212. ISBN 978-0-465-01895-6.

[31] Descartes, R. (1637). *Discours de la méthode (Discourse on Method)* (in French). Imprimerie de Ian Maire. ISBN 0-268-00870-1.

[32] Hooke, R. (1667). *Micrographia: or some physiological descriptions of minute bodies made by magnifying glasses with observations and inquiries thereupon ...* London (UK): Royal Society of London. ISBN 0-486-49564-7.

[33] Huygens, C. (1678). *Traité de la lumière* (in French).. An English translation is available from Project Gutenberg

[34] Newton, I. (1952) [1730]. *Opticks* (4th ed.). Dover (NY): Dover Publications. Book II, Part III, Propositions XII–XX; Queries 25–29. ISBN 0-486-60205-2.

[35] Buchwald, J.Z. (1989). *The Rise of the Wave Theory of Light: Optical Theory and Experiment in the Early Nineteenth Century.* University of Chicago Press. ISBN 0-226-07886-8. OCLC 18069573.

[36] Maxwell, J.C. (1865). "A Dynamical Theory of the Electromagnetic Field". *Philosophical Transactions of the Royal Society.* **155**: 459–512. Bibcode:1865RSPT..155..459C. doi:10.1098/rstl.1865.0008. This article followed a presentation by Maxwell on 8 December 1864 to the Royal Society.

[37] Hertz, H. (1888). "Über Strahlen elektrischer Kraft". *Sitzungsberichte der Preussischen Akademie der Wissenschaften (Berlin)* (in German). **1888**: 1297–1307.

[38] Frequency-dependence of luminiscence p. 276f., photoelectric effect section 1.4 in Alonso, M.; Finn, E.J. (1968). *Fundamental University Physics Volume III: Quantum and Statistical Physics.* Addison-Wesley. ISBN 0-201-00262-0.

[39] Wien, W. (1911). "Wilhelm Wien Nobel Lecture".

[40] Planck, M. (1920). "Max Planck's Nobel Lecture".

[41] Einstein, A. (1909). "Über die Entwicklung unserer Anschauungen über das Wesen und die Konstitution der Strahlung" (PDF). *Physikalische Zeitschrift* (in German). **10**: 817–825.. An English translation is available from Wikisource.

[42] Presentation speech by Svante Arrhenius for the 1921 Nobel Prize in Physics, December 10, 1922. Online text from [nobelprize.org], The Nobel Foundation 2008. Access date 2008-12-05.

[43] Einstein, A. (1916). "Zur Quantentheorie der Strahlung". *Mitteilungen der Physikalischen Gesellschaft zu Zürich.* **16**: 47. Also *Physikalische Zeitschrift*, **18**, 121–128 (1917). (German)

[44] Compton, A. (1923). "A Quantum Theory of the Scattering of X-rays by Light Elements". *Physical Review.* **21** (5): 483–502. Bibcode:1923PhRv...21..483C. doi:10.1103/PhysRev.21.483.

[45] Pais, A. (1982). *Subtle is the Lord: The Science and the Life of Albert Einstein.* Oxford University Press. ISBN 0-19-853907-X.

[46] *Einstein and the Quantum: The Quest of the Valiant Swabian*, A. Douglas Stone, Princeton University Press, 2013.

[47] Millikan, R.A (1924). "Robert A. Millikan's Nobel Lecture".

[48] Hendry, J. (1980). "The development of attitudes to the wave-particle duality of light and quantum theory, 1900–1920". *Annals of Science.* **37** (1): 59–79. doi:10.1080/00033798000200121.

[49] Bohr, N.; Kramers, H.A.; Slater, J.C. (1924). "The Quantum Theory of Radiation". *Philosophical Magazine.* **47**: 785–802. doi:10.1080/14786442408565262. Also *Zeitschrift für Physik*, **24**, 69 (1924).

[50] Heisenberg, W. (1933). "Heisenberg Nobel lecture".

[51] Mandel, L. (1976). E. Wolf, ed. "The case for and against semiclassical radiation theory". *Progress in Optics.* Progress in Optics. North-Holland. **13**: 27–69. doi:10.1016/S0079-6638(08)70018-0. ISBN 978-0-444-10806-7.

[52] Taylor, G.I. (1909). *Interference fringes with feeble light. Proceedings of the Cambridge Philosophical Society.* **15**. pp. 114–115.

[53] Saleh, B. E. A. & Teich, M. C. (2007). *Fundamentals of Photonics.* Wiley. ISBN 0-471-35832-0.

[54] Heisenberg, W. (1927). "Über den anschaulichen Inhalt der quantentheoretischen Kinematik und Mechanik". *Zeitschrift für Physik* (in German). **43** (3–4): 172–198. Bibcode:1927ZPhy...43..172H. doi:10.1007/BF01397280.

[55] E.g., p. 10f. in Schiff, L.I. (1968). *Quantum Mechanics* (3rd ed.). McGraw-Hill. ASIN B001B3MINM. ISBN 0-07-055287-8.

[56] Kramers, H.A. (1958). *Quantum Mechanics.* Amsterdam: North-Holland. ASIN B0006AUW5C. ISBN 0-486-49533-7.

[57] Bohm, D. (1989) [1954]. *Quantum Theory.* Dover Publications. ISBN 0-486-65969-0.

[58] Newton, T.D.; Wigner, E.P. (1949). "Localized states for elementary particles". *Reviews of Modern Physics.* **21** (3): 400–406. Bibcode:1949RvMP...21..400N. doi:10.1103/RevModPhys.21.400.

[59] Bialynicki-Birula, I. (1994). "On the wave function of the photon" (PDF). *Acta Physica Polonica A.* **86**: 97–116.

[60] Sipe, J.E. (1995). "Photon wave functions". *Physical Review A.* **52** (3): 1875–1883. Bibcode:1995PhRvA..52.1875S. doi:10.1103/PhysRevA.52.1875.

[61] Bialynicki-Birula, I. (1996). "Photon wave function". *Progress in Optics.* Progress in Optics. **36**: 245–294. doi:10.1016/S0079-6638(08)70316-0. ISBN 978-0-444-82530-8.

[62] Scully, M.O.; Zubairy, M.S. (1997). *Quantum Optics.* Cambridge (UK): Cambridge University Press. ISBN 0-521-43595-1.

1.18. REFERENCES

[63] The best illustration is the Couder experiment, demonstrating the behaviour of a mechanical analog, see https://www.youtube.com/watch?v=W9yWv5dqSKk

[64] Bell, J. S., "Speakable and Unspeakable in Quantum Mechanics", Cambridge: Cambridge University Press, 1987.

[65] Bose, S.N. (1924). "Plancks Gesetz und Lichtquantenhypothese". *Zeitschrift für Physik* (in German). **26**: 178–181. Bibcode:1924ZPhy...26..178B. doi:10.1007/BF01327326.

[66] Einstein, A. (1924). "Quantentheorie des einatomigen idealen Gases". *Sitzungsberichte der Preussischen Akademie der Wissenschaften (Berlin), Physikalisch-mathematische Klasse* (in German). **1924**: 261–267.

[67] Einstein, A. (1925). "Quantentheorie des einatomigen idealen Gases, Zweite Abhandlung". *Sitzungsberichte der Preussischen Akademie der Wissenschaften (Berlin), Physikalisch-mathematische Klasse* (in German). **1925**: 3–14. doi:10.1002/3527608958.ch28. ISBN 978-3-527-60895-9.

[68] Anderson, M.H.; Ensher, J.R.; Matthews, M.R.; Wieman, C.E.; Cornell, E.A. (1995). "Observation of Bose–Einstein Condensation in a Dilute Atomic Vapor". *Science*. **269** (5221): 198–201. Bibcode:1995Sci...269..198A. doi:10.1126/science.269.5221.198. JSTOR 2888436. PMID 17789847.

[69] "Physicists Slow Speed of Light". News.harvard.edu (1999-02-18). Retrieved on 2015-05-11.

[70] "Light Changed to Matter, Then Stopped and Moved". photonics.com (February 2007). Retrieved on 2015-05-11.

[71] Streater, R.F.; Wightman, A.S. (1989). *PCT, Spin and Statistics, and All That*. Addison-Wesley. ISBN 0-201-09410-X.

[72] Einstein, A. (1916). "Strahlungs-emission und -absorption nach der Quantentheorie". *Verhandlungen der Deutschen Physikalischen Gesellschaft* (in German). **18**: 318–323. Bibcode:1916DPhyG..18..318E.

[73] Section 1.4 in Wilson, J.; Hawkes, F.J.B. (1987). *Lasers: Principles and Applications*. New York: Prentice Hall. ISBN 0-13-523705-X.

[74] P. 322 in Einstein, A. (1916). "Strahlungs-emission und -absorption nach der Quantentheorie". *Verhandlungen der Deutschen Physikalischen Gesellschaft* (in German). **18**: 318–323. Bibcode:1916DPhyG..18..318E.:

> Die Konstanten A_m^n and B_m^n würden sich direkt berechnen lassen, wenn wir im Besitz einer im Sinne der Quantenhypothese modifizierten Elektrodynamik und Mechanik wären."

[75] Dirac, P.A.M. (1926). "On the Theory of Quantum Mechanics". *Proceedings of the Royal Society A*. **112** (762): 661–677. Bibcode:1926RSPSA.112..661D. doi:10.1098/rspa.1926.0133.

[76] Dirac, P.A.M. (1927). "The Quantum Theory of the Emission and Absorption of Radiation" (PDF). *Proceedings of the Royal Society A*. **114** (767): 243–265. Bibcode:1927RSPSA.114..243D. doi:10.1098/rspa.1927.0039.

[77] Dirac, P.A.M. (1927b). *The Quantum Theory of Dispersion*. Proceedings of the Royal Society A. **114**. pp. 710–728. doi:10.1098/rspa.1927.0071.

[78] Heisenberg, W.; Pauli, W. (1929). "Zur Quantentheorie der Wellenfelder". *Zeitschrift für Physik* (in German). **56**: 1. Bibcode:1929ZPhy...56....1H. doi:10.1007/BF01340129.

[79] Heisenberg, W.; Pauli, W. (1930). "Zur Quantentheorie der Wellenfelder". *Zeitschrift für Physik* (in German). **59** (3–4): 139. Bibcode:1930ZPhy...59..168H. doi:10.1007/BF01341423.

[80] Fermi, E. (1932). "Quantum Theory of Radiation" (PDF). *Reviews of Modern Physics*. **4**: 87. Bibcode:1932RvMP....4...87F. doi:10.1103/RevModPhys.4.87.

[81] Born, M. (1926). "Zur Quantenmechanik der Stossvorgänge" (PDF). *Zeitschrift für Physik* (in German). **37** (12): 863–867. Bibcode:1926ZPhy...37..863B. doi:10.1007/BF01397477.

[82] Born, M. (1926). "Quantenmechanik der Stossvorgänge". *Zeitschrift für Physik* (in German). **38** (11–12): 803. Bibcode:1926ZPhy...38..803B. doi:10.1007/BF01397184.

[83] Pais, A. (1986). *Inward Bound: Of Matter and Forces in the Physical World*. Oxford University Press. p. 260. ISBN 0-19-851997-4. Specifically, Born claimed to have been inspired by Einstein's never-published attempts to develop a "ghost-field" theory, in which point-like photons are guided probabilistically by ghost fields that follow Maxwell's equations.

[84] Debye, P. (1910). "Der Wahrscheinlichkeitsbegriff in der Theorie der Strahlung". *Annalen der Physik* (in German). **33** (16): 1427–1434. Bibcode:1910AnP...338.1427D. doi:10.1002/andp.19103381617.

[85] Born, M.; Heisenberg, W.; Jordan, P. (1925). "Quantenmechanik II". *Zeitschrift für Physik* (in German). **35** (8–9): 557–615. Bibcode:1926ZPhy...35..557B. doi:10.1007/BF01379806.

[86] Photon-photon-scattering section 7-3-1, renormalization chapter 8-2 in Itzykson, C.; Zuber, J.-B. (1980). *Quantum Field Theory*. McGraw-Hill. ISBN 0-07-032071-3.

[87] Weiglein, G. (2008). "Electroweak Physics at the ILC". *Journal of Physics: Conference Series*. **110** (4): 042033. arXiv:0711.3003. Bibcode:2008JPhCS.110d2033W. doi:10.1088/1742-6596/110/4/042033.

[88] Bauer, T. H.; Spital, R. D.; Yennie, D. R.; Pipkin, F. M. (1978). "The hadronic properties of the photon in high-energy interactions". *Reviews of Modern Physics*. **50** (2): 261. Bibcode:1978RvMP...50..261B. doi:10.1103/RevModPhys.50.261.

[89] Sakurai, J. J. (1960). "Theory of strong interactions". *Annals of Physics*. **11**: 1. Bibcode:1960AnPhy..11....1S. doi:10.1016/0003-4916(60)90126-3.

[90] Walsh, T. F.; Zerwas, P. (1973). "Two-photon processes in the parton model". *Physics Letters B*. **44** (2): 195. Bibcode:1973PhLB...44..195W. doi:10.1016/0370-2693(73)90520-0.

[91] Witten, E. (1977). "Anomalous cross section for photon-photon scattering in gauge theories". *Nuclear Physics B*. **120** (2): 189. Bibcode:1977NuPhB.120..189W. doi:10.1016/0550-3213(77)90038-4.

[92] Nisius, R. (2000). "The photon structure from deep inelastic electron–photon scattering". *Physics Reports*. **332** (4–6): 165. arXiv:hep-ex/9912049. Bibcode:2000PhR...332..165N. doi:10.1016/S0370-1573(99)00115-5.

[93] Ryder, L.H. (1996). *Quantum field theory* (2nd ed.). Cambridge University Press. ISBN 0-521-47814-6.

[94] Sheldon Glashow Nobel lecture, delivered 8 December 1979.

[95] Abdus Salam Nobel lecture, delivered 8 December 1979.

[96] Steven Weinberg Nobel lecture, delivered 8 December 1979.

[97] E.g., chapter 14 in Hughes, I. S. (1985). *Elementary particles* (2nd ed.). Cambridge University Press. ISBN 0-521-26092-2.

[98] E.g., section 10.1 in Dunlap, R.A. (2004). *An Introduction to the Physics of Nuclei and Particles*. Brooks/Cole. ISBN 0-534-39294-6.

[99] Radiative correction to electron mass section 7-1-2, anomalous magnetic moments section 7-2-1, Lamb shift section 7-3-2 and hyperfine splitting in positronium section 10-3 in Itzykson, C.; Zuber, J.-B. (1980). *Quantum Field Theory*. McGraw-Hill. ISBN 0-07-032071-3.

[100] E. g. sections 9.1 (gravitational contribution of photons) and 10.5 (influence of gravity on light) in Stephani, H.; Stewart, J. (1990). *General Relativity: An Introduction to the Theory of Gravitational Field*. Cambridge University Press. pp. 86 ff, 108 ff. ISBN 0-521-37941-5.

[101] Naeye, R. (1998). *Through the Eyes of Hubble: Birth, Life and Violent Death of Stars*. CRC Press. ISBN 0-750-30484-7. OCLC 40180195.

[102] Polaritons section 10.10.1, Raman and Brillouin scattering section 10.11.3 in Patterson, J.D.; Bailey, B.C. (2007). *Solid-State Physics: Introduction to the Theory*. Springer. ISBN 3-540-24115-9.

[103] E.g. section 11-5 C in Pine, S.H.; Hendrickson, J.B.; Cram, D.J.; Hammond, G.S. (1980). *Organic Chemistry* (4th ed.). McGraw-Hill. ISBN 0-07-050115-7.

[104] Nobel lecture given by G. Wald on December 12, 1967, online at nobelprize.org: The Molecular Basis of Visual Excitation.

[105] Photomultiplier section 1.1.10, CCDs section 1.1.8, Geiger counters section 1.3.2.1 in Kitchin, C.R. (2008). *Astrophysical Techniques*. Boca Raton (FL): CRC Press. ISBN 1-4200-8243-4.

[106] Denk, W.; Svoboda, K. (1997). "Photon upmanship: Why multiphoton imaging is more than a gimmick". *Neuron*. **18** (3): 351–357. doi:10.1016/S0896-6273(00)81237-4. PMID 9115730.

[107] Lakowicz, J.R. (2006). *Principles of Fluorescence Spectroscopy*. Springer. pp. 529 ff. ISBN 0-387-31278-1.

[108] Jennewein, T.; Achleitner, U.; Weihs, G.; Weinfurter, H.; Zeilinger, A. (2000). "A fast and compact quantum random number generator". *Review of Scientific Instruments*. **71** (4): 1675–1680. arXiv:quant-ph/9912118. Bibcode:2000RScI...71.1675J. doi:10.1063/1.1150518.

[109] Stefanov, A.; Gisin, N.; Guinnard, O.; Guinnard, L.; Zbiden, H. (2000). "Optical quantum random number generator". *Journal of Modern Optics*. **47** (4): 595–598. doi:10.1080/095003400147908.

1.19 Additional references

By date of publication:

- Clauser, J.F. (1974). "Experimental distinction between the quantum and classical field-theoretic predictions for the photoelectric effect". *Physical Review D*. **9** (4): 853–860. Bibcode:1974PhRvD...9..853C. doi:10.1103/PhysRevD.9.853.

- Kimble, H.J.; Dagenais, M.; Mandel, L. (1977). "Photon Anti-bunching in Resonance Fluorescence". *Physical Review Letters*. **39** (11): 691–695. Bibcode:1977PhRvL..39..691K. doi:10.1103/PhysRevLett.39.691.

- Pais, A. (1982). *Subtle is the Lord: The Science and the Life of Albert Einstein*. Oxford University Press.

- Feynman, Richard (1985). *QED: The Strange Theory of Light and Matter*. Princeton University Press. ISBN 978-0-691-12575-6.

- Grangier, P.; Roger, G.; Aspect, A. (1986). "Experimental Evidence for a Photon Anticorrelation Effect on a Beam Splitter: A New Light on

Single-Photon Interferences". *Europhysics Letters*. **1** (4): 173–179. Bibcode:1986EL......1..173G. doi:10.1209/0295-5075/1/4/004.

- Lamb, W.E. (1995). "Anti-photon". *Applied Physics B*. **60** (2–3): 77–84. Bibcode:1995ApPhB..60...77L. doi:10.1007/BF01135846.

- Special supplemental issue of *Optics and Photonics News* (vol. 14, October 2003) article web link

  - Roychoudhuri, C.; Rajarshi, R. (2003). "The nature of light: what is a photon?". *Optics and Photonics News*. **14**: S1 (Supplement).
  - Zajonc, A. "Light reconsidered". *Optics and Photonics News*. **14**: S2–S5 (Supplement).
  - Loudon, R. "What is a photon?". *Optics and Photonics News*. **14**: S6–S11 (Supplement).
  - Finkelstein, D. "What is a photon?". *Optics and Photonics News*. **14**: S12–S17 (Supplement).
  - Muthukrishnan, A.; Scully, M.O.; Zubairy, M.S. "The concept of the photon—revisited". *Optics and Photonics News*. **14**: S18–S27 (Supplement).
  - Mack, H.; Schleich, W.P. "A photon viewed from Wigner phase space". *Optics and Photonics News*. **14**: S28–S35 (Supplement).

- Glauber, R. (2005). "One Hundred Years of Light Quanta" (PDF). *2005 Physics Nobel Prize Lecture*.

- Hentschel, K. (2007). "Light quanta: The maturing of a concept by the stepwise accretion of meaning". *Physics and Philosophy*. **1** (2): 1–20.

Education with single photons:

- Thorn, J.J.; Neel, M.S.; Donato, V.W.; Bergreen, G.S.; Davies, R.E.; Beck, M. (2004). "Observing the quantum behavior of light in an undergraduate laboratory" (PDF). *American Journal of Physics*. **72** (9): 1210–1219. Bibcode:2004AmJPh..72.1210T. doi:10.1119/1.1737397.

- Bronner, P.; Strunz, Andreas; Silberhorn, Christine; Meyn, Jan-Peter (2009). "Interactive screen experiments with single photons". *European Journal of Physics*. **30** (2): 345–353. Bibcode:2009EJPh...30..345B. doi:10.1088/0143-0807/30/2/014.

1.20 External links

- The dictionary definition of photon at Wiktionary
- Media related to Photon at Wikimedia Commons

Chapter 2

Light

"Visible light" redirects here. For light that cannot be seen with human eye, see Electromagnetic radiation. For other uses, see Light (disambiguation) and Visible light (disambiguation).

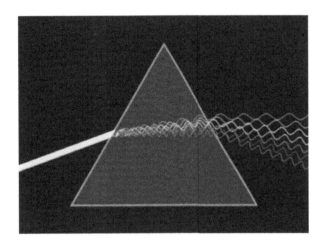

A triangular prism dispersing a beam of white light. The longer wavelengths (red) and the shorter wavelengths (blue) get separated.

Light is electromagnetic radiation within a certain portion of the electromagnetic spectrum. The word usually refers to **visible light**, which is visible to the human eye and is responsible for the sense of sight.[1] Visible light is usually defined as having wavelengths in the range of 400–700 nanometres (nm), or 4.00×10^{-7} to 7.00×10^{-7} m, between the infrared (with longer wavelengths) and the ultraviolet (with shorter wavelengths).[2][3] This wavelength means a frequency range of roughly 430–750 terahertz (THz).

The main source of light on Earth is the Sun. Sunlight provides the energy that green plants use to create sugars mostly in the form of starches, which release energy into the living things that digest them. This process of photosynthesis provides virtually all the energy used by living things. Historically, another important source of light for humans has been fire, from ancient campfires to modern kerosene lamps. With the development of electric lights and power systems, electric lighting has effectively replaced firelight. Some species of animals generate their own light, a process called bioluminescence. For example, fireflies use light to locate mates, and vampire squids use it to hide themselves from prey.

The primary properties of visible light are intensity, propagation direction, frequency or wavelength spectrum, and polarization, while its speed in a vacuum, 299,792,458 metres per second, is one of the fundamental constants of nature. Visible light, as with all types of electromagnetic radiation (EMR), is experimentally found to always move at this speed in a vacuum.

In physics, the term *light* sometimes refers to electromagnetic radiation of any wavelength, whether visible or not.[4][5] In this sense, gamma rays, X-rays, microwaves and radio waves are also light. Like all types of light, visible light is emitted and absorbed in tiny "packets" called photons and exhibits properties of both waves and particles. This property is referred to as the wave–particle duality. The study of light, known as optics, is an important research area in modern physics.

2.1 Electromagnetic spectrum and visible light

Main article: Electromagnetic spectrum

Generally, EM radiation, or EMR (the designation "radiation" excludes static electric and magnetic and near fields), is classified by wavelength into radio, microwave, infrared, the **visible region** that we perceive as light, ultraviolet, X-rays and gamma rays.

The behavior of EMR depends on its wavelength. Higher frequencies have shorter wavelengths, and lower frequencies have longer wavelengths. When EMR interacts with single atoms and molecules, its behavior depends on the amount of energy per quantum it carries.

EMR in the visible light region consists of quanta (called

2.2 Speed of light

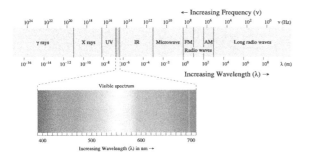

Electromagnetic spectrum with light highlighted

Main article: Speed of light

The speed of light in a vacuum is defined to be exactly 299,792,458 m/s (approx. 186,282 miles per second). The fixed value of the speed of light in SI units results from the fact that the metre is now defined in terms of the speed of light. All forms of electromagnetic radiation move at exactly this same speed in vacuum.

Different physicists have attempted to measure the speed of light throughout history. Galileo attempted to measure the speed of light in the seventeenth century. An early experiment to measure the speed of light was conducted by Ole Rømer, a Danish physicist, in 1676. Using a telescope, Rømer observed the motions of Jupiter and one of its moons, Io. Noting discrepancies in the apparent period of Io's orbit, he calculated that light takes about 22 minutes to traverse the diameter of Earth's orbit.[14] However, its size was not known at that time. If Rømer had known the diameter of the Earth's orbit, he would have calculated a speed of 227,000,000 m/s.

Another, more accurate, measurement of the speed of light was performed in Europe by Hippolyte Fizeau in 1849. Fizeau directed a beam of light at a mirror several kilometers away. A rotating cog wheel was placed in the path of the light beam as it traveled from the source, to the mirror and then returned to its origin. Fizeau found that at a certain rate of rotation, the beam would pass through one gap in the wheel on the way out and the next gap on the way back. Knowing the distance to the mirror, the number of teeth on the wheel, and the rate of rotation, Fizeau was able to calculate the speed of light as 313,000,000 m/s.

Léon Foucault used an experiment which used rotating mirrors to obtain a value of 298,000,000 m/s in 1862. Albert A. Michelson conducted experiments on the speed of light from 1877 until his death in 1931. He refined Foucault's methods in 1926 using improved rotating mirrors to measure the time it took light to make a round trip from Mount Wilson to Mount San Antonio in California. The precise measurements yielded a speed of 299,796,000 m/s.[15]

The effective velocity of light in various transparent substances containing ordinary matter, is less than in vacuum. For example, the speed of light in water is about 3/4 of that in vacuum.

Two independent teams of physicists were said to bring light to a "complete standstill" by passing it through a Bose–Einstein condensate of the element rubidium, one team at Harvard University and the Rowland Institute for Science in Cambridge, Mass., and the other at the Harvard–Smithsonian Center for Astrophysics, also in

photons) that are at the lower end of the energies that are capable of causing electronic excitation within molecules, which leads to changes in the bonding or chemistry of the molecule. At the lower end of the visible light spectrum, EMR becomes invisible to humans (infrared) because its photons no longer have enough individual energy to cause a lasting molecular change (a change in conformation) in the visual molecule retinal in the human retina, which change triggers the sensation of vision.

There exist animals that are sensitive to various types of infrared, but not by means of quantum-absorption. Infrared sensing in snakes depends on a kind of natural thermal imaging, in which tiny packets of cellular water are raised in temperature by the infrared radiation. EMR in this range causes molecular vibration and heating effects, which is how these animals detect it.

Above the range of visible light, ultraviolet light becomes invisible to humans, mostly because it is absorbed by the cornea below 360 nanometers and the internal lens below 400. Furthermore, the rods and cones located in the retina of the human eye cannot detect the very short (below 360 nm) ultraviolet wavelengths and are in fact damaged by ultraviolet. Many animals with eyes that do not require lenses (such as insects and shrimp) are able to detect ultraviolet, by quantum photon-absorption mechanisms, in much the same chemical way that humans detect visible light.

Various sources define visible light as narrowly as 420 to 680[6][7] to as broadly as 380 to 800 nm.[8][9] Under ideal laboratory conditions, people can see infrared up to at least 1050 nm;[10] children and young adults may perceive ultraviolet wavelengths down to about 310 to 313 nm.[11][12][13]

Plant growth is also affected by the color spectrum of light, a process known as photomorphogenesis.

Cambridge.[16] However, the popular description of light being "stopped" in these experiments refers only to light being stored in the excited states of atoms, then re-emitted at an arbitrary later time, as stimulated by a second laser pulse. During the time it had "stopped" it had ceased to be light.

2.3 Optics

Main article: Optics

The study of light and the interaction of light and matter is termed optics. The observation and study of optical phenomena such as rainbows and the aurora borealis offer many clues as to the nature of light.

2.3.1 Refraction

Main article: Refraction
Refraction is the bending of light rays when passing through

An example of refraction of light. The straw appears bent, because of refraction of light as it enters liquid from air.

a surface between one transparent material and another. It is described by Snell's Law:

$$n_1 \sin \theta_1 = n_2 \sin \theta_2 .$$

A cloud illuminated by sunlight

where θ_1 is the angle between the ray and the surface normal in the first medium, θ_2 is the angle between the ray and the surface normal in the second medium, and n_1 and n_2 are the indices of refraction, $n = 1$ in a vacuum and $n > 1$ in a transparent substance.

When a beam of light crosses the boundary between a vacuum and another medium, or between two different media, the wavelength of the light changes, but the frequency remains constant. If the beam of light is not orthogonal (or rather normal) to the boundary, the change in wavelength results in a change in the direction of the beam. This change of direction is known as refraction.

The refractive quality of lenses is frequently used to manipulate light in order to change the apparent size of images. Magnifying glasses, spectacles, contact lenses, microscopes and refracting telescopes are all examples of this manipulation.

2.4 Light sources

Further information: List of light sources

There are many sources of light. The most common light sources are thermal: a body at a given temperature emits a characteristic spectrum of black-body radiation. A simple thermal source is sunlight, the radiation emitted by the chromosphere of the Sun at around 6,000 kelvins (5,730 degrees Celsius; 10,340 degrees Fahrenheit) peaks in the visible region of the electromagnetic spectrum when plotted in wavelength units[17] and roughly 44% of sunlight energy that reaches the ground is visible.[18] Another example is incandescent light bulbs, which emit only around 10% of their energy as visible light and the remainder as infrared. A common thermal light source in history is the glowing solid particles in flames, but these also emit most of their radiation in the infrared, and only a fraction in the visible spectrum. The peak of the blackbody spectrum is in the deep infrared, at about 10 micrometre wavelength, for relatively cool objects like human beings. As the temperature increases, the peak shifts to shorter wavelengths, producing first a red glow, then a white one, and finally a blue-white colour as the peak moves out of the visible part of the spectrum and into the ultraviolet. These colours can be seen when metal is heated to "red hot" or "white hot". Blue-white thermal emission is not often seen, except in stars (the commonly seen pure-blue colour in a gas flame or a welder's torch is in fact due to molecular emission, notably by CH radicals (emitting a wavelength band around 425 nm, and is not seen in stars or pure thermal radiation).

Atoms emit and absorb light at characteristic energies. This produces "emission lines" in the spectrum of each atom. Emission can be spontaneous, as in light-emitting diodes, gas discharge lamps (such as neon lamps and neon signs, mercury-vapor lamps, etc.), and flames (light from the hot gas itself—so, for example, sodium in a gas flame emits characteristic yellow light). Emission can also be stimulated, as in a laser or a microwave maser.

Deceleration of a free charged particle, such as an electron, can produce visible radiation: cyclotron radiation, synchrotron radiation, and bremsstrahlung radiation are all examples of this. Particles moving through a medium faster than the speed of light in that medium can produce visible Cherenkov radiation.

Certain chemicals produce visible radiation by chemoluminescence. In living things, this process is called bioluminescence. For example, fireflies produce light by this means, and boats moving through water can disturb plankton which produce a glowing wake.

Certain substances produce light when they are illuminated by more energetic radiation, a process known as fluorescence. Some substances emit light slowly after excitation by more energetic radiation. This is known as phosphorescence.

Phosphorescent materials can also be excited by bombarding them with subatomic particles. Cathodoluminescence is one example. This mechanism is used in cathode ray tube television sets and computer monitors.

Certain other mechanisms can produce light:

- Bioluminescence
- Cherenkov radiation
- Electroluminescence
- Scintillation

A city illuminated by colorful artificial lighting

- Sonoluminescence
- Triboluminescence

When the concept of light is intended to include very-high-energy photons (gamma rays), additional generation mechanisms include:

- Particle–antiparticle annihilation
- Radioactive decay

2.5 Units and measures

Main articles: Photometry (optics) and Radiometry

Light is measured with two main alternative sets of units: radiometry consists of measurements of light power at all wavelengths, while photometry measures light with wavelength weighted with respect to a standardised model of human brightness perception. Photometry is useful, for example, to quantify Illumination (lighting) intended for human use. The SI units for both systems are summarised in the following tables.

The photometry units are different from most systems of physical units in that they take into account how the human eye responds to light. The cone cells in the human eye are of three types which respond differently across the visible spectrum, and the cumulative response peaks at a wavelength of around 555 nm. Therefore, two sources of light which produce the same intensity (W/m^2) of visible light do not necessarily appear equally bright. The photometry units are designed to take this into account, and therefore are a better representation of how "bright" a light appears to be than raw intensity. They relate to raw power by a quantity called luminous efficacy, and are used for purposes like determining how to best achieve sufficient illumination for various tasks in indoor and outdoor settings. The illumination measured by a photocell sensor does not necessarily correspond to what is perceived by the human eye, and without filters which may be costly, photocells and charge-coupled devices (CCD) tend to respond to some infrared, ultraviolet or both.

2.6 Light pressure

Main article: Radiation pressure

Light exerts physical pressure on objects in its path, a phenomenon which can be deduced by Maxwell's equations, but can be more easily explained by the particle nature of light: photons strike and transfer their momentum. Light pressure is equal to the power of the light beam divided by c, the speed of light. Due to the magnitude of c, the effect of light pressure is negligible for everyday objects. For example, a one-milliwatt laser pointer exerts a force of about 3.3 piconewtons on the object being illuminated; thus, one could lift a U.S. penny with laser pointers, but doing so would require about 30 billion 1-mW laser pointers.[19] However, in nanometre-scale applications such as nanoelectromechanical systems (NEMS), the effect of light pressure is more significant, and exploiting light pressure to drive NEMS mechanisms and to flip nanometre-scale physical switches in integrated circuits is an active area of research.[20]

At larger scales, light pressure can cause asteroids to spin faster,[21] acting on their irregular shapes as on the vanes of a windmill. The possibility of making solar sails that would accelerate spaceships in space is also under investigation.[22][23]

Although the motion of the Crookes radiometer was originally attributed to light pressure, this interpretation is incorrect; the characteristic Crookes rotation is the result of a partial vacuum.[24] This should not be confused with the Nichols radiometer, in which the (slight) motion caused by torque (though not enough for full rotation against friction) *is* directly caused by light pressure.[25]

2.7 Historical theories about light, in chronological order

2.7.1 Classical Greece and Hellenism

In the fifth century BC, Empedocles postulated that everything was composed of four elements; fire, air, earth and water. He believed that Aphrodite made the human eye out of the four elements and that she lit the fire in the eye which shone out from the eye making sight possible. If this were true, then one could see during the night just as well as during the day, so Empedocles postulated an interaction between rays from the eyes and rays from a source such as the sun.

In about 300 BC, Euclid wrote *Optica*, in which he studied the properties of light. Euclid postulated that light travelled in straight lines and he described the laws of reflection and studied them mathematically. He questioned that sight is the result of a beam from the eye, for he asks how one sees the stars immediately, if one closes one's eyes, then opens them at night. Of course if the beam from the eye travels infinitely fast this is not a problem.

In 55 BC, Lucretius, a Roman who carried on the ideas of earlier Greek atomists, wrote:

"*The light & heat of the sun; these are composed of minute atoms which, when they are shoved off, lose no time in shooting right across the interspace of air in the direction imparted by the shove.*" – On the nature of the Universe

Despite being similar to later particle theories, Lucretius's views were not generally accepted.

Ptolemy (c. 2nd century) wrote about the refraction of light in his book *Optics*.[26]

2.7.2 Classical India

In ancient India, the Hindu schools of Samkhya and Vaisheshika, from around the early centuries AD developed theories on light. According to the Samkhya school, light is one of the five fundamental "subtle" elements (*tanmatra*) out of which emerge the gross elements. The atomicity of these elements is not specifically mentioned and it appears that they were actually taken to be continuous.[27] On the other hand, the Vaisheshika school gives an atomic theory of the physical world on the non-atomic ground of ether, space and time. (See *Indian atomism*.) The basic atoms are those of earth (*prthivi*), water (*pani*), fire (*agni*), and air (*vayu*) Light rays are taken to be a stream of high velocity of *tejas* (fire) atoms. The particles of light can exhibit different characteristics depending on the speed and the arrangements of the *tejas* atoms. The *Vishnu Purana* refers to sunlight as "the seven rays of the sun".[27]

The Indian Buddhists, such as Dignāga in the 5th century and Dharmakirti in the 7th century, developed a type of atomism that is a philosophy about reality being composed of atomic entities that are momentary flashes of light or energy. They viewed light as being an atomic entity equivalent to energy.[27]

2.7.3 Descartes

René Descartes (1596–1650) held that light was a mechanical property of the luminous body, rejecting the "forms" of Ibn al-Haytham and Witelo as well as the "species" of Bacon, Grosseteste, and Kepler.[28] In 1637 he published a theory of the refraction of light that assumed, incorrectly, that light travelled faster in a denser medium than in a less dense medium. Descartes arrived at this conclusion by analogy with the behaviour of sound waves. Although Descartes was incorrect about the relative speeds, he was correct in assuming that light behaved like a wave and in concluding that refraction could be explained by the speed of light in different media.

Descartes is not the first to use the mechanical analogies but because he clearly asserts that light is only a mechanical property of the luminous body and the transmitting medium, Descartes' theory of light is regarded as the start of modern physical optics.[28]

Pierre Gassendi.

2.7.4 Particle theory

Main article: Corpuscular theory of light

Pierre Gassendi (1592–1655), an atomist, proposed a particle theory of light which was published posthumously in the 1660s. Isaac Newton studied Gassendi's work at an early age, and preferred his view to Descartes' theory of the *plenum*. He stated in his *Hypothesis of Light* of 1675 that light was composed of corpuscles (particles of matter) which were emitted in all directions from a source. One of Newton's arguments against the wave nature of light was that waves were known to bend around obstacles, while light travelled only in straight lines. He did, however, explain the phenomenon of the diffraction of light (which had been observed by Francesco Grimaldi) by allowing that a light particle could create a localised wave in the aether.

Newton's theory could be used to predict the reflection of light, but could only explain refraction by incorrectly assuming that light accelerated upon entering a denser medium because the gravitational pull was greater. Newton published the final version of his theory in his *Opticks* of 1704. His reputation helped the particle theory of light to hold sway during the 18th century. The particle theory of light led Laplace to argue that a body could be so massive that light could not escape from it. In other words, it would become what is now called a black hole. Laplace withdrew his suggestion later, after a wave theory of light became firmly established as the model for light (as has been explained, neither a particle or wave theory is fully correct). A translation of Newton's essay on light appears in *The large scale structure of space-time,* by Stephen Hawking and George F. R. Ellis.

The fact that light could be polarized was for the first time qualitatively explained by Newton using the particle theory. Étienne-Louis Malus in 1810 created a mathematical particle theory of polarization. Jean-Baptiste Biot in 1812 showed that this theory explained all known phenomena of light polarization. At that time the polarization was considered as the proof of the particle theory.

2.7.5 Wave theory

To explain the origin of colors, Robert Hooke (1635-1703) developed a "pulse theory" and compared the spreading of light to that of waves in water in his 1665 work *Micrographia* ("Observation IX"). In 1672 Hooke suggested that light's vibrations could be perpendicular to the direction of propagation. Christiaan Huygens (1629-1695) worked out a mathematical wave theory of light in 1678, and published it in his *Treatise on light* in 1690. He proposed that light was emitted in all directions as a series of waves in a medium called the *Luminiferous ether*. As waves are not affected by gravity, it was assumed that they slowed down upon entering a denser medium.[29]

The wave theory predicted that light waves could interfere

Christiaan Huygens.

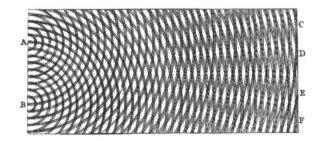

Thomas Young's sketch of a double-slit experiment showing diffraction. Young's experiments supported the theory that light consists of waves.

with each other like sound waves (as noted around 1800 by Thomas Young). Young showed by means of a diffraction experiment that light behaved as waves. He also proposed that different colours were caused by different wavelengths of light, and explained colour vision in terms of three-coloured receptors in the eye.

Another supporter of the wave theory was Leonhard Euler. He argued in *Nova theoria lucis et colorum* (1746) that diffraction could more easily be explained by a wave theory.

In 1816 André-Marie Ampère gave Augustin-Jean Fresnel an idea that the polarization of light can be explained by the wave theory if light were a transverse wave.[30]

Later, Fresnel independently worked out his own wave theory of light, and presented it to the Académie des Sciences in 1817. Siméon Denis Poisson added to Fresnel's mathematical work to produce a convincing argument in favour of the wave theory, helping to overturn Newton's corpuscular theory. By the year 1821, Fresnel was able to show via mathematical methods that polarisation could be explained by the wave theory of light and only if light was entirely transverse, with no longitudinal vibration whatsoever.

The weakness of the wave theory was that light waves, like sound waves, would need a medium for transmission. The existence of the hypothetical substance *luminiferous aether* proposed by Huygens in 1678 was cast into strong doubt in the late nineteenth century by the Michelson–Morley experiment.

Newton's corpuscular theory implied that light would travel faster in a denser medium, while the wave theory of Huygens and others implied the opposite. At that time, the speed of light could not be measured accurately enough to decide which theory was correct. The first to make a sufficiently accurate measurement was Léon Foucault, in 1850.[31] His result supported the wave theory, and the classical particle theory was finally abandoned, only to partly re-emerge in the 20th century.

2.7.6 Electromagnetic theory as explanation for all types of visible light and all EM radiation

Main article: Electromagnetic radiation

In 1845, Michael Faraday discovered that the plane of po-

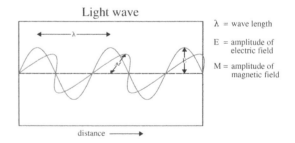

A 3–dimensional rendering of linearly polarised light wave frozen in time and showing the two oscillating components of light; an electric field and a magnetic field perpendicular to each other and to the direction of motion (a transverse wave).

larisation of linearly polarised light is rotated when the light rays travel along the magnetic field direction in the presence of a transparent dielectric, an effect now known as Faraday rotation.[32] This was the first evidence that light was related to electromagnetism. In 1846 he speculated that light might be some form of disturbance propagating along magnetic field lines.[32] Faraday proposed in 1847 that light was a high-frequency electromagnetic vibration, which could propagate even in the absence of a medium such as the ether.

Faraday's work inspired James Clerk Maxwell to study electromagnetic radiation and light. Maxwell discovered that self-propagating electromagnetic waves would travel through space at a constant speed, which happened to be equal to the previously measured speed of light. From this, Maxwell concluded that light was a form of electromagnetic radiation: he first stated this result in 1862 in *On Physical Lines of Force*. In 1873, he published *A Treatise on Electricity and Magnetism*, which contained a full mathematical description of the behaviour of electric and magnetic fields, still known as Maxwell's equations. Soon after, Heinrich Hertz confirmed Maxwell's theory experimentally by generating and detecting radio waves in the laboratory, and demonstrating that these waves behaved exactly like visible light, exhibiting properties such as reflection, refraction, diffraction, and interference. Maxwell's theory and Hertz's experiments led directly to the development of modern radio, radar, television, electromagnetic imaging, and wireless communications.

In the quantum theory, photons are seen as wave packets of the waves described in the classical theory of Maxwell. The quantum theory was needed to explain effects even with visual light that Maxwell's classical theory could not (such as spectral lines).

2.7.7 Quantum theory

In 1900 Max Planck, attempting to explain black body radiation suggested that although light was a wave, these waves could gain or lose energy only in finite amounts related to their frequency. Planck called these "lumps" of light energy "quanta" (from a Latin word for "how much"). In 1905, Albert Einstein used the idea of light quanta to explain the photoelectric effect, and suggested that these light quanta had a "real" existence. In 1923 Arthur Holly Compton showed that the wavelength shift seen when low intensity X-rays scattered from electrons (so called Compton scattering) could be explained by a particle-theory of X-rays, but not a wave theory. In 1926 Gilbert N. Lewis named these light quanta particles photons.

Eventually the modern theory of quantum mechanics came to picture light as (in some sense) *both* a particle and a wave, and (in another sense), as a phenomenon which is *neither* a particle nor a wave (which actually are macroscopic phenomena, such as baseballs or ocean waves). Instead, modern physics sees light as something that can be described sometimes with mathematics appropriate to one type of macroscopic metaphor (particles), and sometimes another macroscopic metaphor (water waves), but is actually something that cannot be fully imagined. As in the case for radio waves and the X-rays involved in Compton scattering, physicists have noted that electromagnetic radiation tends to behave more like a classical wave at lower frequencies, but more like a classical particle at higher frequencies, but never completely loses all qualities of one or the other. Visible light, which occupies a middle ground in frequency, can easily be shown in experiments to be describable using either a wave or particle model, or sometimes both.

2.8 See also

- Automotive lighting
- Ballistic photon
- Color temperature
- Fermat's principle
- Huygens' principle
- Incandescent light bulb
- International Commission on Illumination
- *Journal of Luminescence*
- Light art
- Light beam – in particular about light beams visible from the side
- Light Fantastic (TV series)
- Light mill
- Light Painting
- Light pollution
- Light therapy
- Lighting
- List of light sources
- *Luminescence: The Journal of Biological and Chemical Luminescence*
- Photic sneeze reflex
- Photometry
- Photon
- Rights of Light

- Risks and benefits of sun exposure
- Spectroscopy
- Visible spectrum
- Wave–particle duality

2.9 Notes

[1] Standards organizations recommend that radiometric quantities should be denoted with suffix "e" (for "energetic") to avoid confusion with photometric or photon quantities.

[2] Alternative symbols sometimes seen: *W* or *E* for radiant energy, *P* or *F* for radiant flux, *I* for irradiance, *W* for radiant exitance.

[3] Spectral quantities given per unit frequency are denoted with suffix "ν" (Greek)—not to be confused with suffix "v" (for "visual") indicating a photometric quantity.

[4] Spectral quantities given per unit wavelength are denoted with suffix "λ" (Greek).

[5] Directional quantities are denoted with suffix "Ω" (Greek).

[6] Standards organizations recommend that photometric quantities be denoted with a suffix "v" (for "visual") to avoid confusion with radiometric or photon quantities. For example: *USA Standard Letter Symbols for Illuminating Engineering* USAS Z7.1-1967, Y10.18-1967

[7] Alternative symbols sometimes seen: *W* for luminous energy, *P* or *F* for luminous flux, and ϱ or *K* for luminous efficacy.

[8] "J" here is the symbol for the dimension of luminous intensity, not the symbol for the unit joules.

2.10 References

[1] CIE (1987). *International Lighting Vocabulary*. Number 17.4. CIE, 4th edition. ISBN 978-3-900734-07-7.
By the *International Lighting Vocabulary*, the definition of *light* is: "Any radiation capable of causing a visual sensation directly."

[2] Pal, G. K.; Pal, Pravati (2001). "chapter 52". *Textbook of Practical Physiology* (1st ed.). Chennai: Orient Blackswan. p. 387. ISBN 978-81-250-2021-9. Retrieved 11 October 2013. The human eye has the ability to respond to all the wavelengths of light from 400–700 nm. This is called the visible part of the spectrum.

[3] Buser, Pierre A.; Imbert, Michel (1992). *Vision*. MIT Press. p. 50. ISBN 978-0-262-02336-8. Retrieved 11 October 2013. Light is a special class of radiant energy embracing wavelengths between 400 and 700 nm (or mμ), or 4000 to 7000 Å.

[4] Gregory Hallock Smith (2006). *Camera lenses: from box camera to digital*. SPIE Press. p. 4. ISBN 978-0-8194-6093-6.

[5] Narinder Kumar (2008). *Comprehensive Physics XII*. Laxmi Publications. p. 1416. ISBN 978-81-7008-592-8.

[6] Laufer, Gabriel (13 July 1996). *Introduction to Optics and Lasers in Engineering*. Cambridge University Press. p. 11. ISBN 978-0-521-45233-5. Retrieved 20 October 2013.

[7] Bradt, Hale (2004). *Astronomy Methods: A Physical Approach to Astronomical Observations*. Cambridge University Press. p. 26. ISBN 978-0-521-53551-9. Retrieved 20 October 2013.

[8] Ohannesian, Lena; Streeter, Anthony (9 November 2001). *Handbook of Pharmaceutical Analysis*. CRC Press. p. 187. ISBN 978-0-8247-4194-5. Retrieved 20 October 2013.

[9] Ahluwalia, V. K.; Goyal, Madhuri (1 January 2000). *A Textbook of Organic Chemistry*. Narosa. p. 110. ISBN 978-81-7319-159-6. Retrieved 20 October 2013.

[10] Sliney, David H.; Wangemann, Robert T.; Franks, James K.; Wolbarsht, Myron L. (1976). "Visual sensitivity of the eye to infrared laser radiation". *Journal of the Optical Society of America*. **66** (4): 339–341. doi:10.1364/JOSA.66.000339. (subscription required (help)). The foveal sensitivity to several near-infrared laser wavelengths was measured. It was found that the eye could respond to radiation at wavelengths at least as far as 1064 nm. A continuous 1064 nm laser source appeared red, but a 1060 nm pulsed laser source appeared green, which suggests the presence of second harmonic generation in the retina.

[11] Lynch, David K.; Livingston, William Charles (2001). *Color and Light in Nature* (2nd ed.). Cambridge, UK: Cambridge University Press. p. 231. ISBN 978-0-521-77504-5. Retrieved 12 October 2013. Limits of the eye's overall range of sensitivity extends from about 310 to 1050 nanometers

[12] Dash, Madhab Chandra; Dash, Satya Prakash (2009). *Fundamentals Of Ecology 3E*. Tata McGraw-Hill Education. p. 213. ISBN 978-1-259-08109-5. Retrieved 18 October 2013. Normally the human eye responds to light rays from 390 to 760 nm. This can be extended to a range of 310 to 1,050 nm under artificial conditions.

[13] Saidman, Jean (15 May 1933). "Sur la visibilité de l'ultraviolet jusqu'à la longueur d'onde 3130" [The visibility of the ultraviolet to the wave length of 3130]. *Comptes rendus de l'Académie des sciences* (in French). **196**: 1537–9.

[14] "Scientific Method, Statistical Method and the Speed of Light". *Statistical Science*. **15** (3): 254–278. 2000. doi:10.1214/ss/1009212817. MR 1847825.

[15] Michelson,, A. A. (January 1927). "Measurements of the velocity of light between Mount Wilson and Mount San Antonio". *Astrophysical Journal.* **65**: 1. Bibcode:1927ApJ....65....1M. doi:10.1086/143021. Retrieved 12 March 2014.

[16] Harvard News Office (2001-01-24). "Harvard Gazette: Researchers now able to stop, restart light". News.harvard.edu. Retrieved 2011-11-08.

[17] http://thulescientific.com/LYNCH%20&%20Soffer%20OPN%201999.pdf

[18] "Reference Solar Spectral Irradiance: Air Mass 1.5". Retrieved 2009-11-12.

[19] Tang, Hong (1 October 2009). "May The Force of Light Be With You". *IEEE Spectrum.* **46** (10): 46–51. doi:10.1109/MSPEC.2009.5268000.

[20] See, for example, nano-opto-mechanical systems research at Yale University.

[21] Kathy A. (2004-02-05). "Asteroids Get Spun By the Sun". *Discover Magazine.*

[22] "Solar Sails Could Send Spacecraft 'Sailing' Through Space". *NASA.* 2004-08-31.

[23] "NASA team successfully deploys two solar sail systems". *NASA.* 2004-08-09.

[24] P. Lebedev, Untersuchungen über die Druckkräfte des Lichtes, Ann. Phys. 6, 433 (1901).

[25] Nichols, E.F; Hull, G.F. (1903). "The Pressure due to Radiation". *The Astrophysical Journal.* **17** (5): 315–351. Bibcode:1903ApJ....17..315N. doi:10.1086/141035.

[26] Ptolemy and A. Mark Smith (1996). *Ptolemy's Theory of Visual Perception: An English Translation of the Optics with Introduction and Commentary.* Diane Publishing. p. 23. ISBN 0-87169-862-5.

[27] http://www.sifuae.com/sif/wp-content/uploads/2015/04/Shastra-Pratibha-2015-Seniors-Booklet.pdf

[28] *Theories of light, from Descartes to Newton* A. I. Sabra CUP Archive,1981 pg 48 ISBN 0-521-28436-8, ISBN 978-0-521-28436-3

[29] Fokko Jan Dijksterhuis, Lenses and Waves: Christiaan Huygens and the Mathematical Science of Optics in the 17th Century, Kluwer Academic Publishers, 2004, ISBN 1-4020-2697-8

[30] James R. Hofmann, *André-Marie Ampère: Enlightenment and Electrodynamics*, Cambridge University Press, 1996, p. 222.

[31] David Cassidy; Gerald Holton; James Rutherford (2002). *Understanding Physics.* Birkhäuser. ISBN 0-387-98756-8.

[32] Longair, Malcolm (2003). *Theoretical Concepts in Physics.* p. 87.

2.11 External links

- Media related to Light at Wikimedia Commons
- The dictionary definition of light at Wiktionary
- Quotations related to Light at Wikiquote

Chapter 3

Photon energy

The **photon energy** is the energy carried by a single photon with a certain electromagnetic wavelength and frequency. The higher the photon's frequency, the higher its energy. Equally, the longer the photon's wavelength, the lower its energy.

Photon energy is solely a function of the photon's wavelength. Other factors, such as the intensity of the radiation, do not affect photon energy. In other words, two photons of light with the same color (and, therefore, same frequency) will have the same photon energy, even if one was emitted from a wax candle and the other from the Sun.

The photon energy can be represented by any unit of energy. Among the units commonly used to denote photon energy are the electronvolt (eV) and the joule (as well as its multiples, such as the microjoule). As one joule equals 6.24 × 10^{18} eV, the larger units may be more useful in denoting the energy of photons with higher frequency and higher energy, such as gamma rays, as opposed to lower energy photons, such as those in the radiofrequency region of the electromagnetic spectrum.

Photons being massless, the notion of "photon energy" is not related to mass through the equivalence $E = mc^2$.

3.1 Formula

The equation for photon energy[1] is

$E = \frac{hc}{\lambda}$

Where E is the photon energy, h is the Planck constant, c is the speed of light in vacuum and λ is the photon's wavelength. As h and c are both constants, the photon energy changes with direct relation to wavelength λ.

To find the photon energy in electronvolts, using the wavelength in micrometres, the equation is approximately

$E(eV) = \frac{1.2398}{\lambda(\mu m)}$

Therefore, the photon energy at 1 μm wavelength (the wavelength of near infrared radiation) is approximately 1.2398 eV.

Since $\frac{c}{\lambda} = f$, where f is frequency, the photon energy equation can be simplified to

$E = hf$

This equation is known as the Planck-Einstein relation. Substituting h with its value in J·s and f with its value in hertz gives the photon energy in joules. Therefore, the photon energy at 1 Hz frequency is 6.62606957 × 10^{-34} joules or 4.135667516 × 10^{-15} eV.

3.2 Examples

An FM radio station transmitting at 100 MHz emits photons with an energy of about 4.1357 × 10^{-7} eV. This minuscule amount of energy is approximately 8 × 10^{-13} times the electron's mass (via the mass-energy equivalence).

The highest energy gamma rays detected to date, very-high-energy gamma rays, have a photon energy of 100 GeV to 100 TeV (10^{11} to 10^{14} electronvolts) or 0.016 microjoules to 0.016 millijoules. This corresponds to frequencies of 2.42 × 10^{25} to 2.42 × 10^{28} Hz.

A photon with a wavelength equal to the Planck length would have an energy of about 7.671 × 10^{28} eV or 1.229 × 10^{10} joules (12.29 gigajoules). This is roughly the amount of energy produced by the world's most powerful coal-fired power station, the Taichung Power Plant, during a period of 2.25 seconds.

3.3 See also

- Photon
- Electromagnetic radiation
- Electromagnetic spectrum
- Planck constant and Planck units

- Planck-Einstein relation

3.4 References

[1] "Energy of Photon". Photovoltaic Education Network, pveducation.org. Retrieved 2015-06-21.

Chapter 4

Photon polarization

Photon polarization is the quantum mechanical description of the classical polarized sinusoidal plane electromagnetic wave. Individual photon eigenstates have either right or left circular polarization. A photon that is in a superposition of eigenstates can have linear, circular, or elliptical polarization.

The description of photon polarization contains many of the physical concepts and much of the mathematical machinery of more involved quantum descriptions, such as the quantum mechanics of an electron in a potential well, and forms a fundamental basis for an understanding of more complicated quantum phenomena. Much of the mathematical machinery of quantum mechanics, such as state vectors, probability amplitudes, unitary operators, and Hermitian operators, emerge naturally from the classical Maxwell's equations in the description. The quantum polarization state vector for the photon, for instance, is identical with the Jones vector, usually used to describe the polarization of a classical wave. Unitary operators emerge from the classical requirement of the conservation of energy of a classical wave propagating through lossless media that alter the polarization state of the wave. Hermitian operators then follow for infinitesimal transformations of a classical polarization state.

Many of the implications of the mathematical machinery are easily verified experimentally. In fact, many of the experiments can be performed with two pairs (or one broken pair) of polaroid sunglasses.

The connection with quantum mechanics is made through the identification of a minimum packet size, called a photon, for energy in the electromagnetic field. The identification is based on the theories of Planck and the interpretation of those theories by Einstein. The correspondence principle then allows the identification of momentum and angular momentum (called spin), as well as energy, with the photon.

4.1 Polarization of classical electromagnetic waves

Main article: Polarization (waves)

4.1.1 Polarization states

Linear polarization

Main article: Linear polarization
The wave is linearly polarized (or plane polarized) when

Effect of a polarizer on reflection from mud flats. In the first picture, the polarizer is rotated to minimize the effect; in the second it is rotated 90° to maximize it: almost all reflected sunlight is eliminated.

the phase angles α_x, α_y are equal,

$$\alpha_x = \alpha_y \stackrel{\text{def}}{=} \alpha.$$

This represents a wave with phase α polarized at an angle θ with respect to the x axis. In that case the Jones vector can be written

$$|\psi\rangle = \begin{pmatrix} \cos\theta \\ \sin\theta \end{pmatrix} \exp(i\alpha).$$

The state vectors for linear polarization in x or y are special cases of this state vector.

If unit vectors are defined such that

$$|x\rangle \stackrel{\text{def}}{=} \begin{pmatrix} 1 \\ 0 \end{pmatrix}$$

and

$$|y\rangle \stackrel{\text{def}}{=} \begin{pmatrix} 0 \\ 1 \end{pmatrix}$$

then the linearly polarized polarization state can be written in the "x-y basis" as

$$|\psi\rangle = \cos\theta \exp(i\alpha) |x\rangle + \sin\theta \exp(i\alpha) |y\rangle = \psi_x |x\rangle + \psi_y |y\rangle$$

Circular polarization

Main article: Circular polarization

If the phase angles α_x and α_y differ by exactly $\pi/2$ and the x amplitude equals the y amplitude the wave is circularly polarized. The Jones vector then becomes

$$|\psi\rangle = \frac{1}{\sqrt{2}} \begin{pmatrix} 1 \\ \pm i \end{pmatrix} \exp(i\alpha_x)$$

where the plus sign indicates right circular polarization and the minus sign indicates left circular polarization. In the case of circular polarization, the electric field vector of constant magnitude rotates in the x-y plane.

If unit vectors are defined such that

$$|R\rangle \stackrel{\text{def}}{=} \frac{1}{\sqrt{2}} \begin{pmatrix} 1 \\ i \end{pmatrix}$$

and

$$|L\rangle \stackrel{\text{def}}{=} \frac{1}{\sqrt{2}} \begin{pmatrix} 1 \\ -i \end{pmatrix}$$

then an arbitrary polarization state can be written in the "R-L basis" as

$$|\psi\rangle = \psi_R |R\rangle + \psi_L |L\rangle$$

where

$$\psi_R = \langle R|\psi\rangle = \frac{1}{\sqrt{2}}(\cos\theta \exp(i\alpha_x) - i\sin\theta \exp(i\alpha_y))$$

and

$$\psi_L = \langle L|\psi\rangle = \frac{1}{\sqrt{2}}(\cos\theta \exp(i\alpha_x) + i\sin\theta \exp(i\alpha_y)).$$

We can see that

$$1 = |\psi_R|^2 + |\psi_L|^2.$$

Elliptical polarization

Main article: Elliptical polarization

The general case in which the electric field rotates in the x-y plane and has variable magnitude is called elliptical polarization. The state vector is given by

$$|\psi\rangle \stackrel{\text{def}}{=} \begin{pmatrix} \psi_x \\ \psi_y \end{pmatrix} = \begin{pmatrix} \cos\theta \exp(i\alpha_x) \\ \sin\theta \exp(i\alpha_y) \end{pmatrix}.$$

Geometric visualization of an arbitrary polarization state

To get an understanding of what a polarization state looks like, one can observe the orbit that is made if the polarization state is multiplied by a phase factor of $e^{i\omega t}$ and then having the real parts of its components interpreted as x and y coordinates respectively. That is:

$$\begin{pmatrix} x(t) \\ y(t) \end{pmatrix} = \begin{pmatrix} \Re(e^{i\omega t}\psi_x) \\ \Re(e^{i\omega t}\psi_y) \end{pmatrix} = \Re\left[e^{i\omega t}\begin{pmatrix} \psi_x \\ \psi_y \end{pmatrix}\right] = \Re\left(e^{i\omega t}|\psi\rangle\right).$$

If only the traced out shape and the direction of the rotation of (x(t), y(t)) is considered when interpreting the polarization state, i.e. only

$$M(|\psi\rangle) = \left\{\big(x(t), y(t)\big) \,\big|\, \forall t\right\}$$

(where x(t) and y(t) are defined as above) and whether it is overall more right circularly or left circularly polarized (i.e. whether |ψR| > |ψL| or vice versa), it can be seen that the physical interpretation will be the same even if the state is multiplied by an arbitrary phase factor, since

$$M(e^{i\alpha}|\psi\rangle) = M(|\psi\rangle),\ \alpha \in \mathbb{R}$$

and the direction of rotation will remain the same. In other words, there is no physical difference between two polarization states $|\psi\rangle$ and $e^{i\alpha}|\psi\rangle$, between which only a phase factor differs.

It can be seen that for a linearly polarized state, M will be a line in the xy plane, with length 2 and its middle in the origin, and whose slope equals to $\tan(\theta)$. For a circularly polarized state, M will be a circle with radius $1/\sqrt{2}$ and with the middle in the origin.

4.2 Energy, momentum, and angular momentum of a classical electromagnetic wave

4.2.1 Energy density of classical electromagnetic waves

Energy in a plane wave

Main article: Energy density

The energy per unit volume in classical electromagnetic fields is (cgs units)

$$\mathcal{E}_c = \frac{1}{8\pi}\left[\mathbf{E}^2(\mathbf{r},t) + \mathbf{B}^2(\mathbf{r},t)\right].$$

For a plane wave, this becomes

$$\mathcal{E}_c = \frac{|\mathbf{E}|^2}{8\pi}$$

where the energy has been averaged over a wavelength of the wave.

Fraction of energy in each component

The fraction of energy in the x component of the plane wave is

$$f_x = \frac{|\mathbf{E}|^2 \cos^2\theta}{|\mathbf{E}|^2} = \psi_x^*\psi_x = \cos^2\theta$$

with a similar expression for the y component resulting in $f_y = \sin^2\theta$.

The fraction in both components is

$$\psi_x^*\psi_x + \psi_y^*\psi_y = \langle\psi|\psi\rangle = 1.$$

4.2.2 Momentum density of classical electromagnetic waves

The momentum density is given by the Poynting vector

$$\mathcal{P} = \frac{1}{4\pi c}\mathbf{E}(\mathbf{r},t) \times \mathbf{B}(\mathbf{r},t).$$

For a sinusoidal plane wave traveling in the z direction, the momentum is in the z direction and is related to the energy density:

$$\mathcal{P}_z c = \mathcal{E}_c.$$

The momentum density has been averaged over a wavelength.

4.2.3 Angular momentum density of classical electromagnetic waves

Electromagnetic waves can have both orbital and spin angular momentum.[1] The total angular momentum density is

$$\mathcal{L} = \mathbf{r} \times \mathcal{P} = \frac{1}{4\pi c}\mathbf{r} \times [\mathbf{E}(\mathbf{r},t) \times \mathbf{B}(\mathbf{r},t)].$$

For a sinusoidal plane wave propagating along z axis the orbital angular momentum density vanishes. The spin angular momentum density is in the z direction and is given by

$$\mathcal{L} = \frac{|\mathbf{E}|^2}{8\pi\omega}\left(|\langle R|\psi\rangle|^2 - |\langle L|\psi\rangle|^2\right) = \frac{1}{\omega}\mathcal{E}_c\left(|\psi_R|^2 - |\psi_L|^2\right)$$

where again the density is averaged over a wavelength.

4.3 Optical filters and crystals

4.3.1 Passage of a classical wave through a polaroid filter

A linear filter transmits one component of a plane wave and absorbs the perpendicular component. In that case, if the

4.3. OPTICAL FILTERS AND CRYSTALS

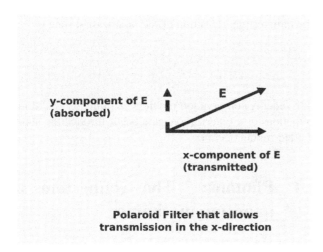

Linear polarization

filter is polarized in the x direction, the fraction of energy passing through the filter is

$$f_x = \psi_x^* \psi_x = \cos^2 \theta.$$

4.3.2 Example of energy conservation: Passage of a classical wave through a birefringent crystal

An ideal birefringent crystal transforms the polarization state of an electromagnetic wave without loss of wave energy. Birefringent crystals therefore provide an ideal test bed for examining the conservative transformation of polarization states. Even though this treatment is still purely classical, standard quantum tools such as unitary and Hermitian operators that evolve the state in time naturally emerge.

Initial and final states

A birefringent crystal is a material that has an **optic axis** with the property that the light has a different index of refraction for light polarized parallel to the axis than it has for light polarized perpendicular to the axis. Light polarized parallel to the axis are called "*extraordinary rays*" or "*extraordinary photons*", while light polarized perpendicular to the axis are called "*ordinary rays*" or "*ordinary photons*". If a linearly polarized wave impinges on the crystal, the extraordinary component of the wave will emerge from the crystal with a different phase than the ordinary component. In mathematical language, if the incident wave is linearly polarized at an angle θ with respect to the optic axis, the incident state vector can be written

$$|\psi\rangle = \begin{pmatrix} \cos \theta \\ \sin \theta \end{pmatrix}$$

and the state vector for the emerging wave can be written

$$|\psi'\rangle = \begin{pmatrix} \cos \theta \exp(i\alpha_x) \\ \sin \theta \exp(i\alpha_y) \end{pmatrix} = \begin{pmatrix} \exp(i\alpha_x) & 0 \\ 0 & \exp(i\alpha_y) \end{pmatrix} \begin{pmatrix} \cos \theta \\ \sin \theta \end{pmatrix} \stackrel{\text{def}}{=} \hat{U} |\psi\rangle.$$

While the initial state was linearly polarized, the final state is elliptically polarized. The birefringent crystal alters the character of the polarization.

Dual of the final state

A calcite crystal laid upon a paper with some letters showing the double refraction

The initial polarization state is transformed into the final state with the operator U. The dual of the final state is given by

$$\langle \psi' | = \langle \psi | \hat{U}^\dagger$$

where $U^\dagger$ is the adjoint of U, the complex conjugate transpose of the matrix.

Unitary operators and energy conservation

The fraction of energy that emerges from the crystal is

$$\langle \psi' | \psi' \rangle = \langle \psi | \hat{U}^\dagger \hat{U} | \psi \rangle = \langle \psi | \psi \rangle = 1.$$

In this ideal case, all the energy impinging on the crystal emerges from the crystal. An operator U with the property that

$$\hat{U}^\dagger \hat{U} = I,$$

where I is the identity operator and U is called a unitary operator. The unitary property is necessary to ensure energy conservation in state transformations.

Hermitian operators and energy conservation

Doubly refracting Calcite from Iceberg claim, Dixon, New Mexico. This 35 pound (16 kg) crystal, on display at the National Museum of Natural History, is one of the largest single crystals in the United States.

If the crystal is very thin, the final state will be only slightly different from the initial state. The unitary operator will be close to the identity operator. We can define the operator H by

$$\hat{U} \approx I + i\hat{H}$$

and the adjoint by

$$\hat{U}^\dagger \approx I - i\hat{H}^\dagger.$$

Energy conservation then requires

$$I = \hat{U}^\dagger \hat{U} \approx \left(I - i\hat{H}^\dagger\right)\left(I + i\hat{H}\right) \approx I - i\hat{H}^\dagger + i\hat{H}.$$

This requires that

$$\hat{H} = \hat{H}^\dagger.$$

Operators like this that are equal to their adjoints are called Hermitian or self-adjoint.

The infinitesimal transition of the polarization state is

$$|\psi'\rangle - |\psi\rangle = i\hat{H}|\psi\rangle.$$

Thus, energy conservation requires that infinitesimal transformations of a polarization state occur through the action of a Hermitian operator.

4.4 Photons: The connection to quantum mechanics

Main article: Photon

4.4.1 Energy, momentum, and angular momentum of photons

Energy

The treatment to this point has been classical. It is a testament, however, to the generality of Maxwell's equations for electrodynamics that the treatment can be made quantum mechanical with only a reinterpretation of classical quantities. The reinterpretation is based on the theories of Max Planck and the interpretation by Albert Einstein of those theories and of other experiments.

Einstein's conclusion from early experiments on the photoelectric effect is that electromagnetic radiation is composed of irreducible packets of energy, known as photons. The energy of each packet is related to the angular frequency of the wave by the relation

$$\epsilon = \hbar\omega$$

where $\hbar$ is an experimentally determined quantity known as Planck's constant. If there are N photons in a box of volume V, the energy in the electromagnetic field is

$$N\hbar\omega$$

and the energy density is

$$\frac{N\hbar\omega}{V}$$

The energy of a photon can be related to classical fields through the correspondence principle which states that for

4.4. PHOTONS: THE CONNECTION TO QUANTUM MECHANICS

a large number of photons, the quantum and classical treatments must agree. Thus, for very large N, the quantum energy density must be the same as the classical energy density

$$\frac{N\hbar\omega}{V} = \mathcal{E}_c = \frac{|\mathbf{E}|^2}{8\pi}.$$

The number of photons in the box is then

$$N = \frac{V}{8\pi\hbar\omega} |\mathbf{E}|^2.$$

Momentum

The correspondence principle also determines the momentum and angular momentum of the photon. For momentum

$$\mathcal{P}_z = \frac{N\hbar\omega}{cV} = \frac{N\hbar k_z}{V}$$

where k_z is the wave number. This implies that the momentum of a photon is

$$p_z = \hbar k_z.$$

Angular momentum and spin

Similarly for the spin angular momentum

$$\mathcal{L} = \frac{1}{\omega}\mathcal{E}_c \left(|\psi_R|^2 - |\psi_L|^2 \right) = \frac{N\hbar}{V} \left(|\psi_R|^2 - |\psi_L|^2 \right)$$

where E_c is field strength. This implies that the spin angular momentum of the photon is

$$l_z = \hbar \left(|\psi_R|^2 - |\psi_L|^2 \right).$$

the quantum interpretation of this expression is that the photon has a probability of $|\psi_R|^2$ of having a spin angular momentum of $\hbar$ and a probability of $|\psi_L|^2$ of having a spin angular momentum of $-\hbar$. We can therefore think of the spin angular momentum of the photon being quantized as well as the energy. The angular momentum of classical light has been verified.[2] Photons have only been observed to have spin angular momenta of $\pm\hbar$.

Spin operator The spin of the photon is defined as the coefficient of $\hbar$ in the spin angular momentum calculation. A photon has spin 1 if it is in the $|R\rangle$ state and -1 if it is in the $|L\rangle$ state. The spin operator is defined as the outer product

$$\hat{S} \stackrel{\text{def}}{=} |R\rangle\langle R| - |L\rangle\langle L| = \begin{pmatrix} 0 & -i \\ i & 0 \end{pmatrix}.$$

The eigenvectors of the spin operator are $|R\rangle$ and $|L\rangle$ with eigenvalues 1 and -1, respectively.

The expected value of a spin measurement on a photon is then

$$\langle\psi|\hat{S}|\psi\rangle = |\psi_R|^2 - |\psi_L|^2.$$

An operator S has been associated with an observable quantity, the spin angular momentum. The eigenvalues of the operator are the allowed observable values. This has been demonstrated for spin angular momentum, but it is in general true for any observable quantity.

Spin states See also: Circular polarization § Quantum mechanics

We can write the circularly polarized states as

$$|s\rangle$$

where $s=1$ for $|R\rangle$ and $s=-1$ for $|L\rangle$. An arbitrary state can be written

$$|\psi\rangle = \sum_{s=-1,1} a_s \exp(i\alpha_x - is\theta) |s\rangle$$

where

$$\sum_{s=-1,1} |a_s|^2 = 1.$$

Spin and angular momentum operators in differential form When the state is written in spin notation, the spin operator can be written

$$\hat{S}_d \to i\frac{\partial}{\partial\theta}$$

$$\hat{S}_d^\dagger \rightarrow -i\frac{\partial}{\partial \theta}.$$

The eigenvectors of the differential spin operator are

$$\exp(i\alpha_x - is\theta)|s\rangle.$$

To see this note

$$\hat{S}_d \exp(i\alpha_x - is\theta)|s\rangle \rightarrow i\frac{\partial}{\partial \theta}\exp(i\alpha_x - is\theta)|s\rangle = s\left[\exp(i\alpha_x - is\theta)|s\rangle\right].$$

The spin angular momentum operator is

$$\hat{l}_z = \hbar \hat{S}_d.$$

4.4.2 The nature of probability in quantum mechanics

Probability for a single photon

There are two ways in which probability can be applied to the behavior of photons; probability can be used to calculate the probable number of photons in a particular state, or probability can be used to calculate the likelihood of a single photon to be in a particular state. The former interpretation violates energy conservation. The latter interpretation is the viable, if nonintuitive, option. Dirac explains this in the context of the double-slit experiment:

> Some time before the discovery of quantum mechanics people realized that the connection between light waves and photons must be of a statistical character. What they did not clearly realize, however, was that the wave function gives information about the probability of **one** photon being in a particular place and not the probable number of photons in that place. The importance of the distinction can be made clear in the following way. Suppose we have a beam of light consisting of a large number of photons split up into two components of equal intensity. On the assumption that the beam is connected with the probable number of photons in it, we should have half the total number going into each component. If the two components are now made to interfere, we should require a photon in one component to be able to interfere with one in the other. Sometimes these two photons would have to annihilate one another and other times they would have to produce four photons. This would contradict the conservation of energy. The new theory, which connects the wave function with probabilities for one photon gets over the difficulty by making each photon go partly into each of the two components. Each photon then interferes only with itself. Interference between two different photons never occurs.
>
> —Paul Dirac, **The Principles of Quantum Mechanics,** Fourth Edition, Chapter 1

Probability amplitudes

The probability for a photon to be in a particular polarization state depends on the fields as calculated by the classical Maxwell's equations. The polarization state of the photon is proportional to the field. The probability itself is quadratic in the fields and consequently is also quadratic in the quantum state of polarization. In quantum mechanics, therefore, the state or probability amplitude contains the basic probability information. In general, the rules for combining probability amplitudes look very much like the classical rules for composition of probabilities: [The following quote is from Baym, Chapter 1]

1. The probability amplitude for two successive probabilities is the product of amplitudes for the individual possibilities. For example, the amplitude for the x polarized photon to be right circularly polarized **and** for the right circularly polarized photon to pass through the y-polaroid is $\langle R|x\rangle\langle y|R\rangle$, the product of the individual amplitudes.

2. The amplitude for a process that can take place in one of several **indistinguishable** ways is the sum of amplitudes for each of the individual ways. For example, the total amplitude for the x polarized photon to pass through the y-polaroid is the sum of the amplitudes for it to pass as a right circularly polarized photon, $\langle y|R\rangle\langle R|x\rangle$, plus the amplitude for it to pass as a left circularly polarized photon, $\langle y|L\rangle\langle L|x\rangle\ldots$

3. The total probability for the process to occur is the absolute value squared of the total amplitude calculated by 1 and 2.

4.4.3 Uncertainty principle

Main article: Uncertainty principle

4.4. PHOTONS: THE CONNECTION TO QUANTUM MECHANICS

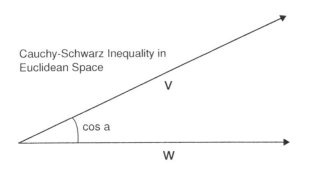

Cauchy-Schwarz inequality in Euclidean space. $\mathbf{V} \cdot \mathbf{W} = \|\mathbf{V}\|\|\mathbf{W}\| \cos a$. *This implies* $\mathbf{V} \cdot \mathbf{W} \leq \|\mathbf{V}\|\|\mathbf{W}\|$.

Mathematical preparation

For any legal operators the following inequality, a consequence of the Cauchy-Schwarz inequality, is true.

$$\frac{1}{4}|\langle(\hat{A}\hat{B} - \hat{B}\hat{A})x|x\rangle|^2 \leq \|\hat{A}x\|^2 \|\hat{B}x\|^2.$$

If $B A \psi$ and $A B \psi$ are defined, then by subtracting the means and re-inserting in the above formula, we deduce

$$\Delta_\psi \hat{A} \, \Delta_\psi \hat{B} \geq \frac{1}{2} \left| \left\langle \left[\hat{A}, \hat{B}\right] \right\rangle_\psi \right|$$

where

$$\left\langle \hat{X} \right\rangle_\psi = \left\langle \psi | \hat{X} \psi \right\rangle$$

is the operator mean of observable X in the system state ψ and

$$\Delta_\psi \hat{X} = \sqrt{\langle \hat{X}^2 \rangle_\psi - \langle \hat{X} \rangle_\psi^2}.$$

Here

$$\left[\hat{A}, \hat{B}\right] \stackrel{\text{def}}{=} \hat{A}\hat{B} - \hat{B}\hat{A}$$

is called the commutator of A and B.

This is a purely mathematical result. No reference has been made to any physical quantity or principle. It simply states that the uncertainty of one operator times the uncertainty of another operator has a lower bound.

Application to angular momentum

The connection to physics can be made if we identify the operators with physical operators such as the angular momentum and the polarization angle. We have then

$$\Delta_\psi \hat{l}_z \, \Delta_\psi \theta \geq \frac{\hbar}{2},$$

which simply states that angular momentum **and** the polarization angle cannot be measured simultaneously with infinite accuracy.

4.4.4 States, probability amplitudes, unitary and Hermitian operators, and eigenvectors

Much of the mathematical apparatus of quantum mechanics appears in the classical description of a polarized sinusoidal electromagnetic wave. The Jones vector for a classical wave, for instance, is identical with the quantum polarization state vector for a photon. The right and left circular components of the Jones vector can be interpreted as probability amplitudes of spin states of the photon. Energy conservation requires that the states be transformed with a unitary operation. This implies that infinitesimal transformations are transformed with a Hermitian operator. These conclusions are a natural consequence of the structure of Maxwell's equations for classical waves.

Quantum mechanics enters the picture when observed quantities are measured and found to be discrete rather than continuous. The allowed observable values are determined by the eigenvalues of the operators associated with the observable. In the case angular momentum, for instance, the allowed observable values are the eigenvalues of the spin operator.

These concepts have emerged naturally from Maxwell's equations and Planck's and Einstein's theories. They have been found to be true for many other physical systems. In fact, the typical program is to assume the concepts of this section and then to infer the unknown dynamics of a physical system. This was done, for instance, with the dynamics of electrons. In that case, working back from the principles in this section, the quantum dynamics of particles were inferred, leading to Schrödinger's equation, a departure from Newtonian mechanics. The solution of this equation for atoms led to the explanation of the Balmer series for atomic spectra and consequently formed a basis for all of atomic physics and chemistry.

This is not the only occasion in which Maxwell's equations have forced a restructuring of Newtonian mechanics. Maxwell's equations are relativistically consistent. Special relativity resulted from attempts to make classical mechanics consistent with Maxwell's equations (see, for example, Moving magnet and conductor problem).

4.5 See also

- Angular momentum of light
  - Spin angular momentum of light
  - Orbital angular momentum of light
- Quantum decoherence
- Stern–Gerlach experiment
- Wave–particle duality
- Double-slit experiment
- Theoretical and experimental justification for the Schrödinger equation
- Spin polarization

4.6 References

[1] Allen, L.; Beijersbergen, M.W.; Spreeuw, R.J.C.; Woerdman, J.P. (June 1992). "Orbital angular momentum of light and the transformation of Laguerre-Gaussian laser modes". *Physical Review A.* **45** (11): 8186–9. Bibcode:1992PhRvA..45.8185A. doi:10.1103/PhysRevA.45.8185. PMID 9906912.

[2] Beth, R.A. (1935). "Direct detection of the angular momentum of light". *Phys. Rev.* **48** (5): 471. Bibcode:1935PhRv...48..471B. doi:10.1103/PhysRev.48.471.

4.7 Further reading

- Jackson, John D. (1998). *Classical Electrodynamics* (3rd ed.). Wiley. ISBN 0-471-30932-X.
- Baym, Gordon (1969). *Lectures on Quantum Mechanics*. W. A. Benjamin. ISBN 0-8053-0667-6.
- Dirac, P. A. M. (1958). *The Principles of Quantum Mechanics* (Fourth ed.). Oxford. ISBN 0-19-851208-2.

Chapter 5

Photon counting

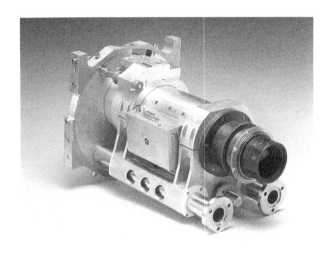

A prototype single-photon detector that was used on the 200-inch Hale Telescope. The Hubble Space Telescope has a similar detector.

Photon counting is a technique in which individual photons are counted using some **single-photon detector** (SPD). The counting efficiency is determined by the quantum efficiency and any electronic losses that are present in the system.

Many photodetectors can be configured to detect individual photons, each with relative advantages and disadvantages,[1] including a photomultiplier, geiger counter, single-photon avalanche diode, superconducting nanowire single-photon detector, transition edge sensor, or scintillation counter. Charge-coupled devices can also sometimes be used.

Single-photon detection is useful in many fields including interplanetary communications,[2] fiber-optic communication, quantum information science, quantum encryption, medical imaging, light detection and ranging, DNA sequencing, astrophysics, and materials science.[1]

5.1 Measured quantities

The number of photons observed per unit time is the photon flux. The photon flux per unit area is the photon irradiance if the photons are incident on a surface, or photon exitance if the emission of photons from a broad-area source is being considered. The flux per unit solid angle is the photon intensity. The flux per unit source area per unit solid angle is photon radiance. SI units for these quantities are summarized in the table below.

[1] Standards organizations recommend that photon quantities be denoted with a suffix "q" (for "quantum") to avoid confusion with radiometric and photometric quantities.

[2] The energy of a single photon at wavelength λ is $Q_p = h \cdot c / \lambda$ with h = Planck's constant and c = velocity of light.

5.2 See also

- Single-photon source
- Shot noise
- Visible-light photon counter
- Transition edge sensor
- Superconducting nanowire single-photon detector
- Time-correlated single photon counting
- Oversampled binary image sensor

5.3 References

[1] Francesco Marsili. "High Efficiency in the Fastest Single-Photon Detector System". 2013.

[2] V. B. Verma, R. Horansky, F. Marsili, J. A. Stern, M. D. Shaw, A. E. Lita, R. P. Mirin, S. W. Nam. "A four-pixel single-photon pulse-position camera fabricated from WSi superconducting nanowire single-photon detectors". arXiv: 1311.1553 [physics.ins-det] . 2013.

Chapter 6

Photonics

Dispersion of light (photons) by a prism.

Photonics is the science of light (photon) generation, detection, and manipulation through emission, transmission, modulation, signal processing, switching, amplification, and detection/sensing.[1][2] Though covering all light's technical applications over the whole spectrum, most photonic applications are in the range of visible and near-infrared light. The term photonics developed as an outgrowth of the first practical semiconductor light emitters invented in the early 1960s and optical fibers developed in the 1970s.

6.1 History of photonics

The word 'photonics' is derived from the Greek word "photos" meaning light; it appeared in the late 1960s to describe a research field whose goal was to use light to perform functions, that traditionally fell within the typical domain of electronics, such as telecommunications, information processing, etc.

Photonics as a field began with the invention of the laser in 1960. Other developments followed: the laser diode in the 1970s, optical fibers for transmitting information, and the erbium-doped fiber amplifier. These inventions formed the basis for the telecommunications revolution of the late 20th century and provided the infrastructure for the Internet.

Though coined earlier, the term photonics came into common use in the 1980s as fiber-optic data transmission was adopted by telecommunications network operators. At that time, the term was used widely at Bell Laboratories. Its use was confirmed when the IEEE Lasers and Electro-Optics Society established an archival journal named Photonics Technology Letters at the end of the 1980s.

During the period leading up to the dot-com crash circa 2001, photonics as a field focused largely on optical telecommunications. However, photonics covers a huge range of science and technology applications, including laser manufacturing, biological and chemical sensing, medical diagnostics and therapy, display technology, and optical computing. Further growth of photonics is likely if current silicon photonics developments are successful.

6.2 Relationship to other fields

6.2.1 Classical optics

Photonics is closely related to optics. Classical optics long preceded the discovery that light is quantized, when Albert Einstein famously explained the photoelectric effect in 1905. Optics tools include the refracting lens, the reflecting mirror, and various optical components and instruments developed throughout the 15th to 19th centuries. Key tenets of classical optics, such as Huygens Principle, developed in the 17th century, Maxwell's Equations and the wave equations, developed in the 19th, do not depend on quantum properties of light.

6.2.2 Modern optics

Photonics is related to quantum optics, optomechanics, electro-optics, optoelectronics and quantum electronics. However, each area has slightly different connotations by scientific and government communities and in the marketplace. Quantum optics often connotes fundamental research, whereas photonics is used to connote applied research and development.

The term *photonics* more specifically connotes:

- The particle properties of light,
- The potential of creating signal processing device technologies using photons,
- The practical application of optics, and
- An analogy to electronics.

The term optoelectronics connotes devices or circuits that comprise both electrical and optical functions, i.e., a thin-film semiconductor device. The term electro-optics came into earlier use and specifically encompasses nonlinear electrical-optical interactions applied, e.g., as bulk crystal modulators such as the Pockels cell, but also includes advanced imaging sensors typically used for surveillance by civilian or government organizations.

6.2.3 Emerging fields

Photonics also relates to the emerging science of quantum information and quantum optics, in those cases where it employs photonic methods. Other emerging fields include optomechanics, which involves the study of the interaction between light and mechanical vibrations of mesoscopic or macroscopic objects; opto-atomics, in which devices integrate both photonic and atomic devices for applications such as precision timekeeping, navigation, and metrology; polaritonics, which differs from photonics in that the fundamental information carrier is a polariton, which is a mixture of photons and phonons, and operates in the range of frequencies from 300 gigahertz to approximately 10 terahertz.

6.3 Applications

Applications of photonics are ubiquitous. Included are all areas from everyday life to the most advanced science, e.g. light detection, telecommunications, information processing, lighting, metrology, spectroscopy, holography, medicine (surgery, vision correction, endoscopy, health monitoring), military technology, laser material processing, visual art, biophotonics, agriculture, and robotics.

Just as applications of electronics have expanded dramatically since the first transistor was invented in 1948, the unique applications of photonics continue to emerge. Economically important applications for semiconductor photonic devices include optical data recording, fiber optic telecommunications, laser printing (based on xerography),

A *sea mouse* (Aphrodita aculeata),[3] *showing colorful spines, a remarkable example of photonic engineering by a living organism*

displays, and optical pumping of high-power lasers. The potential applications of photonics are virtually unlimited and include chemical synthesis, medical diagnostics, on-chip data communication, laser defense, and fusion energy, to name several interesting additional examples.

- Consumer equipment: barcode scanner, printer, CD/DVD/Blu-ray devices, remote control devices
- Telecommunications: optical fiber communications, optical down converter to microwave
- Medicine: correction of poor eyesight, laser surgery, surgical endoscopy, tattoo removal
- Industrial manufacturing: the use of lasers for welding, drilling, cutting, and various methods of surface modification
- Construction: laser leveling, laser rangefinding, smart structures
- Aviation: photonic gyroscopes lacking mobile parts
- Military: IR sensors, command and control, navigation, search and rescue, mine laying and detection
- Entertainment: laser shows, beam effects, holographic art
- Information processing
- Metrology: time and frequency measurements, rangefinding
- Photonic computing: clock distribution and communication between computers, printed circuit boards, or within optoelectronic integrated circuits; in the future: quantum computing

Microphotonics and nanophotonics usually includes photonic crystals and solid state devices.[4]

6.4 Overview of photonics research

The science of photonics includes investigation of the emission, transmission, amplification, detection, and modulation of light.

6.4.1 Light sources

Light sources used in photonics are usually far more sophisticated than light bulbs. Photonics commonly uses semiconductor light sources like light-emitting diodes (LEDs), superluminescent diodes, and lasers. Other light sources include single photon sources, fluorescent lamps, cathode ray tubes (CRTs), and plasma screens. Note that while CRTs, plasma screens, and organic light-emitting diode displays generate their own light, liquid crystal displays (LCDs) like TFT screens require a backlight of either cold cathode fluorescent lamps or, more often today, LEDs.

Characteristic for research on semiconductor light sources is the frequent use of III-V semiconductors instead of the classical semiconductors like silicon and germanium. This is due to the special properties of III-V semiconductors that allow for the implementation of light emitting devices. Examples for material systems used are gallium arsenide (GaAs) and aluminium gallium arsenide (AlGaAs) or other compound semiconductors. They are also used in conjunction with silicon to produce hybrid silicon lasers.

6.4.2 Transmission media

Light can be transmitted through any transparent medium. Glass fiber or plastic optical fiber can be used to guide the light along a desired path. In optical communications optical fibers allow for transmission distances of more than 100 km without amplification depending on the bit rate and modulation format used for transmission. A very advanced research topic within photonics is the investigation and fabrication of special structures and "materials" with engineered optical properties. These include photonic crystals, photonic crystal fibers and metamaterials.

6.4.3 Amplifiers

Optical amplifiers are used to amplify an optical signal. Optical amplifiers used in optical communications are erbium-doped fiber amplifiers, semiconductor optical amplifiers, Raman amplifiers and optical parametric amplifiers. A very advanced research topic on optical amplifiers is the research on quantum dot semiconductor optical amplifiers.

6.4.4 Detection

Photodetectors detect light. Photodetectors range from very fast photodiodes for communications applications over medium speed charge coupled devices (CCDs) for digital cameras to very slow solar cells that are used for energy harvesting from sunlight. There are also many other photodetectors based on thermal, chemical, quantum, photoelectric and other effects.

6.4.5 Modulation

Modulation of a light source is used to encode information on a light source. Modulation can be achieved by the light source directly. One of the simplest examples is to use a flashlight to send Morse code. Another method is to take the light from a light source and modulate it in an external optical modulator.

An additional topic covered by modulation research is the modulation format. On-off keying has been the commonly used modulation format in optical communications. In the last years more advanced modulation formats like phase-shift keying or even orthogonal frequency-division multiplexing have been investigated to counteract effects like dispersion that degrade the quality of the transmitted signal.

6.4.6 Photonic systems

Photonics also includes research on photonic systems. This term is often used for optical communication systems. This area of research focuses on the implementation of photonic systems like high speed photonic networks. This also includes research on optical regenerators, which improve optical signal quality.

6.4.7 Photonic integrated circuits

Photonic integrated circuits (PICs) are optically active integrated semiconductor photonic devices which consist of at least two different functional blocks, (gain region and a grating based mirror in a laser...). These devices are responsible for commercial successes of optical communications and the ability to increase the available bandwidth without significant cost increases to the end user, through improved performance and cost reduction that they provide. The most widely deployed PICs are based on Indium phosphide material system. Silicon photonics is an active area of research.

6.5 See also

- Nano-optics
- Optronics/optoelectronics
- Organic photonics
- Photonics mast (on submarines)
- European Photonics Industry Consortium
- Photonics21 - a voluntary association of industrial enterprises and other stakeholders in the field of photonics in Europe

6.6 References

[1] Chai Yeh (2 December 2012). *Applied Photonics*. Elsevier. pp. 1–. ISBN 978-0-08-049926-0.

[2] Richard S. Quimby (14 April 2006). *Photonics and Lasers: An Introduction*. John Wiley & Sons. ISBN 978-0-471-79158-4.

[3] "Sea mouse promises bright future". BBC News. 2001-01-03. Retrieved 2013-05-05.

[4] Hervé Rigneault; Jean-Michel Lourtioz; Claude Delalande; Ariel Levenson (5 January 2010). *Nanophotonics*. John Wiley & Sons. pp. 5–. ISBN 978-0-470-39459-5. Cite uses deprecated parameter |coauthors= (help)

Chapter 7

Electromagnetic radiation

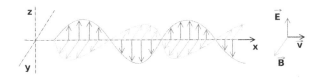

The electromagnetic waves that compose electromagnetic radiation can be imagined as a self-propagating transverse oscillating wave of electric and magnetic fields. This diagram shows a plane linearly polarized EMR wave propagating from left to right. The electric field is in a vertical plane and the magnetic field in a horizontal plane. The electric and magnetic fields in EMR waves are always in phase and at 90 degrees to each other.

Electromagnetic radiation (**EM radiation** or **EMR**) is the radiant energy released by certain electromagnetic processes. Visible light is an electromagnetic radiation. Other familiar electromagnetic radiations are invisible to the human eye, such as radio waves, infrared light and X-rays.

Classically, electromagnetic radiation consists of **electromagnetic waves**, which are synchronized oscillations of electric and magnetic fields that propagate at the speed of light through a vacuum. The oscillations of the two fields are perpendicular to each other and perpendicular to the direction of energy and wave propagation, forming a transverse wave. Electromagnetic waves can be characterized by either the frequency or wavelength of their oscillations to form the electromagnetic spectrum, which includes, in order of increasing frequency and decreasing wavelength: radio waves, microwaves, infrared radiation, visible light, ultraviolet radiation, X-rays and gamma rays.

Electromagnetic waves are produced whenever charged particles are accelerated, and these waves can subsequently interact with any charged particles. EM waves carry energy, momentum and angular momentum away from their source particle and can impart those quantities to matter with which they interact. Quanta of EM waves are called photons, which are massless, but they are still affected by gravity. Electromagnetic radiation is associated with those EM waves that are free to propagate themselves ("radiate") without the continuing influence of the moving charges that produced them, because they have achieved sufficient distance from those charges. Thus, EMR is sometimes referred to as the far field. In this language, the *near field* refers to EM fields near the charges and current that directly produced them, specifically, electromagnetic induction and electrostatic induction phenomena.

In the quantum theory of electromagnetism, EMR consists of photons, the elementary particles responsible for all electromagnetic interactions. Quantum effects provide additional sources of EMR, such as the transition of electrons to lower energy levels in an atom and black-body radiation. The energy of an individual photon is quantized and is greater for photons of higher frequency. This relationship is given by Planck's equation $E = h\nu$, where E is the energy per photon, ν is the frequency of the photon, and h is Planck's constant. A single gamma ray photon, for example, might carry ~100,000 times the energy of a single photon of visible light.

The effects of EMR upon biological systems (and also to many other chemical systems, under standard conditions) depend both upon the radiation's power and its frequency. For EMR of visible frequencies or lower (i.e., radio, microwave, infrared), the damage done to cells and other materials is determined mainly by power and caused primarily by heating effects from the combined energy transfer of many photons. By contrast, for ultraviolet and higher frequencies (i.e., X-rays and gamma rays), chemical materials and living cells can be further damaged beyond that done by simple heating, since individual photons of such high frequency have enough energy to cause direct molecular damage.

7.1 Physics

7.1.1 Theory

Main articles: Maxwell's equations and Near and far field

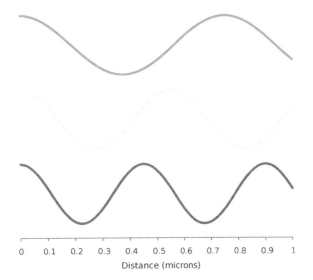

Shows the relative wavelengths of the electromagnetic waves of three different colours of light (blue, green, and red) with a distance scale in micrometers along the x-axis.

Maxwell's equations

Maxwell derived a wave form of the electric and magnetic equations, thus uncovering the wave-like nature of electric and magnetic fields and their symmetry. Because the speed of EM waves predicted by the wave equation coincided with the measured speed of light, Maxwell concluded that light itself is an EM wave. Maxwell's equations were confirmed by Heinrich Hertz through experiments with radio waves.

According to Maxwell's equations, a spatially varying electric field is always associated with a magnetic field that changes over time. Likewise, a spatially varying magnetic field is associated with specific changes over time in the electric field. In an electromagnetic wave, the changes in the electric field are always accompanied by a wave in the magnetic field in one direction, and vice versa. This relationship between the two occurs without either type field causing the other; rather, they occur together in the same way that time and space changes occur together and are interlinked in special relativity. In fact, magnetic fields may be viewed as relativistic distortions of electric fields, so the close relationship between space and time changes here is more than an analogy. Together, these fields form a propagating electromagnetic wave, which moves out into space and need never again affect the source. The distant EM field formed in this way by the acceleration of a charge carries energy with it that "radiates" away through space, hence the term.

Near and far fields

Main articles: Near and far field and Liénard–Wiechert potential

Maxwell's equations established that some charges and cur-

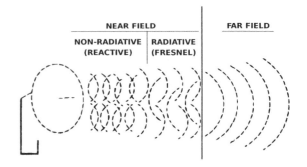

In electromagnetic radiation (such as microwaves from an antenna, shown here) the term applies only to the parts of the electromagnetic field that radiate into infinite space and decrease in intensity by an inverse-square law of power, so that the total radiation energy that crosses through an imaginary spherical surface is the same, no matter how far away from the antenna the spherical surface is drawn. Electromagnetic radiation thus includes the far field part of the electromagnetic field around a transmitter. A part of the "near-field" close to the transmitter, forms part of the changing electromagnetic field, but does not count as electromagnetic radiation.

rents ("sources") produce a local type of electromagnetic field near them that does *not* have the behaviour of EMR. Currents directly produce a magnetic field, but it is of a magnetic dipole type that dies out with distance from the current. In a similar manner, moving charges pushed apart in a conductor by a changing electrical potential (such as in an antenna) produce an electric dipole type electrical field, but this also declines with distance. These fields make up the near-field near the EMR source. Neither of these behaviours are responsible for EM radiation. Instead, they cause electromagnetic field behaviour that only efficiently transfers power to a receiver very close to the source, such as the magnetic induction inside a transformer, or the feedback behaviour that happens close to the coil of a metal detector. Typically, near-fields have a powerful effect on their own sources, causing an increased "load" (decreased electrical reactance) in the source or transmitter, whenever energy is withdrawn from the EM field by a receiver. Otherwise, these fields do not "propagate" freely out into space, carrying their energy away without distance-limit, but rather oscillate, returning their energy to the transmitter if it is not received by a receiver.

By contrast, the EM far-field is composed of *radiation* that is free of the transmitter in the sense that (unlike the case in an electrical transformer) the transmitter requires the same power to send these changes in the fields out, whether the signal is immediately picked up or not. This distant part of the electromagnetic field *is* "electromagnetic radiation" (also called the far-field). The far-fields propagate (radiate) without allowing the transmitter to affect them. This causes them to be independent in the sense that their existence and their energy, after they have left the transmitter, is completely independent of both transmitter and receiver. Because such waves conserve the amount of energy they transmit through any spherical boundary surface drawn around their source, and because such surfaces have an area that is defined by the square of the distance from the source, the power of EM radiation always varies according to an inverse-square law. This is in contrast to dipole parts of the EM field close to the source (the near-field), which varies in power according to an inverse cube power law, and thus does *not* transport a conserved amount of energy over distances, but instead fades with distance, with its energy (as noted) rapidly returning to the transmitter or absorbed by a nearby receiver (such as a transformer secondary coil).

The far-field (EMR) depends on a different mechanism for its production than the near-field, and upon different terms in Maxwell's equations. Whereas the magnetic part of the near-field is due to currents in the source, the magnetic field in EMR is due only to the local change in the electric field. In a similar way, while the electric field in the near-field is due directly to the charges and charge-separation in the source, the electric field in EMR is due to a change in the local magnetic field. Both processes for producing electric and magnetic EMR fields have a different dependence on distance than do near-field dipole electric and magnetic fields. That is why the EMR type of EM field becomes dominant in power "far" from sources. The term "far from sources" refers to how far from the source (moving at the speed of light) any portion of the outward-moving EM field is located, by the time that source currents are changed by the varying source potential, and the source has therefore begun to generate an outwardly moving EM field of a different phase.

A more compact view of EMR is that the far-field that composes EMR is generally that part of the EM field that has traveled sufficient distance from the source, that it has become completely disconnected from any feedback to the charges and currents that were originally responsible for it. Now independent of the source charges, the EM field, as it moves farther away, is dependent only upon the accelerations of the charges that produced it. It no longer has a strong connection to the direct fields of the charges, or to the velocity of the charges (currents).

In the Liénard–Wiechert potential formulation of the electric and magnetic fields due to motion of a single particle (according to Maxwell's equations), the terms associated with acceleration of the particle are those that are responsible for the part of the field that is regarded as electromagnetic radiation. By contrast, the term associated with the changing static electric field of the particle and the magnetic term that results from the particle's uniform velocity, are both associated with the electromagnetic near-field, and do not comprise EM radiation.

7.1.2 Properties

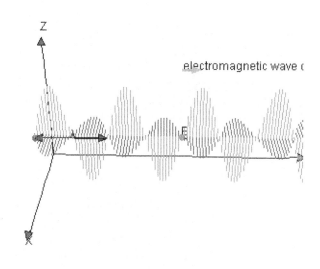

Electromagnetic waves can be imagined as a self-propagating transverse oscillating wave of electric and magnetic fields. This 3D animation shows a plane linearly polarized wave propagating from left to right. Note that the electric and magnetic fields in such a wave are in-phase with each other, reaching minima and maxima together

Electrodynamics is the physics of electromagnetic radiaton, and electromagnetism is the physical phenomenon associated with the theory of electrodynamics. Electric and magnetic fields obey the properties of superposition. Thus, a field due to any particular particle or time-varying electric or magnetic field contributes to the fields present in the same space due to other causes. Further, as they are vector fields, all magnetic and electric field vectors add together according to vector addition. For example, in optics two or more coherent lightwaves may interact and by constructive or destructive interference yield a resultant irradiance deviating from the sum of the component irradiances of the individual lightwaves.

Since light is an oscillation it is not affected by travelling through static electric or magnetic fields in a linear medium such as a vacuum. However, in nonlinear media, such as some crystals, interactions can occur between light and static electric and magnetic fields — these interactions include the Faraday effect and the Kerr effect.

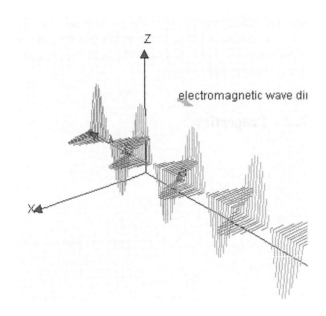

An alternate view of the wave shown above.

In refraction, a wave crossing from one medium to another of different density alters its speed and direction upon entering the new medium. The ratio of the refractive indices of the media determines the degree of refraction, and is summarized by Snell's law. Light of composite wavelengths (natural sunlight) disperses into a visible spectrum passing through a prism, because of the wavelength-dependent refractive index of the prism material (dispersion); that is, each component wave within the composite light is bent a different amount.

EM radiation exhibits both wave properties and particle properties at the same time (see wave-particle duality). Both wave and particle characteristics have been confirmed in many experiments. Wave characteristics are more apparent when EM radiation is measured over relatively large timescales and over large distances while particle characteristics are more evident when measuring small timescales and distances. For example, when electromagnetic radiation is absorbed by matter, particle-like properties will be more obvious when the average number of photons in the cube of the relevant wavelength is much smaller than 1. It is not too difficult to experimentally observe non-uniform deposition of energy when light is absorbed, however this alone is not evidence of "particulate" behavior. Rather, it reflects the quantum nature of *matter*.[1] Demonstrating that the light itself is quantized, not merely its interaction with matter, is a more subtle affair.

Some experiments display both the wave and particle natures of electromagnetic waves, such as the self-interference of a single photon.[2] When a single photon is sent through an interferometer, it passes through both paths, interfering with itself, as waves do, yet is detected by a photomultiplier or other sensitive detector only once.

A quantum theory of the interaction between electromagnetic radiation and matter such as electrons is described by the theory of quantum electrodynamics.

Electromagnetic waves can be polarized, reflected, refracted, diffracted or interfere with each other.

7.1.3 Wave model

Electromagnetic radiation is a transverse wave, meaning that its oscillations are perpendicular to the direction of energy transfer and travel. The electric and magnetic parts of the field stand in a fixed ratio of strengths in order to satisfy the two Maxwell equations that specify how one is produced from the other. These **E** and **B** fields are also in phase, with both reaching maxima and minima at the same points in space (see illustrations). A common misconception is that the **E** and **B** fields in electromagnetic radiation are out of phase because a change in one produces the other, and this would produce a phase difference between them as sinusoidal functions (as indeed happens in electromagnetic induction, and in the near-field close to antennas). However, in the far-field EM radiation which is described by the two source-free Maxwell curl operator equations, a more correct description is that a time-change in one type of field is proportional to a space-change in the other. These derivatives require that the **E** and **B** fields in EMR are in-phase (see math section below).

An important aspect of light's nature is its frequency. The frequency of a wave is its rate of oscillation and is measured in hertz, the SI unit of frequency, where one hertz is equal to one oscillation per second. Light usually has multiple frequencies that sum to form the resultant wave. Different frequencies undergo different angles of refraction, a phenomenon known as dispersion.

A wave consists of successive troughs and crests, and the distance between two adjacent crests or troughs is called the wavelength. Waves of the electromagnetic spectrum vary in size, from very long radio waves the size of buildings to very short gamma rays smaller than atom nuclei. Frequency is inversely proportional to wavelength, according to the equation:

$$v = f\lambda$$

where v is the speed of the wave (c in a vacuum, or less in other media), f is the frequency and λ is the wavelength. As waves cross boundaries between different media, their speeds change but their frequencies remain constant.

Electromagnetic waves in free space must be solutions of Maxwell's electromagnetic wave equation. Two main classes of solutions are known, namely plane waves and spherical waves. The plane waves may be viewed as the limiting case of spherical waves at a very large (ideally infinite) distance from the source. Both types of waves can have a waveform which is an arbitrary time function (so long as it is sufficiently differentiable to conform to the wave equation). As with any time function, this can be decomposed by means of Fourier analysis into its frequency spectrum, or individual sinusoidal components, each of which contains a single frequency, amplitude and phase. Such a component wave is said to be *monochromatic*. A monochromatic electromagnetic wave can be characterized by its frequency or wavelength, its peak amplitude, its phase relative to some reference phase, its direction of propagation and its polarization.

Interference is the superposition of two or more waves resulting in a new wave pattern. If the fields have components in the same direction, they constructively interfere, while opposite directions cause destructive interference. An example of interference caused by EMR is electromagnetic interference (EMI) or as it is more commonly known as, radio-frequency interference (RFI). Additionally, multiple polarization signals can be combined (i.e. interfered) to form new states of polarization, which is known as parallel polarization state generation. [3]

The energy in electromagnetic waves is sometimes called radiant energy.

7.1.4 Particle model and quantum theory

See also: Quantization (physics) and Quantum optics

An anomaly arose in the late 19th century involving a contradiction between the wave theory of light and measurements of the electromagnetic spectra that were being emitted by thermal radiators known as black bodies. Physicists struggled with this problem, which later became known as the ultraviolet catastrophe, unsuccessfully for many years. In 1900, Max Planck developed a new theory of black-body radiation that explained the observed spectrum. Planck's theory was based on the idea that black bodies emit light (and other electromagnetic radiation) only as discrete bundles or packets of energy. These packets were called quanta. Later, Albert Einstein proposed that light quanta be regarded as real particles. Later the particle of light was given the name photon, to correspond with other particles being described around this time, such as the electron and proton. A photon has an energy, E, proportional to its frequency, f, by

$$E = hf = \frac{hc}{\lambda}$$

where h is Planck's constant, λ is the wavelength and c is the speed of light. This is sometimes known as the Planck–Einstein equation.[4] In quantum theory (see first quantization) the energy of the photons is thus directly proportional to the frequency of the EMR wave.[5]

Likewise, the momentum p of a photon is also proportional to its frequency and inversely proportional to its wavelength:

$$p = \frac{E}{c} = \frac{hf}{c} = \frac{h}{\lambda}.$$

The source of Einstein's proposal that light was composed of particles (or could act as particles in some circumstances) was an experimental anomaly not explained by the wave theory: the photoelectric effect, in which light striking a metal surface ejected electrons from the surface, causing an electric current to flow across an applied voltage. Experimental measurements demonstrated that the energy of individual ejected electrons was proportional to the *frequency*, rather than the *intensity*, of the light. Furthermore, below a certain minimum frequency, which depended on the particular metal, no current would flow regardless of the intensity. These observations appeared to contradict the wave theory, and for years physicists tried in vain to find an explanation. In 1905, Einstein explained this puzzle by resurrecting the particle theory of light to explain the observed effect. Because of the preponderance of evidence in favor of the wave theory, however, Einstein's ideas were met initially with great skepticism among established physicists. Eventually Einstein's explanation was accepted as new particle-like behavior of light was observed, such as the Compton effect.

As a photon is absorbed by an atom, it excites the atom, elevating an electron to a higher energy level (one that is on average farther from the nucleus). When an electron in an excited molecule or atom descends to a lower energy level, it emits a photon of light at a frequency corresponding to the energy difference. Since the energy levels of electrons in atoms are discrete, each element and each molecule emits and absorbs its own characteristic frequencies. Immediate photon emission is called fluorescence, a type of photoluminescence. An example is visible light emitted from fluorescent paints, in response to ultraviolet (blacklight). Many other fluorescent emissions are known in spectral bands other than visible light. Delayed emission is called phosphorescence.

7.1.5 Wave–particle duality

Main article: Wave-particle duality

The modern theory that explains the nature of light includes the notion of wave–particle duality. More generally, the theory states that everything has both a particle nature and a wave nature, and various experiments can be done to bring out one or the other. The particle nature is more easily discerned using an object with a large mass. A bold proposition by Louis de Broglie in 1924 led the scientific community to realize that electrons also exhibited wave–particle duality.

7.1.6 Wave and particle effects of electromagnetic radiation

Together, wave and particle effects fully explain the emission and absorption spectra of EM radiation. The matter-composition of the medium through which the light travels determines the nature of the absorption and emission spectrum. These bands correspond to the allowed energy levels in the atoms. Dark bands in the absorption spectrum are due to the atoms in an intervening medium between source and observer. The atoms absorb certain frequencies of the light between emitter and detector/eye, then emit them in all directions. A dark band appears to the detector, due to the radiation scattered out of the beam. For instance, dark bands in the light emitted by a distant star are due to the atoms in the star's atmosphere. A similar phenomenon occurs for emission, which is seen when an emitting gas glows due to excitation of the atoms from any mechanism, including heat. As electrons descend to lower energy levels, a spectrum is emitted that represents the jumps between the energy levels of the electrons, but lines are seen because again emission happens only at particular energies after excitation. An example is the emission spectrum of nebulae. Rapidly moving electrons are most sharply accelerated when they encounter a region of force, so they are responsible for producing much of the highest frequency electromagnetic radiation observed in nature.

These phenomena can aid various chemical determinations for the composition of gases lit from behind (absorption spectra) and for glowing gases (emission spectra). Spectroscopy (for example) determines what chemical elements comprise a particular star. Spectroscopy is also used in the determination of the distance of a star, using the red shift.

7.1.7 Propagation speed

Main article: Speed of light

Any electric charge that accelerates, or any changing magnetic field, produces electromagnetic radiation. Electromagnetic information about the charge travels at the speed of light. Accurate treatment thus incorporates a concept known as retarded time, which adds to the expressions for the electrodynamic electric field and magnetic field. These extra terms are responsible for electromagnetic radiation.

When any wire (or other conducting object such as an antenna) conducts alternating current, electromagnetic radiation is propagated at the same frequency as the current. In many such situations it is possible to identify an electrical dipole moment that arises from separation of charges due to the exciting electrical potential, and this dipole moment oscillates in time, as the charges move back and forth. This oscillation at a given frequency gives rise to changing electric and magnetic fields, which then set the electromagnetic radiation in motion.

At the quantum level, electromagnetic radiation is produced when the wavepacket of a charged particle oscillates or otherwise accelerates. Charged particles in a stationary state do not move, but a superposition of such states may result in a transition state that has an electric dipole moment that oscillates in time. This oscillating dipole moment is responsible for the phenomenon of radiative transition between quantum states of a charged particle. Such states occur (for example) in atoms when photons are radiated as the atom shifts from one stationary state to another.

As a wave, light is characterized by a velocity (the speed of light), wavelength, and frequency. As particles, light is a stream of photons. Each has an energy related to the frequency of the wave given by Planck's relation $E = hf$, where E is the energy of the photon, $h = 6.626 \times 10^{-34}$ J·s is Planck's constant, and f is the frequency of the wave.

One rule is obeyed regardless of circumstances: EM radiation in a vacuum travels at the speed of light, *relative to the observer*, regardless of the observer's velocity. (This observation led to Einstein's development of the theory of special relativity.)

In a medium (other than vacuum), velocity factor or refractive index are considered, depending on frequency and application. Both of these are ratios of the speed in a medium to speed in a vacuum.

7.1.8 Special theory of relativity

Main article: Special theory of relativity

By the late nineteenth century, various experimental anomalies could not be explained by the simple wave theory. One of these anomalies involved a controversy over the speed of light. The speed of light and other EMR predicted by Maxwell's equations did not appear unless the equations were modified in a way first suggested by FitzGerald and Lorentz (see history of special relativity), or else otherwise that speed would depend on the speed of observer relative to the "medium" (called luminiferous aether) which supposedly "carried" the electromagnetic wave (in a manner analogous to the way air carries sound waves). Experiments failed to find any observer effect. In 1905, Einstein proposed that space and time appeared to be velocity-changeable entities for light propagation and all other processes and laws. These changes accounted for the constancy of the speed of light and all electromagnetic radiation, from the viewpoints of all observers—even those in relative motion.

7.2 History of discovery

See also: History of electromagnetic theory and Timeline of electromagnetic theory

Electromagnetic radiation of wavelengths other than those of visible light were discovered in the early 19th century. The discovery of infrared radiation is ascribed to astronomer William Herschel, who published his results in 1800 before the Royal Society of London.[6] Herschel used a glass prism to refract light from the Sun and detected invisible rays that caused heating beyond the red part of the spectrum, through an increase in the temperature recorded with a thermometer. These "calorific rays" were later termed infrared.

In 1801, German physicist Johann Wilhelm Ritter discovered ultraviolet in an experiment similar to Hershel's, using sunlight and a glass prism. Ritter noted that invisible rays near the violet edge of a solar spectrum dispersed by a triangular prism darkened silver chloride preparations more quickly than did the nearby violet light. Ritter's experiments were an early precursor to what would become photography. Ritter noted that the ultraviolet rays (which at first were called "chemical rays") were capable of causing chemical reactions.

In 1862-4 James Clerk Maxwell developed equations for the electromagnetic field which suggested that waves in the field would travel with a speed that was very close to the known speed of light. Maxwell therefore suggested that visible light (as well as invisible infrared and ultraviolet rays by inference) all consisted of propagating disturbances (or radiation) in the electromagnetic field. Radio waves were first produced deliberately by Heinrich Hertz in 1887, using electrical circuits calculated to produce oscillations at a much lower frequency than that of visible light, following recipes for producing oscillating charges and currents suggested by Maxwell's equations. Hertz also developed ways to detect these waves, and produced and characterized what were later termed radio waves and microwaves.[7]:286,7

Wilhelm Röntgen discovered and named X-rays. After experimenting with high voltages applied to an evacuated tube on 8 November 1895, he noticed a fluorescence on a nearby plate of coated glass. In one month, he discovered X-rays' main properties.[7]:307

The last portion of the EM spectrum to be discovered was associated with radioactivity. Henri Becquerel found that uranium salts caused fogging of an unexposed photographic plate through a covering paper in a manner similar to X-rays, and Marie Curie discovered that only certain elements gave off these rays of energy, soon discovering the intense radiation of radium. The radiation from pitchblende was differentiated into alpha rays (alpha particles) and beta rays (beta particles) by Ernest Rutherford through simple experimentation in 1899, but these proved to be charged particulate types of radiation. However, in 1900 the French scientist Paul Villard discovered a third neutrally charged and especially penetrating type of radiation from radium, and after he described it, Rutherford realized it must be yet a third type of radiation, which in 1903 Rutherford named gamma rays. In 1910 British physicist William Henry Bragg demonstrated that gamma rays are electromagnetic radiation, not particles, and in 1914 Rutherford and Edward Andrade measured their wavelengths, finding that they were similar to X-rays but with shorter wavelengths and higher frequency, although a 'cross-over' between X and gamma rays makes it possible to have X-rays with a higher energy (and hence shorter wavelength) than gamma rays and vice versa. The origin of the ray differentiates them, gamma rays tend to be a natural phenomena originating from the unstable nucleus of an atom and X-rays are electrically generated (and hence man-made) unless they are as a result of bremsstrahlung X-radiation caused by the interaction of fast moving particles (such as beta particles) colliding with certain materials, usually of higher atomic numbers.[7]:308,9

7.3 Electromagnetic spectrum

Main article: Electromagnetic spectrum

EM radiation (the designation 'radiation' excludes static electric and magnetic and near fields) is classified by wavelength into radio, microwave, infrared, visible, ultraviolet, X-rays and gamma rays. Arbitrary electromagnetic waves can be expressed by Fourier analysis in terms of sinusoidal monochromatic waves, which in turn can each be classified

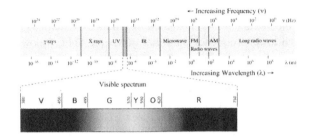

Electromagnetic spectrum with visible light highlighted

CLASS	FREQUENCY	WAVELENGTH	ENERGY
γ	300 EHz	1 pm	1.24 MeV
HX	30 EHz	10 pm	124 keV
HX	3 EHz	100 pm	12.4 keV
SX	300 PHz	1 nm	1.24 keV
EUV	30 PHz	10 nm	124 eV
NUV	3 PHz	100 nm	12.4 eV
NIR	300 THz	1 μm	1.24 eV
NIR	30 THz	10 μm	124 meV
MIR	3 THz	100 μm	12.4 meV
FIR	300 GHz	1 mm	1.24 meV
EHF	30 GHz	1 cm	124 μeV
SHF	3 GHz	1 dm	12.4 μeV
UHF	300 MHz	1 m	1.24 μeV
VHF	30 MHz	10 m	124 neV
HF	3 MHz	100 m	12.4 neV
MF	300 kHz	1 km	1.24 neV
LF	30 kHz	10 km	124 peV
VLF	3 kHz	100 km	12.4 peV
VF/ULF	300 Hz	1 Mm	1.24 peV
SLF	30 Hz	10 Mm	124 feV
ELF	3 Hz	100 Mm	12.4 feV

Legend:
γ = *Gamma rays*
HX = *Hard X-rays*
SX = *Soft X-Rays*
EUV = *Extreme-ultraviolet*
NUV = *Near-ultraviolet*
Visible light *(colored bands)*
NIR = *Near-infrared*
MIR = *Mid-infrared*
FIR = *Far-infrared*
EHF = *Extremely high frequency (microwaves)*
SHF = *Super-high frequency (microwaves)*
UHF = *Ultrahigh frequency (radio waves)*
VHF = *Very high frequency (radio)*
HF = *High frequency (radio)*
MF = *Medium frequency (radio)*
LF = *Low frequency (radio)*
VLF = *Very low frequency (radio)*
VF = *Voice frequency*
ULF = *Ultra-low frequency (radio)*
SLF = *Super-low frequency (radio)*
ELF = *Extremely low frequency(radio)*

into these regions of the EMR spectrum.

For certain classes of EM waves, the waveform is most usefully treated as *random*, and then spectral analysis must be done by slightly different mathematical techniques appropriate to random or stochastic processes. In such cases, the individual frequency components are represented in terms of their *power* content, and the phase information is not preserved. Such a representation is called the power spectral density of the random process. Random electromagnetic radiation requiring this kind of analysis is, for example, encountered in the interior of stars, and in certain other very wideband forms of radiation such as the Zero point wave field of the electromagnetic vacuum.

The behavior of EM radiation depends on its frequency. Lower frequencies have longer wavelengths, and higher frequencies have shorter wavelengths, and are associated with photons of higher energy. There is no fundamental limit known to these wavelengths or energies, at either end of the spectrum, although photons with energies near the Planck energy or exceeding it (far too high to have ever been observed) will require new physical theories to describe.

Soundwaves are not electromagnetic radiation. At the lower end of the electromagnetic spectrum, about 20 Hz to about 20 kHz, are frequencies that might be considered in the audio range. However, electromagnetic waves cannot be directly perceived by human ears. Sound waves are instead the oscillating compression of molecules. To be heard, electromagnetic radiation must be converted to pressure waves of the fluid in which the ear is located (whether the fluid is air, water or something else).

7.3.1 Interactions as a function of frequency

When EM radiation interacts with matter, its behavior changes qualitatively as its frequency changes.

Radio and microwave

At radio and microwave frequencies, EMR interacts with matter largely as a bulk collection of charges which are spread out over large numbers of affected atoms. In electrical conductors, such induced bulk movement of charges (electric currents) results in absorption of the EMR, or else separations of charges that cause generation of new EMR (effective reflection of the EMR). An example is absorption or emission of radio waves by antennas, or absorption of microwaves by water or other molecules with an electric dipole moment, as for example inside a microwave oven. These interactions produce either electric currents or heat, or both.

Infrared

Like radio and microwave, infrared also is reflected by metals (as is most EMR into the ultraviolet). However, unlike lower-frequency radio and microwave radiation, Infrared EMR commonly interacts with dipoles present in single molecules, which change as atoms vibrate at the ends of a single chemical bond. It is consequently absorbed by a wide range of substances, causing them to increase in temperature as the vibrations dissipate as heat. The same process, run in reverse, causes bulk substances to radiate in the infrared spontaneously (see thermal radiation section below).

Visible light

As frequency increases into the visible range, photons have enough energy to change the bond structure of some individual molecules. It is not a coincidence that this happens in the "visible range," as the mechanism of vision involves the change in bonding of a single molecule (retinal) which absorbs light in the rhodopsin in the retina of the human eye. Photosynthesis becomes possible in this range as well, for similar reasons, as a single molecule of chlorophyll is excited by a single photon. Animals that detect infrared make use of small packets of water that change temperature, in an essentially thermal process that involves many photons (see infrared sensing in snakes). For this reason, infrared, microwaves and radio waves are thought to damage molecules and biological tissue only by bulk heating, not excitation from single photons of the radiation.

Visible light is able to affect a few molecules with single photons, but usually not in a permanent or damaging way, in the absence of power high enough to increase temperature to damaging levels. However, in plant tissues that conduct photosynthesis, carotenoids act to quench electronically excited chlorophyll produced by visible light in a process called non-photochemical quenching, in order to prevent reactions that would otherwise interfere with photosynthesis at high light levels. Limited evidence indicate that some reactive oxygen species are created by visible light in skin, and that these may have some role in photoaging, in the same manner as ultraviolet A.[8]

Ultraviolet

As frequency increases into the ultraviolet, photons now carry enough energy (about three electron volts or more) to excite certain doubly bonded molecules into permanent chemical rearrangement. In DNA, this causes lasting damage. DNA is also indirectly damaged by reactive oxygen species produced by ultraviolet A (UVA), which has energy too low to damage DNA directly. This is why ultraviolet at all wavelengths can damage DNA, and is capable of causing cancer, and (for UVB) skin burns (sunburn) that are far worse than would be produced by simple heating (temperature increase) effects. This property of causing molecular damage that is out of proportion to heating effects, is characteristic of all EMR with frequencies at the visible light range and above. These properties of high-frequency EMR are due to quantum effects that permanently damage materials and tissues at the molecular level.

At the higher end of the ultraviolet range, the energy of photons becomes large enough to impart enough energy to electrons to cause them to be liberated from the atom, in a process called photoionisation. The energy required for this is always larger than about 10 electron volts (eV) corresponding with wavelengths smaller than 124 nm (some sources suggest a more realistic cutoff of 33 eV, which is the energy required to ionize water). This high end of the ultraviolet spectrum with energies in the approximate ionization range, is sometimes called "extreme UV." Ionizing UV is strongly filtered by the Earth's atmosphere).

X-rays and gamma rays

Electromagnetic radiation composed of photons that carry minimum-ionization energy, or more, (which includes the entire spectrum with shorter wavelengths), is therefore termed ionizing radiation. (Many other kinds of ionizing radiation are made of non-EM particles). Electromagnetic-type ionizing radiation extends from the extreme ultraviolet to all higher frequencies and shorter wavelengths, which means that all X-rays and gamma rays qualify. These are capable of the most severe types of molecular damage, which can happen in biology to any type of biomolecule, including mutation and cancer, and often at great depths below the skin, since the higher end of the X-ray spectrum, and all of the gamma ray spectrum, penetrate matter.

7.4 Atmosphere and magnetosphere

Main articles: ozone layer, shortwave radio, skywave, and ionosphere

Most UV and X-rays are blocked by absorption first from

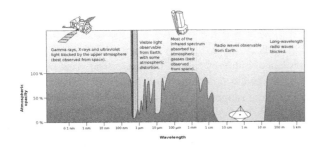

Rough plot of Earth's atmospheric absorption and scattering (or opacity) of various wavelengths of electromagnetic radiation

molecular nitrogen, and then (for wavelengths in the upper UV) from the electronic excitation of dioxygen and finally ozone at the mid-range of UV. Only 30% of the Sun's ultraviolet light reaches the ground, and almost all of this is well transmitted.

Visible light is well transmitted in air, as it is not energetic enough to excite nitrogen, oxygen, or ozone, but too energetic to excite molecular vibrational frequencies of water vapor.

Absorption bands in the infrared are due to modes of vibrational excitation in water vapor. However, at energies too low to excite water vapor, the atmosphere becomes transparent again, allowing free transmission of most microwave and radio waves.

Finally, at radio wavelengths longer than 10 meters or so (about 30 MHz), the air in the lower atmosphere remains transparent to radio, but plasma in certain layers of the ionosphere begins to interact with radio waves (see skywave). This property allows some longer wavelengths (100 meters or 3 MHz) to be reflected and results in shortwave radio beyond line-of-sight. However, certain ionospheric effects begin to block incoming radiowaves from space, when their frequency is less than about 10 MHz (wavelength longer than about 30 meters).

7.5 Types and sources, classed by spectral band

See electromagnetic spectrum

7.5.1 Radio waves

Main article: Radio waves

When radio waves impinge upon a conductor, they couple to the conductor, travel along it and induce an electric current on the conductor surface by moving the electrons of the conducting material in correlated bunches of charge. Such effects can cover macroscopic distances in conductors (such as radio antennas), since the wavelength of radiowaves is long.

7.5.2 Microwaves

Main article: Microwaves

Microwaves are a form of electromagnetic radiation with wavelengths ranging from as long as one meter to as short as one millimeter; with frequencies between 300 MHz (0.3 GHz) and 300 GHz.

7.5.3 Infrared

Main article: Infrared

7.5.4 Visible light

Main article: Light

Natural sources produce EM radiation across the spectrum. EM radiation with a wavelength between approximately 400 nm and 700 nm is directly detected by the human eye and perceived as visible light. Other wavelengths, especially nearby infrared (longer than 700 nm) and ultraviolet (shorter than 400 nm) are also sometimes referred to as light.

7.5.5 Ultraviolet

Main article: Ultraviolet

7.5.6 X-rays

Main article: X-rays

7.5.7 Gamma rays

Main article: Gamma rays

7.5.8 Thermal radiation and electromagnetic radiation as a form of heat

Main articles: Thermal radiation and Planck's law

The basic structure of matter involves charged particles bound together. When electromagnetic radiation impinges on matter, it causes the charged particles to oscillate and gain energy. The ultimate fate of this energy depends on the context. It could be immediately re-radiated and appear as scattered, reflected, or transmitted radiation. It may get dissipated into other microscopic motions within the matter, coming to thermal equilibrium and manifesting itself as thermal energy in the material. With a few exceptions related to high-energy photons (such as fluorescence, harmonic generation, photochemical reactions, the photovoltaic effect for ionizing radiations at far ultraviolet, X-ray and gamma radiation), absorbed electromagnetic radiation simply deposits its energy by heating the material. This happens for infrared, microwave and radio wave radiation. Intense radio waves can thermally burn living tissue and can cook food. In addition to infrared lasers, sufficiently intense visible and ultraviolet lasers can easily set paper afire.

Ionizing radiation creates high-speed electrons in a material and breaks chemical bonds, but after these electrons collide many times with other atoms eventually most of the energy becomes thermal energy all in a tiny fraction of a second. This process makes ionizing radiation far more dangerous per unit of energy than non-ionizing radiation. This caveat also applies to UV, even though almost all of it is not ionizing, because UV can damage molecules due to electronic excitation, which is far greater per unit energy than heating effects.

Infrared radiation in the spectral distribution of a black body is usually considered a form of heat, since it has an equivalent temperature and is associated with an entropy change per unit of thermal energy. However, "heat" is a technical term in physics and thermodynamics and is often confused with thermal energy. Any type of electromagnetic energy can be transformed into thermal energy in interaction with matter. Thus, *any* electromagnetic radiation can "heat" (in the sense of increase the thermal energy termperature of) a material, when it is absorbed.

The inverse or time-reversed process of absorption is thermal radiation. Much of the thermal energy in matter consists of random motion of charged particles, and this energy can be radiated away from the matter. The resulting radiation may subsequently be absorbed by another piece of matter, with the deposited energy heating the material.

The electromagnetic radiation in an opaque cavity at thermal equilibrium is effectively a form of thermal energy, having maximum radiation entropy.

7.6 Biological effects

Main articles: Electromagnetic radiation and health and Mobile phone radiation and health

Bioelectromagnetics is the study of the interactions and effects of EM radiation on living organisms. The effects of electromagnetic radiation upon living cells, including those in humans, depends upon the radiation's power and frequency. For low-frequency radiation (radio waves to visible light) the best-understood effects are those due to radiation power alone, acting through heating when radiation is absorbed. For these thermal effects, frequency is important only as it affects penetration into the organism (for example, microwaves penetrate better than infrared). Initially, it was believed that low frequency fields that were too weak to cause significant heating could not possibly have any biological effect.[9]

Despite this opinion among researchers, evidence has accumulated that supports the existence of complex biological effects of weaker *non-thermal* electromagnetic fields, (including weak ELF magnetic fields, although the latter does not strictly qualify as EM radiation[9][10][11]), and modulated RF and microwave fields.[12][13][14] Fundamental mechanisms of the interaction between biological material and electromagnetic fields at non-thermal levels are not fully understood.[9]

The World Health Organization has classified radio frequency electromagnetic radiation as Group 2B - possibly carcinogenic.[15][16] This group contains possible carcinogens that have weaker evidence, at the same level as coffee and automobile exhaust. For example, epidemiological studies looking for a relationship between cell phone use and brain cancer development, have been largely inconclusive, save to demonstrate that the effect, if it exists, cannot be a large one.

At higher frequencies (visible and beyond), the effects of individual photons begin to become important, as these now have enough energy individually to directly or indirectly damage biological molecules.[17] All UV frequencies have been classed as Group 1 carcinogens by the World Health Organization. Ultraviolet radiation from sun exposure is the primary cause of skin cancer.[18][19]

Thus, at UV frequencies and higher (and probably somewhat also in the visible range),[8] electromagnetic radiation does more damage to biological systems than simple heating predicts. This is most obvious in the "far" (or "extreme") ultraviolet. UV, with X-ray and gamma radiation, are referred to as ionizing radiation due to the ability of photons of this radiation to produce ions and free radicals in materials (including living tissue). Since such radiation can severely damage life at energy levels that produce little heating, it is considered far more dangerous (in terms of damage-produced per unit of energy, or power) than the rest of the electromagnetic spectrum.

7.7 Derivation from electromagnetic theory

Main article: Electromagnetic wave equation

Electromagnetic waves were predicted by the classical laws of electricity and magnetism, known as Maxwell's equations. Inspection of Maxwell's equations without sources (charges or currents) results in nontrivial solutions of changing electric and magnetic fields. Beginning with Maxwell's equations in free space:

$$\nabla \cdot \mathbf{E} = 0 \qquad (1)$$

$$\nabla \times \mathbf{E} = -\frac{\partial \mathbf{B}}{\partial t} \qquad (2)$$

$$\nabla \cdot \mathbf{B} = 0 \qquad (3)$$

$$\nabla \times \mathbf{B} = \mu_0 \epsilon_0 \frac{\partial \mathbf{E}}{\partial t} \qquad (4)$$

where

∇ is a vector differential operator (see Del).

One solution,

$$\mathbf{E} = \mathbf{B} = \mathbf{0},$$

is trivial.

For a more useful solution, we utilize vector identities, which work for any vector, as follows:

$$\nabla \times (\nabla \times \mathbf{A}) = \nabla (\nabla \cdot \mathbf{A}) - \nabla^2 \mathbf{A}$$

The curl of equation (2):

$$\nabla \times (\nabla \times \mathbf{E}) = \nabla \times \left(-\frac{\partial \mathbf{B}}{\partial t}\right) \qquad (5)$$

Evaluating the left hand side:

$$\nabla \times (\nabla \times \mathbf{E}) = \nabla (\nabla \cdot \mathbf{E}) - \nabla^2 \mathbf{E} = -\nabla^2 \mathbf{E} \qquad (6)$$

simplifying the above by using equation (1).

Evaluating the right hand side:

$$\nabla \times \left(-\frac{\partial \mathbf{B}}{\partial t}\right) = -\frac{\partial}{\partial t}(\nabla \times \mathbf{B}) = -\mu_0 \epsilon_0 \frac{\partial^2 \mathbf{E}}{\partial t^2} \qquad (7)$$

Equations (6) and (7) are equal, so this results in a vector-valued differential equation for the electric field, namely

Applying a similar pattern results in similar differential equation for the magnetic field:

These differential equations are equivalent to the wave equation:

$$\nabla^2 f = \frac{1}{c_0^2} \frac{\partial^2 f}{\partial t^2}$$

where

c_0 is the speed of the wave in free space and

f describes a displacement

Or more simply:

$$\Box f = 0$$

where $\Box$ is d'Alembertian:

$$\Box = \nabla^2 - \frac{1}{c_0^2}\frac{\partial^2}{\partial t^2} = \frac{\partial^2}{\partial x^2} + \frac{\partial^2}{\partial y^2} + \frac{\partial^2}{\partial z^2} - \frac{1}{c_0^2}\frac{\partial^2}{\partial t^2}$$

In the case of the electric and magnetic fields, the speed is:

$$c_0 = \frac{1}{\sqrt{\mu_0 \epsilon_0}}$$

This is the speed of light in vacuum. Maxwell's equations unified the vacuum permittivity ϵ_0, the vacuum permeability μ_0, and the speed of light itself, c_0. This relationship had been discovered by Wilhelm Eduard Weber and Rudolf Kohlrausch prior to the development of Maxwell's electrodynamics, however Maxwell was the first to produce a field theory consistent with waves traveling at the speed of light.

These are only two equations versus the original four, so more information pertains to these waves hidden within Maxwell's equations. A generic vector wave for the electric field.

$$\mathbf{E} = \mathbf{E}_0 f\left(\hat{\mathbf{k}} \cdot \mathbf{x} - c_0 t\right)$$

Here, $\mathbf{E}_0$ is the constant amplitude, f is any second differentiable function, $\hat{\mathbf{k}}$ is a unit vector in the direction of propagation, and $\mathbf{x}$ is a position vector. $f\left(\hat{\mathbf{k}} \cdot \mathbf{x} - c_0 t\right)$ is a generic solution to the wave equation. In other words,

$$\nabla^2 f\left(\hat{\mathbf{k}} \cdot \mathbf{x} - c_0 t\right) = \frac{1}{c_0^2} \frac{\partial^2}{\partial t^2} f\left(\hat{\mathbf{k}} \cdot \mathbf{x} - c_0 t\right),$$

for a generic wave traveling in the $\hat{\mathbf{k}}$ direction.

This form will satisfy the wave equation.

$$\nabla \cdot \mathbf{E} = \hat{\mathbf{k}} \cdot \mathbf{E}_0 f'\left(\hat{\mathbf{k}} \cdot \mathbf{x} - c_0 t\right) = 0$$

$$\mathbf{E} \cdot \hat{\mathbf{k}} = 0$$

The first of Maxwell's equations implies that the electric field is orthogonal to the direction the wave propagates.

$$\nabla \times \mathbf{E} = \hat{\mathbf{k}} \times \mathbf{E}_0 f'\left(\hat{\mathbf{k}} \cdot \mathbf{x} - c_0 t\right) = -\frac{\partial \mathbf{B}}{\partial t}$$

$$\mathbf{B} = \frac{1}{c_0} \hat{\mathbf{k}} \times \mathbf{E}$$

The second of Maxwell's equations yields the magnetic field. The remaining equations will be satisfied by this choice of $\mathbf{E}, \mathbf{B}$.

The electric and magnetic field waves in the far-field travel at the speed of light. They have a special restricted orientation and proportional magnitudes, $E_0 = c_0 B_0$, which can be seen immediately from the Poynting vector. The electric field, magnetic field, and direction of wave propagation are all orthogonal, and the wave propagates in the same direction as $\mathbf{E} \times \mathbf{B}$. Also, $\mathbf{E}$ and $\mathbf{B}$ far-fields in free space, which as wave solutions depend primarily on these two Maxwell equations, are in-phase with each other. This is guaranteed since the generic wave solution is first order in both space and time, and the curl operator on one side of these equations results in first-order spatial derivatives of the wave solution, while the time-derivative on the other side of the equations, which gives the other field, is first order in time, resulting in the same phase shift for both fields in each mathematical operation.

From the viewpoint of an electromagnetic wave traveling forward, the electric field might be oscillating up and down, while the magnetic field oscillates right and left. This picture can be rotated with the electric field oscillating right and left and the magnetic field oscillating down and up. This is a different solution that is traveling in the same direction. This arbitrariness in the orientation with respect to propagation direction is known as polarization. On a quantum level, it is described as photon polarization. The direction of the polarization is defined as the direction of the electric field.

More general forms of the second-order wave equations given above are available, allowing for both non-vacuum propagation media and sources. Many competing derivations exist, all with varying levels of approximation and intended applications. One very general example is a form of the electric field equation,[20] which was factorized into a pair of explicitly directional wave equations, and then efficiently reduced into a single uni-directional wave equation by means of a simple slow-evolution approximation.

7.8 See also

- Antenna (radio)
- Antenna measurement
- Bioelectromagnetism
- Bolometer
- Control of electromagnetic radiation
- Electromagnetic field
- Electromagnetic pulse
- Electromagnetic radiation and health
- Electromagnetic spectrum
- Electromagnetic wave equation

- Evanescent wave coupling
- Finite-difference time-domain method
- Helicon
- Impedance of free space
- Light
- Maxwell's equations
- Near and far field
- Radiant energy
- Radiation reaction
- Risks and benefits of sun exposure
- Sinusoidal plane-wave solutions of the electromagnetic wave equation

7.9 References

[1] Carmichael, H. J. "Einstein and the Photoelectric Effect" (PDF). Quantum Optics Theory Group, University of Auckland. Retrieved 22 December 2009.

[2] Thorn, J. J.; Neel, M. S.; Donato, V. W.; Bergreen, G. S.; Davies, R. E.; Beck, M. (2004). "Observing the quantum behavior of light in an undergraduate laboratory" (PDF). *American Journal of Physics*. **72** (9): 1210. Bibcode:2004AmJPh..72.1210T. doi:10.1119/1.1737397.

[3] She, Alan; Capasso, Federico (17 May 2016). "Parallel Polarization State Generation". *Scientific Reports*. Nature. doi:10.1038/srep26019. Retrieved 27 June 2016.

[4] Paul M. S. Monk (2004). *Physical Chemistry*. John Wiley and Sons. p. 435. ISBN 978-0-471-49180-4.

[5] Weinberg, S. (1995). *The Quantum Theory of Fields*. **1**. Cambridge University Press. pp. 15–17. ISBN 0-521-55001-7.

[6] Herschel, William (1 January 1800). "Experiments on the Refrangibility of the Invisible Rays of the Sun. By William Herschel, LL. D. F. R. S". *Philosophical Transactions of the Royal Society of London*. **90**: 284–292. JSTOR 107057.

[7] James Jeans (1947) The Growth of Physical Science, link from Internet Archive

[8] Liebel, F.; Kaur, S.; Ruvolo, E.; Kollias, N.; Southall, M. D. (2012). "Irradiation of Skin with Visible Light Induces Reactive Oxygen Species and Matrix-Degrading Enzymes". *Journal of Investigative Dermatology*. **132** (7): 1901–1907. doi:10.1038/jid.2011.476. PMID 22318388.

[9] Binhi, Vladimir N; Repiev, A & Edelev, M (translators from Russian) (2002). *Magnetobiology: Underlying Physical Problems*. San Diego: Academic Press. pp. 1–16. ISBN 978-0-12-100071-4. OCLC 49700531. Cite uses deprecated parameter |coauthors= (help)

[10] Delgado, J. M.; Leal, J.; Monteagudo, J. L.; Gracia, M. G. (1982). "Embryological changes induced by weak, extremely low frequency electromagnetic fields". *Journal of Anatomy*. **134** (Pt 3): 533–551. PMC 1167891. PMID 7107514.

[11] Harland, J. D.; Liburdy, R. P. (1997). "Environmental magnetic fields inhibit the antiproliferative action of tamoxifen and melatonin in a human breast cancer cell line". *Bioelectromagnetics*. **18** (8): 555–562. doi:10.1002/(SICI)1521-186X(1997)18:8<555::AID-BEM4>3.0.CO;2-1. PMID 9383244.

[12] Aalto, S.; Haarala, C.; Brück, A.; Sipilä, H.; Hämäläinen, H.; Rinne, J. O. (2006). "Mobile phone affects cerebral blood flow in humans". *Journal of Cerebral Blood Flow & Metabolism*. **26** (7): 885–890. doi:10.1038/sj.jcbfm.9600279. PMID 16495939.

[13] Cleary, S. F.; Liu, L. M.; Merchant, R. E. (1990). "In vitro lymphocyte proliferation induced by radiofrequency electromagnetic radiation under isothermal conditions". *Bioelectromagnetics*. **11** (1): 47–56. doi:10.1002/bem.2250110107. PMID 2346507.

[14] Ramchandani, P. (2004). "Prevalence of childhood psychiatric disorders may be underestimated". *Evidence-based mental health*. **7** (2): 59. doi:10.1136/ebmh.7.2.59. PMID 15107355.

[15] IARC classifies Radiofrequency Electromagnetic Fields as possibly carcinogenic to humans. World Health Organization. 31 May 2011

[16] "Trouble with cell phone radiation standard". *CBS News*.

[17] See Liebel, F; Kaur, S; Ruvolo, E; Kollias, N; Southall, M. D. (July 2012). "Irradiation of skin with visible light induces reactive oxygen species and matrix-degrading enzymes". *J. Invest. Dermatol.* **132** (7): 1901–7. doi:10.1038/jid.2011.476. PMID 22318388. for evidence of quantum damage from visible light via reactive oxygen species generated in skin. This happens also with UVA. With UVB, the damage to DNA becomes direct, with photochemical formation of pyrimidine dimers.

[18] Narayanan, DL; Saladi, RN; Fox, JL (September 2010). "Ultraviolet radiation and skin cancer". *International Journal of Dermatology*. **49** (9): 978–86. doi:10.1111/j.1365-4632.2010.04474.x. PMID 20883261.

[19] Saladi, RN; Persaud, AN (January 2005). "The causes of skin cancer: a comprehensive review". *Drugs of today (Barcelona, Spain : 1998)*. **41** (1): 37–53. doi:10.1358/dot.2005.41.1.875777. PMID 15753968.

[20] Kinsler, P. (2010). "Optical pulse propagation with minimal approximations". *Phys. Rev. A.* **81**: 013819. arXiv:0810.5689. Bibcode:2010PhRvA..81a3819K. doi:10.1103/PhysRevA.81.013819.

7.10 Further reading

- Hecht, Eugene (2001). *Optics* (4th ed.). Pearson Education. ISBN 0-8053-8566-5.

- Serway, Raymond A.; Jewett, John W. (2004). *Physics for Scientists and Engineers* (6th ed.). Brooks Cole. ISBN 0-534-40842-7.

- Tipler, Paul (2004). *Physics for Scientists and Engineers: Electricity, Magnetism, Light, and Elementary Modern Physics* (5th ed.). W. H. Freeman. ISBN 0-7167-0810-8.

- Reitz, John; Milford, Frederick; Christy, Robert (1992). *Foundations of Electromagnetic Theory* (4th ed.). Addison Wesley. ISBN 0-201-52624-7.

- Jackson, John David (1999). *Classical Electrodynamics* (3rd ed.). John Wiley & Sons. ISBN 0-471-30932-X.

- Allen Taflove and Susan C. Hagness (2005). *Computational Electrodynamics: The Finite-Difference Time-Domain Method, 3rd ed.* Artech House Publishers. ISBN 1-58053-832-0.

7.11 External links

- Electromagnetism – a chapter from an online textbook

- *Electromagnetic Waves from Maxwell's Equations* on Project PHYSNET.

- Radiation of atoms? e-m wave, Polarisation, ...

- An Introduction to The Wigner Distribution in Geometric Optics

- The windows of the electromagnetic spectrum, on Astronoo

- Introduction to light and electromagnetic radiation course video from the Khan Academy

- Lectures on electromagnetic waves course video and notes from MIT Professor Walter Lewin

- *Encyclopedia Britannica* Electromagnetic Radiation

- Physics for the 21st Century Early Unification for Electromagnetism Harvard-Smithsonian Center for Astrophysics

Chapter 8

Photoelectric effect

The **photoelectric effect** or *photoemission* is the production of electrons or other free carriers when light is shone onto a material. Electrons emitted in this manner can be called *photoelectrons*. The phenomenon is commonly studied in electronic physics, as well as in fields of chemistry, such as quantum chemistry or electrochemistry.

According to classical electromagnetic theory, this effect can be attributed to the transfer of energy from the light to an electron. From this perspective, an alteration in either the intensity or wavelength of light would induce changes in the rate of emission of electrons from the metal. Furthermore, according to this theory, a sufficiently dim light would be expected to show a time lag between the initial shining of its light and the subsequent emission of an electron. However, the experimental results did not correlate with either of the two predictions made by classical theory.

Instead, electrons are only dislodged by the impingement of photons when those photons reach or exceed a threshold frequency (energy). Below that threshold, no electrons are emitted from the metal regardless of the light intensity or the length of time of exposure to the light. To make sense of the fact that light can eject electrons even if its intensity is low, Albert Einstein proposed that a beam of light is not a wave propagating through space, but rather a collection of discrete wave packets (photons), each with energy hf. This shed light on Max Planck's previous discovery of the Planck relation ($E = hf$) linking energy (E) and frequency (f) as arising from quantization of energy. The factor h is known as the Planck constant.[1][2]

In 1887, Heinrich Hertz[2][3] discovered that electrodes illuminated with ultraviolet light create electric sparks more easily. In 1905 Albert Einstein published a paper that explained experimental data from the photoelectric effect as the result of light energy being carried in discrete quantized packets. This discovery led to the quantum revolution. In 1914, Robert Millikan's experiment confirmed Einstein's law on photoelectric effect. Einstein was awarded the Nobel Prize in 1921 for "his discovery of the law of the photoelectric effect",[4] and Millikan was awarded the Nobel Prize in 1923 for "his work on the elementary charge of electricity and on the photoelectric effect".[5]

The photoelectric effect requires photons with energies approaching zero (in the case of negative electron affinity) to over 1 MeV for core electrons in elements with a high atomic number. Emission of conduction electrons from typical metals usually requires a few electron-volts, corresponding to short-wavelength visible or ultraviolet light. Study of the photoelectric effect led to important steps in understanding the quantum nature of light and electrons and influenced the formation of the concept of wave–particle duality.[1] Other phenomena where light affects the movement of electric charges include the photoconductive effect (also known as photoconductivity or photoresistivity), the photovoltaic effect, and the photoelectrochemical effect.

Photoemission can occur from any material, but it is most easily observable from metals or other conductors because the process produces a charge imbalance, and if this charge imbalance is not neutralized by current flow (enabled by conductivity), the potential barrier to emission increases until the emission current ceases. It is also usual to have the emitting surface in a vacuum, since gases impede the flow of photoelectrons and make them difficult to observe. Additionally, the energy barrier to photoemission is usually increased by thin oxide layers on metal surfaces if the metal has been exposed to oxygen, so most practical experiments and devices based on the photoelectric effect use clean metal surfaces in a vacuum.

When the photoelectron is emitted into a solid rather than into a vacuum, the term *internal photoemission* is often used, and emission into a vacuum distinguished as *external photoemission*.

8.1 Emission mechanism

The photons of a light beam have a characteristic energy proportional to the frequency of the light. In the photoemission process, if an electron within some material absorbs the energy of one photon and acquires more energy than the work function (the electron binding energy) of the material, it is ejected. If the photon energy is too low, the electron is unable to escape the material. Since an increase in the intensity of low-frequency light will only increase the number of low-energy photons sent over a given interval of time, this change in intensity will not create any single photon with enough energy to dislodge an electron. Thus, the energy of the emitted electrons does not depend on the intensity of the incoming light, but only on the energy (equivalently frequency) of the individual photons. It is an interaction between the incident photon and the outermost electrons.

Electrons can absorb energy from photons when irradiated, but they usually follow an "all or nothing" principle. All of the energy from one photon must be absorbed and used to liberate one electron from atomic binding, or else the energy is re-emitted. If the photon energy is absorbed, some of the energy liberates the electron from the atom, and the rest contributes to the electron's kinetic energy as a free particle.[6][7][8]

8.1.1 Experimental observations of photoelectric emission

The theory of the photoelectric effect must explain the experimental observations of the emission of electrons from an illuminated metal surface.

For a given metal, there exists a certain minimum frequency of incident radiation below which no photoelectrons are emitted. This frequency is called the *threshold frequency*. Increasing the frequency of the incident beam, keeping the number of incident photons fixed (this would result in a proportionate increase in energy) increases the maximum kinetic energy of the photoelectrons emitted. Thus the stopping voltage increases. The number of electrons also changes because the probability that each photon results in an emitted electron is a function of photon energy. If the intensity of the incident radiation of a given frequency is increased, there is no effect on the kinetic energy of each photoelectron.

Above the threshold frequency, the maximum kinetic energy of the emitted photoelectron depends on the frequency of the incident light, but is independent of the intensity of the incident light so long as the latter is not too high.[9]

For a given metal and frequency of incident radiation, the rate at which photoelectrons are ejected is directly proportional to the intensity of the incident light. An increase in the intensity of the incident beam (keeping the frequency fixed) increases the magnitude of the photoelectric current, although the stopping voltage remains the same.

The time lag between the incidence of radiation and the emission of a photoelectron is very small, less than 10^{-9} second.

The direction of distribution of emitted electrons peaks in the direction of polarization (the direction of the electric field) of the incident light, if it is linearly polarized.[10]

8.1.2 Mathematical description

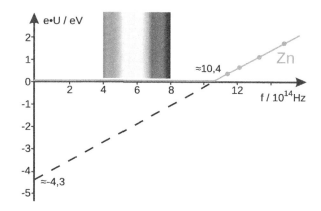

Diagram of the maximum kinetic energy as a function of the frequency of light on zinc

The maximum kinetic energy K_{max} of an ejected electron is given by

$$K_{max} = hf - \varphi,$$

where h is the Planck constant and f is the frequency of the incident photon. The term φ is the work function (sometimes denoted W, or ϕ [11]), which gives the minimum energy required to remove a delocalised electron from the surface of the metal. The work function satisfies

$$\varphi = hf_0,$$

where f_0 is the threshold frequency for the metal. The maximum kinetic energy of an ejected electron is then

$$K_{max} = h(f - f_0).$$

Kinetic energy is positive, so we must have $f > f_0$ for the photoelectric effect to occur.[12]

8.1.3 Stopping potential

The relation between current and applied voltage illustrates the nature of the photoelectric effect. For discussion, a light source illuminates a plate P, and another plate electrode Q collects any emitted electrons. We vary the potential between P and Q and measure the current flowing in the external circuit between the two plates.

$$E = \frac{hc}{\lambda}$$

Work function = E.

Cut-off wavelength = λ

Work function and cut off frequency

If the frequency and the intensity of the incident radiation are fixed, the photoelectric current increases gradually with an increase in the positive potential on the collector electrode until all the photoelectrons emitted are collected. The photoelectric current attains a saturation value and does not increase further for any increase in the positive potential. The saturation current increases with the increase of the light intensity. It also increases with greater frequencies due to a greater probability of electron emission when collisions happen with higher energy photons.

If we apply a negative potential to the collector plate Q with respect to the plate P and gradually increase it, the photoelectric current decreases, becoming zero at a certain negative potential. The negative potential on the collector at which the photoelectric current becomes zero is called the *stopping potential* or *cut off* potential[13]

i. For a given frequency of incident radiation, the stopping potential is independent of its intensity.

ii. For a given frequency of incident radiation, the stopping potential is determined by the maximum kinetic energy K_{max} of the photoelectrons that are emitted. If qe is the charge on the electron and V_0 is the stopping potential, then the work done by the retarding potential in stopping the electron is $q_e V_0$, so we have

$$q_e V_0 = K_{max}.$$

Recalling

$$K_{max} = h(f - f_0),$$

we see that the stopping voltage varies linearly with frequency of light, but depends on the type of material. For any particular material, there is a threshold frequency that must be exceeded, independent of light intensity, to observe any electron emission.

8.1.4 Three-step model

In the X-ray regime, the photoelectric effect in crystalline material is often decomposed into three steps:[14]:50–51

1. Inner photoelectric effect (see photodiode below). The hole left behind can give rise to Auger effect, which is visible even when the electron does not leave the material. In molecular solids phonons are excited in this step and may be visible as lines in the final electron energy. The inner photoeffect has to be dipole allowed. The transition rules for atoms translate via the tight-binding model onto the crystal. They are similar in geometry to plasma oscillations in that they have to be transversal.

2. Ballistic transport of half of the electrons to the surface. Some electrons are scattered.

3. Electrons escape from the material at the surface.

In the three-step model, an electron can take multiple paths through these three steps. All paths can interfere in the sense of the path integral formulation. For surface states and molecules the three-step model does still make some sense as even most atoms have multiple electrons which can scatter the one electron leaving.

8.2 History

When a surface is exposed to electromagnetic radiation above a certain threshold frequency (typically visible light for alkali metals, near ultraviolet for other metals, and extreme ultraviolet for non-metals), the radiation is absorbed and electrons are emitted. Light, and especially ultra-violet light, discharges negatively electrified bodies with the production of rays of the same nature as cathode rays.[15] Under certain circumstances it can directly ionize gases.[15] The first of these phenomena was discovered by Hertz and Hallwachs in 1887.[15] The second was announced first by Philipp Lenard in 1900.[15]

The ultra-violet light to produce these effects may be obtained from an arc lamp, or by burning magnesium, or by sparking with an induction coil between zinc or cadmium terminals, the light from which is very rich in ultra-violet rays. Sunlight is not rich in ultra-violet rays, as these have been absorbed by the atmosphere, and it does not produce nearly so large an effect as the arc-light. Many substances besides metals discharge negative electricity under the action of ultraviolet light: lists of these substances will be found in papers by G. C. Schmidt[16] and O. Knoblauch.[17]

8.2.1 19th century

In 1839, Alexandre Edmond Becquerel discovered the photovoltaic effect while studying the effect of light on electrolytic cells.[18] Though not equivalent to the photoelectric effect, his work on photovoltaics was instrumental in showing a strong relationship between light and electronic properties of materials. In 1873, Willoughby Smith discovered photoconductivity in selenium while testing the metal for its high resistance properties in conjunction with his work involving submarine telegraph cables.[19]

Johann Elster (1854–1920) and Hans Geitel (1855–1923), students in Heidelberg, developed the first practical photoelectric cells that could be used to measure the intensity of light.[20][21]:458 Elster and Geitel had investigated with great success the effects produced by light on electrified bodies.[22]

Heinrich Rudolf Hertz

In 1887, Heinrich Hertz observed the photoelectric effect and the production and reception of electromagnetic waves.[15] He published these observations in the journal Annalen der Physik. His receiver consisted of a coil with a spark gap, where a spark would be seen upon detection of electromagnetic waves. He placed the apparatus in a darkened box to see the spark better. However, he noticed that the maximum spark length was reduced when in the box. A glass panel placed between the source of electromagnetic waves and the receiver absorbed ultraviolet radiation that assisted the electrons in jumping across the gap. When removed, the spark length would increase. He observed no decrease in spark length when he replaced glass with quartz, as quartz does not absorb UV radiation. Hertz concluded his months of investigation and reported the results obtained. He did not further pursue investigation of this effect.

The discovery by Hertz[23] in 1887 that the incidence of ultra-violet light on a spark gap facilitated the passage of the spark, led immediately to a series of investigations by Hallwachs,[24] Hoor,[25] Righi[26] and Stoletow[27][28][29][30][31][32][33] on the effect of light, and especially of ultra-violet light, on charged bodies. It was proved by these investigations that a newly cleaned surface of zinc, if charged with negative electricity, rapidly loses this charge however small it may be when ultra-violet light falls upon the surface; while if the surface is uncharged to begin with, it acquires a positive charge when exposed to the light, the negative electrification going out into the gas by which the metal is surrounded; this positive electrification can be much increased by directing a strong airblast against the surface. If however the zinc surface is positively electrified it suffers no loss of charge when exposed to the light: this result has been questioned, but a very careful examination of the phenomenon by Elster and Geitel[34] has shown that the loss observed under certain circumstances is due to the discharge by the light reflected from the zinc surface of negative electrification on neighbouring conductors induced by the positive charge, the negative electricity under the influence of the electric field moving up to the positively electrified surface.[35]

With regard to the *Hertz effect*, the researches from the start showed a great complexity of the phenomenon of photoelectric fatigue — that is, the progressive diminution of the effect observed upon fresh metallic surfaces. According to an important research by Wilhelm Hallwachs, ozone played an important part in the phenomenon.[36] However, other elements enter such as oxidation, the humidity, the mode of polish of the surface, etc. It was at the time not even sure that the fatigue is absent in a vacuum.

In the period from February 1888 and until 1891, a detailed analysis of photoeffect was performed by Aleksandr Stoletov with results published in 6 works;[37][38][39][40][41][42] four of them in *Comptes Rendus*, one review in *Physikalische Revue* (translated from Russian), and the last work in *Journal de Physique*. First, in these works Stoletov invented a new experimental setup which was more suitable for a quantitative analysis of photoeffect. Using this setup, he discovered the direct proportionality between the intensity of light and the induced photo electric current (the first law of photoeffect or Stoletov's law). One of his other findings resulted from measurements of the dependence of the intensity of the electric photo current on the gas pressure, where he found the existence of an optimal gas pressure P_m corresponding to a maximum photocurrent; this property was used for a creation of solar cells.

In 1899, J. J. Thomson investigated ultraviolet light in Crookes tubes.[43] Thomson deduced that the ejected particles were the same as those previously found in the cathode ray, later called electrons, which he called "corpuscles". In the research, Thomson enclosed a metal plate (a cathode) in a vacuum tube, and exposed it to high frequency radiation.[44] It was thought that the oscillating electromagnetic fields caused the atoms' field to resonate and, after reaching a certain amplitude, caused a subatomic "corpuscle" to be emitted, and current to be detected. The amount of this current varied with the intensity and colour of the radiation. Larger radiation intensity or frequency would produce more current.

8.2.2 20th century

The discovery of the ionization of gases by ultra-violet light was made by Philipp Lenard in 1900. As the effect was produced across several centimeters of air and made very great positive and small negative ions, it was natural to interpret the phenomenon, as did J. J. Thomson, as a *Hertz effect* upon the solid or liquid particles present in the gas.[15]

German physicist Philipp Lenard

In 1902, Lenard observed that the energy of individual emitted electrons increased with the frequency (which is related to the color) of the light.[6]

This appeared to be at odds with Maxwell's wave theory of light, which predicted that the electron energy would be proportional to the intensity of the radiation.

Lenard observed the variation in electron energy with light frequency using a powerful electric arc lamp which enabled him to investigate large changes in intensity, and that had sufficient power to enable him to investigate the variation of potential with light frequency. His experiment directly measured potentials, not electron kinetic energy: he found the electron energy by relating it to the maximum stopping potential (voltage) in a phototube. He found that the calculated maximum electron kinetic energy is determined by the frequency of the light. For example, an increase in frequency results in an increase in the maximum kinetic energy calculated for an electron upon liberation – ultraviolet radiation would require a higher applied stopping potential to stop current in a phototube than blue light. However Lenard's results were qualitative rather than quantitative because of the difficulty in performing the experiments: the experiments needed to be done on freshly cut metal so that the pure metal was observed, but it oxidised in a matter of minutes even in the partial vacuums he used. The current emitted by the surface was determined by the light's intensity, or brightness: doubling the intensity of the light doubled the number of electrons emitted from the surface.

The researches of Langevin and those of Eugene Bloch[45] have shown that the greater part of the Lenard effect is certainly due to this 'Hertz effect'. The Lenard effect upon the gas itself nevertheless does exist. Refound by J. J. Thomson[46] and then more decisively by Frederic Palmer, Jr.,[47][48] it was studied and showed very different characteristics than those at first attributed to it by Lenard.[15]

In 1905, Albert Einstein solved this apparent paradox by describing light as composed of discrete quanta, now called photons, rather than continuous waves. Based upon Max Planck's theory of black-body radiation, Einstein theorized that the energy in each quantum of light was equal to the frequency multiplied by a constant, later called Planck's constant. A photon above a threshold frequency has the required energy to eject a single electron, creating the observed effect. This discovery led to the quantum revolution in physics and earned Einstein the Nobel Prize in Physics in 1921.[49] By wave-particle duality the effect can be analyzed purely in terms of waves though not as conveniently.[50]

Albert Einstein's mathematical description of how the photoelectric effect was caused by absorption of quanta of light was in one of his 1905 papers, named "*On a Heuristic Viewpoint Concerning the Production and Transformation of Light*". This paper proposed the simple description of "light quanta", or photons, and showed how they explained

8.2. HISTORY

Einstein, in 1905, when he wrote the Annus Mirabilis *papers*

such phenomena as the photoelectric effect. His simple explanation in terms of absorption of discrete quanta of light explained the features of the phenomenon and the characteristic frequency.

The idea of light quanta began with Max Planck's published law of black-body radiation (*"On the Law of Distribution of Energy in the Normal Spectrum"*[51]) by assuming that Hertzian oscillators could only exist at energies E proportional to the frequency f of the oscillator by $E = hf$, where h is Planck's constant. By assuming that light actually consisted of discrete energy packets, Einstein wrote an equation for the photoelectric effect that agreed with experimental results. It explained why the energy of photoelectrons was dependent only on the *frequency* of the incident light and not on its *intensity*: a low-intensity, high-frequency source could supply a few high energy photons, whereas a high-intensity, low-frequency source would supply no photons of sufficient individual energy to dislodge any electrons. This was an enormous theoretical leap, but the concept was strongly resisted at first because it contradicted the wave theory of light that followed naturally from James Clerk Maxwell's equations for electromagnetic behavior, and more generally, the assumption of infinite divisibility of energy in physical systems. Even after experiments showed that Einstein's equations for the photoelectric effect were accurate, resistance to the idea of photons continued, since it appeared to contradict Maxwell's equations, which were well-understood and verified.

Robert Millikan (picture around 1923), who first experimentally showed Einstein's prediction on photoelectric effect was correct.

Einstein's work predicted that the energy of individual ejected electrons increases linearly with the frequency of the light. Perhaps surprisingly, the precise relationship had not at that time been tested. By 1905 it was known that the energy of photoelectrons increases with increasing *frequency* of incident light and is independent of the *intensity* of the light. However, the manner of the increase was not experimentally determined until 1914 when Robert Andrews Millikan showed that Einstein's prediction was correct.[7]

The photoelectric effect helped to propel the then-emerging concept of wave–particle duality in the nature of light. Light simultaneously possesses the characteristics of both waves and particles, each being manifested according to the circumstances. The effect was impossible to understand in terms of the classical wave description of light,[52][53][54] as the energy of the emitted electrons did not depend on the intensity of the incident radiation. Classical theory predicted that the electrons would 'gather up' energy over a period of time, and then be emitted.[53][55]

8.3 Uses and effects

8.3.1 Photomultipliers

Main article: Photomultiplier

These are extremely light-sensitive vacuum tubes with a photocathode coated onto part (an end or side) of the inside of the envelope. The photocathode contains combinations of materials such as caesium, rubidium and antimony specially selected to provide a low work function, so when illuminated even by very low levels of light, the photocathode readily releases electrons. By means of a series of electrodes (dynodes) at ever-higher potentials, these electrons are accelerated and substantially increased in number through secondary emission to provide a readily detectable output current. Photomultipliers are still commonly used wherever low levels of light must be detected.[56]

8.3.2 Image sensors

Video camera tubes in the early days of television used the photoelectric effect, for example, Philo Farnsworth's "Image dissector" used a screen charged by the photoelectric effect to transform an optical image into a scanned electronic signal.[57]

8.3.3 Gold-leaf electroscope

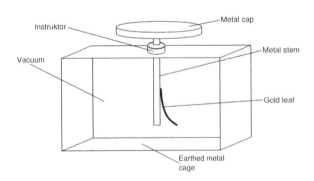

The gold leaf electroscope

Gold-leaf electroscopes are designed to detect static electricity. Charge placed on the metal cap spreads to the stem and the gold leaf of the electroscope. Because they then have the same charge, the stem and leaf repel each other. This will cause the leaf to bend away from the stem.

The electroscope is an important tool in illustrating the photoelectric effect. For example, if the electroscope is negatively charged throughout, there is an excess of electrons and the leaf is separated from the stem. If high-frequency light shines on the cap, the electroscope discharges and the leaf will fall limp. This is because the frequency of the light shining on the cap is above the cap's threshold frequency. The photons in the light have enough energy to liberate electrons from the cap, reducing its negative charge. This will discharge a negatively charged electroscope and further charge a positive electroscope. However, if the electromagnetic radiation hitting the metal cap does not have a high enough frequency (its frequency is below the threshold value for the cap), then the leaf will never discharge, no matter how long one shines the low-frequency light at the cap.[58]:389–390

8.3.4 Photoelectron spectroscopy

Since the energy of the photoelectrons emitted is exactly the energy of the incident photon minus the material's work function or binding energy, the work function of a sample can be determined by bombarding it with a monochromatic X-ray source or UV source, and measuring the kinetic energy distribution of the electrons emitted.[14]:14–20

Photoelectron spectroscopy is usually done in a high-vacuum environment, since the electrons would be scattered by gas molecules if they were present. However, some companies are now selling products that allow photoemission in air. The light source can be a laser, a discharge tube, or a synchrotron radiation source.[59]

The concentric hemispherical analyser (CHA) is a typical electron energy analyzer, and uses an electric field to change the directions of incident electrons, depending on their kinetic energies. For every element and core (atomic orbital) there will be a different binding energy. The many electrons created from each of these combinations will show up as spikes in the analyzer output, and these can be used to determine the elemental composition of the sample.

8.3.5 Spacecraft

The photoelectric effect will cause spacecraft exposed to sunlight to develop a positive charge. This can be a major problem, as other parts of the spacecraft in shadow develop a negative charge from nearby plasma, and the imbalance can discharge through delicate electrical components. The static charge created by the photoelectric effect is self-limiting, though, because a more highly charged object gives up its electrons less easily.[60][61]

8.3.6 Moon dust

Light from the sun hitting lunar dust causes it to become charged through the photoelectric effect. The charged dust then repels itself and lifts off the surface of the Moon by electrostatic levitation.[62][63] This manifests itself almost like an "atmosphere of dust", visible as a thin haze and blurring of distant features, and visible as a dim glow after the sun has set. This was first photographed by the Surveyor program probes in the 1960s. It is thought that the smallest particles are repelled up to kilometers high, and that the particles move in "fountains" as they charge and discharge.

8.3.7 Night vision devices

Photons hitting a thin film of alkali metal or semiconductor material such as gallium arsenide in an image intensifier tube cause the ejection of photoelectrons due to the photoelectric effect. These are accelerated by an electrostatic field where they strike a phosphor coated screen, converting the electrons back into photons. Intensification of the signal is achieved either through acceleration of the electrons or by increasing the number of electrons through secondary emissions, such as with a micro-channel plate. Sometimes a combination of both methods is used. Additional kinetic energy is required to move an electron out of the conduction band and into the vacuum level. This is known as the electron affinity of the photocathode and is another barrier to photoemission other than the forbidden band, explained by the band gap model. Some materials such as Gallium Arsenide have an effective electron affinity that is below the level of the conduction band. In these materials, electrons that move to the conduction band are all of sufficient energy to be emitted from the material and as such, the film that absorbs photons can be quite thick. These materials are known as negative electron affinity materials.

8.4 Cross section

The photoelectric effect is one interaction mechanism between photons and atoms. It is one of 12 theoretically possible interactions.[64]

At the high photon energies comparable to the electron rest energy of 511 keV, Compton scattering, another process, may take place. Above twice this (1.022 MeV) pair production may take place.[65] Compton scattering and pair production are examples of two other competing mechanisms.

Indeed, even if the photoelectric effect is the favoured reaction for a particular single-photon bound-electron interaction, the result is also subject to statistical processes and is not guaranteed, albeit the photon has certainly disappeared and a bound electron has been excited (usually K or L shell electrons at gamma ray energies). The probability of the photoelectric effect occurring is measured by the cross section of interaction, σ. This has been found to be a function of the atomic number of the target atom and photon energy. A crude approximation, for photon energies above the highest atomic binding energy, is given by:[66]

$$\sigma = \text{constant} \cdot \frac{Z^n}{E^3}$$

Here Z is atomic number and n is a number which varies between 4 and 5. (At lower photon energies a characteristic structure with edges appears, K edge, L edges, M edges, etc.) The obvious interpretation follows that the photoelectric effect rapidly decreases in significance, in the gamma ray region of the spectrum, with increasing photon energy, and that photoelectric effect increases steeply with atomic number. The corollary is that high-Z materials make good gamma-ray shields, which is the principal reason that lead ($Z = 82$) is a preferred and ubiquitous gamma radiation shield.[67]

8.5 See also

- Anomalous photovoltaic effect
- Dember effect
- Photo-Dember
- Photomagnetic effect
- Photochemistry
- Timeline of mechanics and physics

8.6 References

[1] Serway, R. A. (1990). *Physics for Scientists & Engineers* (3rd ed.). Saunders. p. 1150. ISBN 0-03-030258-7.

[2] Sears, F. W.; Zemansky, M. W.; Young, H. D. (1983). *University Physics* (6th ed.). Addison-Wesley. pp. 843–844. ISBN 0-201-07195-9.

[3] Hertz, H. (1887). "Ueber den Einfluss des ultravioletten Lichtes auf die electrische Entladung" [On an effect of ultra-violet light upon the electrical discharge]. *Annalen der Physik*. **267** (8): S. 983–1000. Bibcode:1887AnP...267..983H. doi:10.1002/andp.18872670827.

[4] "The Nobel Prize in Physics 1921". Nobel Foundation. Retrieved 2013-03-16.

[5] "The Nobel Prize in Physics 1923". Nobel Foundation. Retrieved 2015-03-29.

[6] Lenard, P. (1902). "Ueber die lichtelektrische Wirkung". *Annalen der Physik*. **313** (5): 149–198. Bibcode:1902AnP...313..149L. doi:10.1002/andp.19023130510.

[7] Millikan, R. (1914). "A Direct Determination of "*h*."". *Physical Review*. **4** (1): 73–75. Bibcode:1914PhRv....4R..73M. doi:10.1103/PhysRev.4.73.2.

[8] Millikan, R. (1916). "A Direct Photoelectric Determination of Planck's "*h*"" (PDF). *Physical Review*. **7** (3): 355–388. Bibcode:1916PhRv....7..355M. doi:10.1103/PhysRev.7.355.

[9] Zhang, Q. (1996). "Intensity dependence of the photoelectric effect induced by a circularly polarized laser beam". *Physics Letters A*. **216** (1–5): 125. Bibcode:1996PhLA..216..125Z. doi:10.1016/0375-9601(96)00259-9.

[10] Bubb, F. (1924). "Direction of Ejection of Photo-Electrons by Polarized X-rays". *Physical Review*. **23** (2): 137–143. Bibcode:1924PhRv...23..137B. doi:10.1103/PhysRev.23.137.

[11] Mee, C.; Crundell, M.; Arnold, B.; Brown, W. (2011). *International A/AS Level Physics*. Hodder Education. p. 241. ISBN 978-0-340-94564-3.

[12] Fromhold, A. T. (1991). *Quantum Mechanics for Applied Physics and Engineering*. Courier Dover Publications. pp. 5–6. ISBN 978-0-486-66741-6.

[13] Gautreau, R.; Savin, W. (1999). *Schaum's Outline of Modern Physics* (2nd ed.). McGraw-Hill. pp. 60–61. ISBN 0-07-024830-3.

[14] Hüfner, S. (2003). *Photoelectron Spectroscopy: Principles and Applications*. Springer. ISBN 3-540-41802-4.

[15] Report of the Board of Regents By Smithsonian Institution. Board of Regents, United States National Museum, Smithsonian Institution. p. 239.

[16] Schmidt, G. C. (1898) Wied. Ann. Uiv. p. 708.

[17] Knoblauch, O. (1899). *Zeitschrift für Physikalische Chemie*. **xxix**. p. 527.

[18] Vesselinka Petrova-Koch; Rudolf Hezel; Adolf Goetzberger (2009). *High-Efficient Low-Cost Photovoltaics: Recent Developments*. Springer. pp. 1–. ISBN 978-3-540-79358-8.

[19] Smith, W. (1873). "Effect of Light on Selenium during the passage of an Electric Current". *Nature*. **7** (173): 303. Bibcode:1873Natur...7R.303.. doi:10.1038/007303e0.

[20] Asimov, A. (1964) *Asimov's Biographical Encyclopedia of Science and Technology*, Doubleday, ISBN ISBN 0-385-04693-6.

[21] Robert Bud; Deborah Jean Warner (1998). *Instruments of Science: An Historical Encyclopedia*. Science Museum, London, and National Museum of American History, Smithsonian Institution. ISBN 978-0-8153-1561-2.

[22] Elster and Geitel arrange the metals in the following order with respect to their power of discharging negative electricity: rubidium, potassium, alloy of potassium and sodium, sodium, lithium, magnesium, thallium and zinc. For copper, platinum, lead, iron, cadmium, carbon, and mercury the effects with ordinary light are too small to be measurable. The order of the metals for this effect is the same as in Volta's series for contact-electricity, the most electropositive metals giving the largest photo-electric effect.

[23] Hertz, Wied. Ann. xxxi. p. 983, 1887.

[24] Hallwachs, Wied. Ann. xxxiii. p. 301, 1888.

[25] Hoor, Repertorium des Physik, xxv. p. 91, 1889.

[26] Bighi, C. R. cvi. p. 1349; cvii. p. 559, 1888

[27] Stoletow. C. R. cvi. pp. 1149, 1593; cvii. p. 91; cviii. p. 1241; Physikalische Revue, Bd. i., 1892.

[28] Stoletow, A. (1888). "Sur une sorte de courants electriques provoques par les rayons ultraviolets". *Comptes Rendus*. **CVI**: 1149. (Reprinted in Stoletow, M.A. (1888). "On a kind of electric current produced by ultra-violet rays". *Philosophical Magazine Series 5*. **26** (160): 317. doi:10.1080/14786448808628270.; abstract in Beibl. Ann. d. Phys. 12, 605, 1888).

[29] Stoletow, A. (1888). "Sur les courants actino-electriqies au travers deTair". *Comptes Rendus*. **CVI**: 1593. (Abstract in Beibl. Ann. d. Phys. 12, 723, 1888).

[30] Stoletow, A. (1888). "Suite des recherches actino-electriques". *Comptes Rendus*. **CVII**: 91. (Abstract in Beibl. Ann. d. Phys. 12, 723, 1888).

[31] Stoletow, A. (1889). "Sur les phénomènes actino-électriques". *Comptes Rendus*. CVIII: 1241.

[32] Stoletow, A. (1889). "Актино-электрические исследования". *Journal of the Russian Physico-chemical Society* (in Russian). **21**: 159.

[33] Stoletow, A. (1890). "Sur les courants actino-électriques dans l'air raréfié". *Journal de Physique*. **9**: 468. doi:10.1051/jphystap:018900090046800.

[34] Elster and Geitel, Wied. Ann. xxxviii. pp. 40, 497, 1889; xli. p. 161, 1890; xlii. p. 564, 1891; xliii. p. 225, 1892; lii. p. 433, 1894 ; lv. p. 684, 1895.

[35] Thomson, J. J. (2005). *Conduction of Electricity Through Gases*. Watchmaker Publishing. ISBN 978-1-929148-49-3. Retrieved 9 July 2011.

8.6. REFERENCES

[36] Hallwachs, W. (1907). "Über die lichtelektrische Ermüdung". *Annalen der Physik*. **328** (8): 459–516. Bibcode:1907AnP...328..459H. doi:10.1002/andp.19073280807.

[37] Stoletow, A. (1888). "Sur une sorte de courants electriques provoques par les rayons ultraviolets". *Comptes Rendus*. **CVI**: 1149. (Reprinted in Stoletow, M.A. (1888). "On a kind of electric current produced by ultra-violet rays". *Philosophical Magazine Series 5*. **26** (160): 317. doi:10.1080/14786448808628270.; abstract in Beibl. Ann. d. Phys. 12, 605, 1888).

[38] Stoletow, A. (1888). "Sur les courants actino-electriques au travers deTair". *Comptes Rendus*. **CVI**: 1593. (Abstract in Beibl. Ann. d. Phys. 12, 723, 1888).

[39] Stoletow, A. (1888). "Suite des recherches actino-électriques". *Comptes Rendus*. **CVII**: 91. (Abstract in Beibl. Ann. d. Phys. 12, 723, 1888).

[40] Stoletow, A. (1889). "Sur les phénomènes actino-électriques". *Comptes Rendus*. CVIII: 1241.

[41] Stoletow, A. (1889). "Актино-электрические исследования". *Journal of the Russian Physico-chemical Society* (in Russian). **21**: 159.

[42] Stoletow, A. (1890). "Sur les courants actino-électriques dans l'air raréfié". *Journal de Physique*. **9**: 468. doi:10.1051/jphystap:018900090046800.

[43] *The International Year Book*. (1900). New York: Dodd, Mead & Company. p. 659.

[44] Buchwald, Jed; Warwick, Andrew, eds. (2004). *Histories of the Electron: The Birth of Microphysics* (PDF) (illustrated, reprint ed.). MIT Press. pp. 21–23. ISBN 978-0-262-52424-7.

[45] Bloch, E. (1908). "L'ionisation de l'air par la lumière ultraviolette". *Le Radium*. **5** (8): 240. doi:10.1051/radium:0190800508024001.

[46] Thomson, J. J. (1907). "On the Ionisation of Gases by Ultra-Violet Light and on the evidence as to the Structure of Light afforded by its Electrical Effects". *Proc. Cambr. Phil. Soc.* **14**: 417.

[47] Palmer, Frederic (1908). "Ionisation of Air by Ultra-violet Light". *Nature*. **77** (2008): 582–582. Bibcode:1908Natur..77..582P. doi:10.1038/077582b0.

[48] Palmer, Frederic (1911). "Volume Ionization Produced by Light of Extremely Short Wave-Length". *Physical Review (Series I)*. **32**: 1–22. Bibcode:1911PhRvI..32....1P. doi:10.1103/PhysRevSeriesI.32.1.

[49] "The Nobel Prize in Physics 1921". Nobel Foundation. Retrieved 2008-10-09.

[50] Lamb, Willis E.; Scully, Marlan O. (1968). "Photoelectric effect without photons, discussing classical field falling on quantized atomic electron" (PDF).

[51] Planck, Max (1901). "Ueber das Gesetz der Energieverteilung im Normalspectrum (On the Law of Distribution of Energy in the Normal Spectrum)". *Annalen der Physik*. **4** (3): 553. Bibcode:1901AnP...309..553P. doi:10.1002/andp.19013090310.

[52] Resnick, Robert (1972) *Basic Concepts in Relativity and Early Quantum Theory*, Wiley, p. 137, ISBN 0-471-71702-9.

[53] Knight, Randall D. (2004) *Physics for Scientists and Engineers With Modern Physics: A Strategic Approach*, Pearson-Addison-Wesley, p. 1224, ISBN 0-8053-8685-8.

[54] Penrose, Roger (2005) *The Road to Reality: A Complete Guide to the Laws of the Universe*, Knopf, p. 502, ISBN 0-679-45443-8

[55] Resnick, Robert (1972) *Basic Concepts in Relativity and Early Quantum Theory*, Wiley, p. 138, ISBN 0-471-71702-9.

[56] Timothy, J. Gethyn (2010) in Huber, Martin C.E. (ed.) *Observing Photons in Space*, ISSI Scientific Report 009, ESA Communications, pp. 365–408, ISBN 978-92-9221-938-3

[57] Burns, R. W. (1998) *Television: An International History of the Formative Years*, IET, p. 358, ISBN 0-85296-914-7.

[58] Tsokos, K. A. (2010). *Cambridge Physics for the IB Diploma* (revised ed.). Cambridge University Press. ISBN 978-0-521-13821-5.

[59] Weaver, J. H.; Margaritondo, G. (1979). "Solid-State Photoelectron Spectroscopy with Synchrotron Radiation". *Science*. **206** (4415): 151–156. Bibcode:1979Sci...206..151W. doi:10.1126/science.206.4415.151. PMID 17801770.

[60] Lai, Shu T. (2011). *Fundamentals of Spacecraft Charging: Spacecraft Interactions with Space Plasmas* (illustrated ed.). Princeton University Press. pp. 1–6. ISBN 978-0-691-12947-1.

[61] "Spacecraft charging". *Arizona State University*.

[62] Bell, Trudy E., "Moon fountains", NASA.gov, 2005-03-30.

[63] Dust gets a charge in a vacuum. spacedaily.com, July 14, 2000.

[64] Evans, R. D. (1955). *The Atomic Nucleus*. Malabar, Fla.: Krieger. p. 673. ISBN 0-89874-414-8.

[65] Evans, R. D. (1955). *The Atomic Nucleus*. Malabar, Fla.: Krieger. p. 712. ISBN 0-89874-414-8.

[66] Davisson, C. M. (1965). *Interaction of gamma-radiation with matter*. pp. 37–78. Bibcode:1965abgs.conf...37D.

[67] Knoll, Glenn F. (1999). *Radiation Detection and Measurement*. New York: Wiley. p. 49. ISBN 0-471-49545-X.

8.7 External links

- AstronomyCast "http://www.astronomycast.com/2014/02/ep-335-photoelectric-effect/". AstronomyCast.

- Nave, R., "*Wave-Particle Duality*". HyperPhysics.

- "*Photoelectric effect*". Physics 2000. University of Colorado, Boulder, Colorado.

- ACEPT W3 Group, "*The Photoelectric Effect*". Department of Physics and Astronomy, Arizona State University, Tempe, AZ.

- Haberkern, Thomas, and N Deepak "*Grains of Mystique: Quantum Physics for the Layman*". Einstein Demystifies Photoelectric Effect, Chapter 3.

- Department of Physics, "*The Photoelectric effect*". Physics 320 Laboratory, Davidson College, Davidson.

- Fowler, Michael, "*The Photoelectric Effect*". Physics 252, University of Virginia.

- Go to "*Concerning an Heuristic Point of View Toward the Emission and Transformation of Light*" to read an English translation of Einstein's 1905 paper. (Retrieved: 2014 Apr 11)

- http://www.chemistryexplained.com/Ru-Sp/Solar-Cells.html

- Photo-electric transducers: http://sensorse.com/page4en.html

Applets

- "*Photoelectric Effect*". The Physics Education Technology (PhET) project. (Java)

- Fendt, Walter, "*The Photoelectric Effect*". (Java)

- "*Applet: Photo Effect*". Open Source Distributed Learning Content Management and Assessment System. (Java)

Chapter 9

Wave–particle duality

Wave–particle duality is the concept that every elementary particle or quantic entity may be partly described in terms not only of particles, but also of waves. It expresses the inability of the classical concepts "particle" or "wave" to fully describe the behavior of quantum-scale objects. As Albert Einstein wrote: *"It seems as though we must use sometimes the one theory and sometimes the other, while at times we may use either. We are faced with a new kind of difficulty. We have two contradictory pictures of reality; separately neither of them fully explains the phenomena of light, but together they do"*.[1]

Through the work of Max Planck, Einstein, Louis de Broglie, Arthur Compton, Niels Bohr and many others, current scientific theory holds that all particles also have a wave nature (and vice versa).[2] This phenomenon has been verified not only for elementary particles, but also for compound particles like atoms and even molecules. For macroscopic particles, because of their extremely short wavelengths, wave properties usually cannot be detected.[3]

Although the use of the wave-particle duality has worked well in physics, the *meaning* or *interpretation* has not been satisfactorily resolved; see Interpretations of quantum mechanics.

Niels Bohr regarded the "duality paradox" as a fundamental or metaphysical fact of nature. A given kind of quantum object will exhibit sometimes wave, sometimes particle, character, in respectively different physical settings. He saw such duality as one aspect of the concept of complementarity.[4] Bohr regarded renunciation of the cause-effect relation, or complementarity, of the space-time picture, as essential to the quantum mechanical account.[5]

Werner Heisenberg considered the question further. He saw the duality as present for all quantic entities, but not quite in the usual quantum mechanical account considered by Bohr. He saw it in what is called second quantization, which generates an entirely new concept of fields which exist in ordinary space-time, causality still being visualizable. Classical field values (e.g. the electric and magnetic field strengths of Maxwell) are replaced by an entirely new kind of field value, as considered in quantum field theory. Turning the reasoning around, ordinary quantum mechanics can be deduced as a specialized consequence of quantum field theory.[6][7]

9.1 Brief history of wave and particle viewpoints

Democritus—the original *atomist*—argued that all things in the universe, including light, are composed of indivisible sub-components (light being some form of solar atom).[8] At the beginning of the 11th Century, the Arabic scientist Alhazen wrote the first comprehensive treatise on optics; describing refraction, reflection, and the operation of a pinhole lens via rays of light traveling from the point of emission to the eye. He asserted that these rays were composed of particles of light. In 1630, René Descartes popularized and accredited the opposing wave description in his treatise on light, showing that the behavior of light could be re-created by modeling wave-like disturbances in a universal medium ("plenum"). Beginning in 1670 and progressing over three decades, Isaac Newton developed and championed his corpuscular hypothesis, arguing that the perfectly straight lines of reflection demonstrated light's particle nature; only particles could travel in such straight lines. He explained refraction by positing that particles of light accelerated laterally upon entering a denser medium. Around the same time, Newton's contemporaries Robert Hooke and Christiaan Huygens—and later Augustin-Jean Fresnel—mathematically refined the wave viewpoint, showing that if light traveled at different speeds in different media (such as water and air), refraction could be easily explained as the medium-dependent propagation of light waves. The resulting Huygens–Fresnel principle was extremely successful at reproducing light's behavior and was subsequently supported by Thomas Young's 1803 discovery of double-slit interference.[9][10] The wave view did not immediately displace the ray and particle view, but began to dominate scientific thinking about light in the mid 19th century, since it could explain polarization phenomena that the alternatives could not.[11]

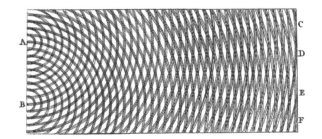

Thomas Young's sketch of two-slit diffraction of waves, 1803

James Clerk Maxwell discovered that he could combine four simple equations, which had been previously discovered, along with a slight modification to describe self-propagating waves of oscillating electric and magnetic fields. When the propagation speed of these electromagnetic waves was calculated, the speed of light fell out. It quickly became apparent that visible light, ultraviolet light, and infrared light (phenomena thought previously to be unrelated) were all electromagnetic waves of differing frequency. The wave theory had prevailed—or at least it seemed to.

While the 19th century had seen the success of the wave theory at describing light, it had also witnessed the rise of the atomic theory at describing matter. Antoine Lavoisier deduced the law of conservation of mass and categorized many new chemical elements and compounds; and Joseph Louis Proust advanced chemistry towards the atom by showing that elements combined in definite proportions. This led John Dalton to propose that elements were invisible sub components; Amedeo Avogadro discovered diatomic gases and completed the basic atomic theory, allowing the correct molecular formulae of most known compounds—as well as the correct weights of atoms—to be deduced and categorized in a consistent manner. Dimitri Mendeleev saw an order in recurring chemical properties, and created a table presenting the elements in unprecedented order and symmetry.

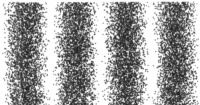 Particle impacts make visible the interference pattern of waves.

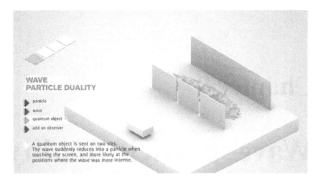

*Animation showing the wave-particle duality with a double slit experiment and effect of an observer. Increase size to see explanations in the video itself. See also **quiz based on this animation**.*

 A quantum particle is represented by a wave packet.

 Interference of a quantum particle with itself.
Click images for animations.

9.2 Turn of the 20th century and the paradigm shift

9.2.1 Particles of electricity

At the close of the 19th century, the reductionism of atomic theory began to advance into the atom itself; determining, through physics, the nature of the atom and the operation of chemical reactions. Electricity, first thought to be a fluid, was now understood to consist of particles called electrons. This was first demonstrated by J. J. Thomson in 1897 when, using a cathode ray tube, he found that an electrical charge would travel across a vacuum (which would possess infinite resistance in classical theory). Since the vacuum offered no medium for an electric fluid to travel, this discovery could only be explained via a particle carrying a negative charge and moving through the vacuum.

This *electron* flew in the face of classical electrodynamics, which had successfully treated electricity as a fluid for many years (leading to the invention of batteries, electric motors, dynamos, and arc lamps). More importantly, the intimate relation between electric charge and electromagnetism had been well documented following the discoveries of Michael Faraday and James Clerk Maxwell. Since electromagnetism was *known* to be a wave generated by a changing electric or magnetic *field* (a continuous, wavelike entity itself) an atomic/particle description of electricity and charge was a non sequitur. Furthermore, classical electrodynamics was not the only classical theory rendered incomplete.

9.2.2 Radiation quantization

Main article: Planck's law

In 1901, Max Planck published an analysis that succeeded in reproducing the observed spectrum of light emitted by a glowing object. To accomplish this, Planck had to make an ad hoc mathematical assumption of quantized energy of the oscillators (atoms of the black body) that emit radiation. It was Einstein who later proposed that it is the electromagnetic radiation itself that is quantized, and not the energy of radiating atoms.

Black-body radiation, the emission of electromagnetic energy due to an object's heat, could not be explained from classical arguments alone. The equipartition theorem of classical mechanics, the basis of all classical thermodynamic theories, stated that an object's energy is partitioned equally among the object's vibrational modes. But applying the same reasoning to the electromagnetic emission of such a thermal object was not so successful. It had been long known that thermal objects emit light. Since light was known to be waves of electromagnetism, physicists hoped to describe this emission via classical laws. This became known as the black body problem. Since the equipartition theorem worked so well in describing the vibrational modes of the thermal object itself, it was natural to assume that it would perform equally well in describing the radiative emission of such objects. But a problem quickly arose: if each mode received an equal partition of energy, the short wavelength modes would consume all the energy. This became clear when plotting the Rayleigh–Jeans law which, while correctly predicting the intensity of long wavelength emissions, predicted infinite total energy as the intensity diverges to infinity for short wavelengths. This became known as the ultraviolet catastrophe.

In 1900, Max Planck hypothesized that the frequency of light emitted by the black body depended on the frequency of the *oscillator* that emitted it, and the energy of these oscillators increased linearly with frequency (according to his constant h, where $E = h\nu$). This was not an unsound proposal considering that macroscopic oscillators operate similarly: when studying five simple harmonic oscillators of equal amplitude but different frequency, the oscillator with the highest frequency possesses the highest energy (though this relationship is not linear like Planck's). By demanding that high-frequency light must be emitted by an oscillator of equal frequency, and further requiring that this oscillator occupy higher energy than one of a lesser frequency, Planck avoided any catastrophe; giving an equal partition to high-frequency oscillators produced successively fewer oscillators and less emitted light. And as in the Maxwell–Boltzmann distribution, the low-frequency, low-energy oscillators were suppressed by the onslaught of thermal jiggling from higher energy oscillators, which necessarily increased their energy and frequency.

The most revolutionary aspect of Planck's treatment of the black body is that it inherently relies on an integer number of oscillators in thermal equilibrium with the electromagnetic field. These oscillators *give* their entire energy to the electromagnetic field, creating a quantum of light, as often as they are *excited* by the electromagnetic field, absorbing a quantum of light and beginning to oscillate at the corresponding frequency. Planck had intentionally created an atomic theory of the black body, but had unintentionally generated an atomic theory of light, where the black body never generates quanta of light at a given frequency with an energy less than **hν**. However, once realizing that he had quantized the electromagnetic field, he denounced particles of light as a limitation of his approximation, not a property of reality.

9.2.3 Photoelectric effect illuminated

While Planck had solved the ultraviolet catastrophe by using atoms and a quantized electromagnetic field, most contemporary physicists agreed that Planck's "light quanta" represented only flaws in his model. A more-complete derivation of black body radiation would yield a fully continuous and 'wave-like' electromagnetic field with no quantization. However, in 1905 Albert Einstein took Planck's black body model to produce his solution to another outstanding problem of the day: the photoelectric effect, wherein electrons are emitted from atoms when they absorb energy from light. Since their discovery eight years previously, electrons had been studied in physics laboratories worldwide.

In 1902 Philipp Lenard discovered that the energy of these ejected electrons did *not* depend on the intensity of the incoming light, but instead on its *frequency*. So if one shines a little low-frequency light upon a metal, a few low energy electrons are ejected. If one now shines a very intense beam of low-frequency light upon the same metal, a whole slew of electrons are ejected; however they possess the same low energy, there are merely *more of them*. The more light there is, the more electrons are ejected. Whereas in order to get high energy electrons, one must illuminate the metal with high-frequency light. Like blackbody radiation, this was at odds with a theory invoking continuous transfer of energy between radiation and matter. However, it can still be explained using a fully classical description of light, as long as matter is quantum mechanical in nature.[12]

If one used Planck's energy quanta, and demanded that electromagnetic radiation at a given frequency could only transfer energy to matter in integer multiples of an energy quantum **hv**, then the photoelectric effect could be explained very simply. Low-frequency light only ejects low-energy electrons because each electron is excited by the absorption of a single photon. Increasing the intensity of the low-frequency light (increasing the number of photons) only increases the number of excited electrons, not their energy, because the energy of each photon remains low. Only by increasing the frequency of the light, and thus increasing the energy of the photons, can one eject electrons with higher energy. Thus, using Planck's constant h to determine the energy of the photons based upon their frequency, the energy of ejected electrons should also increase linearly with frequency; the gradient of the line being Planck's constant. These results were not confirmed until 1915, when Robert Andrews Millikan, who had previously determined the charge of the electron, produced experimental results in perfect accord with Einstein's predictions. While the energy of ejected electrons reflected Planck's constant, the existence of photons was not explicitly proven until the discovery of the photon antibunching effect, of which a modern experiment can be performed in undergraduate-level labs.[13] This phenomenon could only be explained via photons, and not through any semi-classical theory (which could alternatively explain the photoelectric effect). When Einstein received his Nobel Prize in 1921, it was not for his more difficult and mathematically laborious special and general relativity, but for the simple, yet totally revolutionary, suggestion of quantized light. Einstein's "light quanta" would not be called photons until 1925, but even in 1905 they represented the quintessential example of wave-particle duality. Electromagnetic radiation propagates following linear wave equations, but can only be emitted or absorbed as discrete elements, thus acting as a wave and a particle simultaneously.

9.2.4 Einstein's explanation of the photoelectric effect

Main article: Photoelectric effect

In 1905, Albert Einstein provided an explanation of the

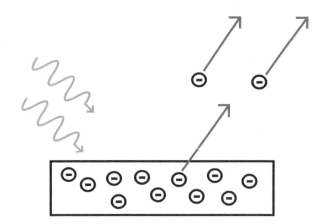

The photoelectric effect. Incoming photons on the left strike a metal plate (bottom), and eject electrons, depicted as flying off to the right.

photoelectric effect, a hitherto troubling experiment that the wave theory of light seemed incapable of explaining. He did so by postulating the existence of photons, quanta of light energy with particulate qualities.

In the photoelectric effect, it was observed that shining a light on certain metals would lead to an electric current in a circuit. Presumably, the light was knocking electrons out of the metal, causing current to flow. However, using the case of potassium as an example, it was also observed that while a dim blue light was enough to cause a current, even the strongest, brightest red light available with the technology of the time caused no current at all. According to the classical theory of light and matter, the strength or amplitude of a light wave was in proportion to its brightness: a bright light should have been easily strong enough to create a large current. Yet, oddly, this was not so.

Einstein explained this conundrum by postulating that the electrons can receive energy from electromagnetic field only in discrete portions (quanta that were called photons): an amount of energy E that was related to the frequency f of the light by

$$E = hf$$

where h is Planck's constant (6.626×10^{-34} J seconds). Only photons of a high enough frequency (above a certain *threshold* value) could knock an electron free. For example, photons of blue light had sufficient energy to free an electron from the metal, but photons of red light did not. One photon of light above the threshold frequency could release only one electron; the higher the frequency of a photon, the higher the kinetic energy of the emitted electron, but no amount of light (using technology available at the time) below the threshold frequency could release an electron. To "violate" this law would require extremely high-intensity lasers which had not yet been invented. Intensity-dependent phenomena have now been studied in detail with such lasers.[14]

Einstein was awarded the Nobel Prize in Physics in 1921 for his discovery of the law of the photoelectric effect.

9.2.5 De Broglie's wavelength

Main article: Matter wave
In 1924, Louis-Victor de Broglie formulated the de Broglie hypothesis, claiming that *all* matter,[15][16] not just light, has a wave-like nature; he related wavelength (denoted as λ), and momentum (denoted as p):

$$\lambda = \frac{h}{p}$$

This is a generalization of Einstein's equation above, since the momentum of a photon is given by $p = \frac{E}{c}$ and the wavelength (in a vacuum) by $\lambda = \frac{c}{f}$, where c is the speed of light in vacuum.

De Broglie's formula was confirmed three years later for electrons (which differ from photons in having a rest mass) with the observation of electron diffraction in two independent experiments. At the University of Aberdeen, George Paget Thomson passed a beam of electrons through a thin metal film and observed the predicted interference patterns. At Bell Labs, Clinton Joseph Davisson and Lester Halbert Germer guided their beam through a crystalline grid.

De Broglie was awarded the Nobel Prize for Physics in 1929 for his hypothesis. Thomson and Davisson shared the Nobel Prize for Physics in 1937 for their experimental work.

9.2.6 Heisenberg's uncertainty principle

Main article: Heisenberg uncertainty principle

In his work on formulating quantum mechanics, Werner Heisenberg postulated his uncertainty principle, which states:

$$\Delta x \Delta p \geq \frac{\hbar}{2}$$

where

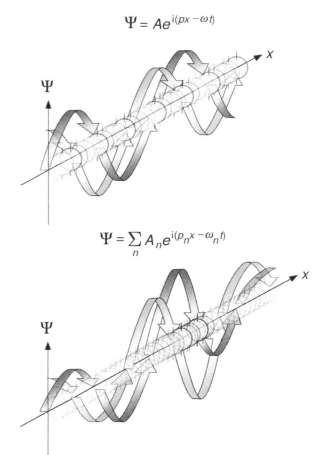

Propagation of de Broglie waves in 1d—real part of the complex amplitude is blue, imaginary part is green. The probability (shown as the colour opacity) of finding the particle at a given point x is spread out like a waveform; there is no definite position of the particle. As the amplitude increases above zero the curvature decreases, so the amplitude decreases again, and vice versa—the result is an alternating amplitude: a wave. Top: Plane wave. Bottom: Wave packet.

Δ here indicates standard deviation, a measure of spread or uncertainty;

x and **p** are a particle's position and linear momentum respectively.

$\hbar$ is the reduced Planck's constant (Planck's constant divided by 2π).

Heisenberg originally explained this as a consequence of the process of measuring: Measuring position accurately would disturb momentum and vice versa, offering an example (the "gamma-ray microscope") that depended crucially on the de Broglie hypothesis. It is now thought, however, that this only partly explains the phenomenon, but that the uncertainty also exists in the particle itself, even before the measurement is made.

In fact, the modern explanation of the uncertainty principle, extending the Copenhagen interpretation first put forward by Bohr and Heisenberg, depends even more centrally on the wave nature of a particle: Just as it is nonsensical to discuss the precise location of a wave on a string, particles do not have perfectly precise positions; likewise, just as it is nonsensical to discuss the wavelength of a "pulse" wave traveling down a string, particles do not have perfectly precise momenta (which corresponds to the inverse of wavelength). Moreover, when position is relatively well defined, the wave is pulse-like and has a very ill-defined wavelength (and thus momentum). And conversely, when momentum (and thus wavelength) is relatively well defined, the wave looks long and sinusoidal, and therefore it has a very ill-defined position.

9.2.7 de Broglie–Bohm theory

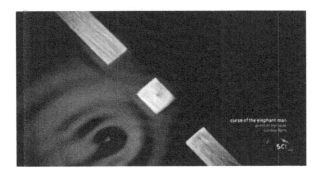

Couder experiments,[17] *"materializing" the* pilot wave *model.*

De Broglie himself had proposed a pilot wave construct to explain the observed wave-particle duality. In this view, each particle has a well-defined position and momentum, but is guided by a wave function derived from Schrödinger's equation. The pilot wave theory was initially rejected because it generated non-local effects when applied to systems involving more than one particle. Non-locality, however, soon became established as an integral feature of quantum theory (see EPR paradox), and David Bohm extended de Broglie's model to explicitly include it.

In the resulting representation, also called the de Broglie–Bohm theory or Bohmian mechanics,[18] the wave-particle duality vanishes, and explains the wave behaviour as a scattering with wave appearance, because the particle's motion is subject to a guiding equation or quantum potential. *"This idea seems to me so natural and simple, to resolve the wave-particle dilemma in such a clear and ordinary way, that it is a great mystery to me that it was so generally ignored"*,[19] J.S.Bell.

The best illustration of the *pilot-wave model* was given by Couder's 2010 "walking droplets" experiments,[20] demonstrating the pilot-wave behaviour in a macroscopic mechanical analog.[17]

9.3 Wave behavior of large objects

Since the demonstrations of wave-like properties in photons and electrons, similar experiments have been conducted with neutrons and protons. Among the most famous experiments are those of Estermann and Otto Stern in 1929.[21] Authors of similar recent experiments with atoms and molecules, described below, claim that these larger particles also act like waves. A wave is basically a group of particles which moves in a particular form of motion, i.e. to and fro. If we break that flow by an object it will convert into radiants.

A dramatic series of experiments emphasizing the action of gravity in relation to wave–particle duality was conducted in the 1970s using the neutron interferometer.[22] Neutrons, one of the components of the atomic nucleus, provide much of the mass of a nucleus and thus of ordinary matter. In the neutron interferometer, they act as quantum-mechanical waves directly subject to the force of gravity. While the results were not surprising since gravity was known to act on everything, including light (see tests of general relativity and the Pound–Rebka falling photon experiment), the self-interference of the quantum mechanical wave of a massive fermion in a gravitational field had never been experimentally confirmed before.

In 1999, the diffraction of C_{60} fullerenes by researchers from the University of Vienna was reported.[23] Fullerenes are comparatively large and massive objects, having an atomic mass of about 720 u. The de Broglie wavelength of the incident beam was about 2.5 pm, whereas the diameter of the molecule is about 1 nm, about 400 times larger. In 2012, these far-field diffraction experiments could be extended to phthalocyanine molecules and their heavier derivatives, which are composed of 58 and 114 atoms respectively. In these experiments the build-up of such interference patterns could be recorded in real time and with single molecule sensitivity.[24][25]

In 2003, the Vienna group also demonstrated the wave nature of tetraphenylporphyrin[26]—a flat biodye with an extension of about 2 nm and a mass of 614 u. For this demonstration they employed a near-field Talbot Lau interferometer.[27][28] In the same interferometer they also found interference fringes for $C_{60}F_{48}$., a fluorinated buckyball with a mass of about 1600 u, composed of 108 atoms.[26] Large molecules are already so complex that they give experimental access to some aspects of the quantum-classical interface, i.e., to certain decoherence mechanisms.[29][30] In 2011, the interference of molecules as heavy as 6910 u could be demonstrated in a Kapitza–Dirac–Talbot–Lau interferometer.[31] In 2013, the interference of molecules beyond 10,000 u has been demonstrated.[32]

Whether objects heavier than the Planck mass (about the weight of a large bacterium) have a de Broglie wavelength is theoretically unclear and experimentally unreachable; above the Planck mass a particle's Compton wavelength would be smaller than the Planck length and its own Schwarzschild radius, a scale at which current theories of physics may break down or need to be replaced by more general ones.[33]

Recently Couder, Fort, *et al.* showed[34] that we can use macroscopic oil droplets on a vibrating surface as a model of wave–particle duality—localized droplet creates periodical waves around and interaction with them leads to quantum-like phenomena: interference in double-slit experiment,[35] unpredictable tunneling[36] (depending in complicated way on practically hidden state of field), orbit quantization[37] (that particle has to 'find a resonance' with field perturbations it creates—after one orbit, its internal phase has to return to the initial state) and Zeeman effect.[38]

9.4 Treatment in modern quantum mechanics

Wave–particle duality is deeply embedded into the foundations of quantum mechanics. In the formalism of the theory, all the information about a particle is encoded in its *wave function*, a complex-valued function roughly analogous to the amplitude of a wave at each point in space. This function evolves according to a differential equation (generically called the Schrödinger equation). For particles with mass this equation has solutions that follow the form of the wave equation. Propagation of such waves leads to wave-like phenomena such as interference and diffraction. Particles without mass, like photons, have no solutions of the Schrödinger equation so have another wave.

The particle-like behavior is most evident due to phenomena associated with measurement in quantum mechanics. Upon measuring the location of the particle, the particle will be forced into a more localized state as given by the uncertainty principle. When viewed through this formalism, the measurement of the wave function will randomly "collapse", or rather "decohere", to a sharply peaked function at some location. For particles with mass the likelihood of detecting the particle at any particular location is equal to the squared amplitude of the wave function there. The measurement will return a well-defined position, (subject to uncertainty), a property traditionally associated with particles. It is important to note that a measurement is only a particular type of interaction where some data is recorded and the measured quantity is forced into a particular eigenstate. The act of measurement is therefore not fundamentally different from any other interaction.

Following the development of quantum field theory the ambiguity disappeared. The field permits solutions that follow the wave equation, which are referred to as the wave functions. The term particle is used to label the irreducible representations of the Lorentz group that are permitted by the field. An interaction as in a Feynman diagram is accepted as a calculationally convenient approximation where the outgoing legs are known to be simplifications of the propagation and the internal lines are for some order in an expansion of the field interaction. Since the field is non-local and quantized, the phenomena which previously were thought of as paradoxes are explained. Within the limits of the wave-particle duality the quantum field theory gives the same results.

9.4.1 Visualization

There are two ways to visualize the wave-particle behaviour: by the "standard model", described below; and by the Broglie–Bohm model, where no duality is perceived.

Below is an illustration of wave–particle duality as it relates to De Broglie's hypothesis and Heisenberg's uncertainty principle (above), in terms of the position and momentum space wavefunctions for one spinless particle with mass in one dimension. These wavefunctions are Fourier transforms of each other.

The more localized the position-space wavefunction, the more likely the particle is to be found with the position coordinates in that region, and correspondingly the momentum-space wavefunction is less localized so the possible momentum components the particle could have are more widespread.

Conversely the more localized the momentum-space wavefunction, the more likely the particle is to be found with those values of momentum components in that region, and correspondingly the less localized the position-space wavefunction, so the position coordinates the particle could occupy are more widespread.

9.5 Alternative views

Wave–particle duality is an ongoing conundrum in modern physics. Most physicists accept wave-particle duality as the best explanation for a broad range of observed phenomena; however, it is not without controversy. Alternative views are also presented here. These views are not generally accepted by mainstream physics, but serve as a basis for valu-

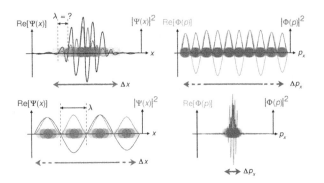

Position x and momentum p wavefunctions corresponding to quantum particles. The colour opacity (%) of the particles corresponds to the probability density of finding the particle with position x or momentum component p.
Top: *If wavelength λ is unknown, so are momentum p, wave-vector k and energy E (de Broglie relations). As the particle is more localized in position space, Δx is smaller than for Δp$_x$.*
Bottom: *If λ is known, so are p, k, and E. As the particle is more localized in momentum space, Δp is smaller than for Δx.*

able discussion within the community.

9.5.1 Both-particle-and-wave view

The pilot wave model, originally developed by Louis de Broglie and further developed by David Bohm into the hidden variable theory proposes that there is no duality, but rather a system exhibits both particle properties and wave properties simultaneously, and particles are guided, in a deterministic fashion, by the pilot wave (or its "quantum potential") which will direct them to areas of constructive interference in preference to areas of destructive interference. This idea is held by a significant minority within the physics community.[39]

At least one physicist considers the "wave-duality" as not being an incomprehensible mystery. L.E. Ballentine, *Quantum Mechanics, A Modern Development*, p. 4, explains:

> When first discovered, particle diffraction was a source of great puzzlement. Are "particles" really "waves?" In the early experiments, the diffraction patterns were detected holistically by means of a photographic plate, which could not detect individual particles. As a result, the notion grew that particle and wave properties were mutually incompatible, or complementary, in the sense that different measurement apparatuses would be required to observe them. That idea, however, was only an unfortunate generalization from a technological limitation. Today it is possible to detect the arrival of individual electrons, and to see the diffraction pattern emerge as a statistical pattern made up of many small spots (Tonomura et al., 1989). Evidently, quantum particles are indeed particles, but whose behaviour is very different from classical physics would have us to expect.

It has been claimed that the Afshar experiment[40] (2007) shows that it is possible to simultaneously observe both wave and particle properties of photons. This claim is, however, rejected by other scientists.

9.5.2 Wave-only view

At least one scientist proposes that the duality can be replaced by a "wave-only" view. In his book *Collective Electrodynamics: Quantum Foundations of Electromagnetism* (2000), Carver Mead purports to analyze the behavior of electrons and photons purely in terms of electron wave functions, and attributes the apparent particle-like behavior to quantization effects and eigenstates. According to reviewer David Haddon:[41]

> Mead has cut the Gordian knot of quantum complementarity. He claims that atoms, with their neutrons, protons, and electrons, are not particles at all but pure waves of matter. Mead cites as the gross evidence of the exclusively wave nature of both light and matter the discovery between 1933 and 1996 of ten examples of pure wave phenomena, including the ubiquitous laser of CD players, the self-propagating electrical currents of superconductors, and the Bose–Einstein condensate of atoms.

Albert Einstein, who, in his search for a Unified Field Theory, did not accept wave-particle duality, wrote:[42]

> This double nature of radiation (and of material corpuscles)...has been interpreted by quantum-mechanics in an ingenious and amazingly successful fashion. This interpretation...appears to me as only a temporary way out...

The many-worlds interpretation (MWI) is sometimes presented as a waves-only theory, including by its originator, Hugh Everett who referred to MWI as "the wave interpretation".[43]

The *Three Wave Hypothesis* of R. Horodecki relates the particle to wave.[44][45] The hypothesis implies that a massive particle is an intrinsically spatially as well as temporally extended wave phenomenon by a nonlinear law.

9.5.3 Particle-only view

Still in the days of the old quantum theory, a pre-quantum-mechanical version of wave–particle duality was pioneered by William Duane,[46] and developed by others including Alfred Landé.[47] Duane explained diffraction of x-rays by a crystal in terms solely of their particle aspect. The deflection of the trajectory of each diffracted photon was explained as due to quantized momentum transfer from the spatially regular structure of the diffracting crystal.[48]

9.5.4 Neither-wave-nor-particle view

It has been argued that there are never exact particles or waves, but only some compromise or intermediate between them. For this reason, in 1928 Arthur Eddington[49] coined the name "wavicle" to describe the objects although it is not regularly used today. One consideration is that zero-dimensional mathematical points cannot be observed. Another is that the formal representation of such points, the Dirac delta function is unphysical, because it cannot be normalized. Parallel arguments apply to pure wave states. Roger Penrose states:[50]

> "Such 'position states' are idealized wavefunctions in the opposite sense from the momentum states. Whereas the momentum states are infinitely spread out, the position states are infinitely concentrated. Neither is normalizable [...]."

9.5.5 Relational approach to wave–particle duality

Relational quantum mechanics is developed which regards the detection event as establishing a relationship between the quantized field and the detector. The inherent ambiguity associated with applying Heisenberg's uncertainty principle and thus wave–particle duality is subsequently avoided.[51]

9.6 Applications

Although it is difficult to draw a line separating wave–particle duality from the rest of quantum mechanics, it is nevertheless possible to list some applications of this basic idea.

- Wave–particle duality is exploited in electron microscopy, where the small wavelengths associated with the electron can be used to view objects much smaller than what is visible using visible light.

- Similarly, neutron diffraction uses neutrons with a wavelength of about 0.1 nm, the typical spacing of atoms in a solid, to determine the structure of solids.

- Photos are now able to show this dual nature, perhaps this will lead to new ways of examining and recording this behaviour.[52]

9.7 See also

- Arago spot
- Afshar experiment
- Basic concepts of quantum mechanics
- Complementarity (physics)
- Englert–Greenberger–Yasin duality relation
- Kapitsa–Dirac effect
- Electron wave-packet interference
- Faraday wave
- Hanbury Brown and Twiss effect
- Photon polarization
- Scattering theory
- Wavelet
- Wheeler's delayed choice experiment

9.8 Notes and references

[1] Harrison, David (2002). "Complementarity and the Copenhagen Interpretation of Quantum Mechanics". *UPSCALE*. Dept. of Physics, U. of Toronto. Retrieved 2008-06-21.

[2] Walter Greiner (2001). *Quantum Mechanics: An Introduction*. Springer. ISBN 3-540-67458-6.

[3] R. Eisberg & R. Resnick (1985). *Quantum Physics of Atoms, Molecules, Solids, Nuclei, and Particles* (2nd ed.). John Wiley & Sons. pp. 59–60. ISBN 047187373X. For both large and small wavelengths, both matter and radiation have both particle and wave aspects.... But the wave aspects of their motion become more difficult to observe as their wavelengths become shorter.... For ordinary macroscopic particles the mass is so large that the momentum is always sufficiently large to make the de Broglie wavelength small enough to be beyond the range of experimental detection, and classical mechanics reigns supreme.

[4] Kumar, Manjit (2011). *Quantum: Einstein, Bohr, and the Great Debate about the Nature of Reality* (Reprint ed.). W. W. Norton & Company. pp. 242, 375–376. ISBN 978-0393339888.

[5] Bohr, N. (1927/1928). The quantum postulate and the recent development of atomic theory, *Nature* Supplement April 14 1928, **121**: 580–590.

[6] Camilleri, K. (2009). *Heisenberg and the Interpretation of Quantum Mechanics: the Physicist as Philosopher*, Cambridge University Press, Cambridge UK, ISBN 978-0-521-88484-6.

[7] Preparata, G. (2002). *An Introduction to a Realistic Quantum Physics*, World Scientific, River Edge NJ, ISBN 978-981-238-176-7.

[8] Nathaniel Page Stites, M.A./M.S. "Light I: Particle or Wave?," Visionlearning Vol. PHY-1 (3), 2005. http://www.visionlearning.com/library/module_viewer.php?mid=132

[9] Young, Thomas (1804). "Bakerian Lecture: Experiments and calculations relative to physical optics". *Philosophical Transactions of the Royal Society*. **94**: 1–16. Bibcode:1804RSPT...94....1Y. doi:10.1098/rstl.1804.0001.

[10] Thomas Young: The Double Slit Experiment

[11] Buchwald, Jed (1989). *The Rise of the Wave Theory of Light: Optical Theory and Experiment in the Early Nineteenth Century*. Chicago: University of Chicago Press. ISBN 0-226-07886-8. OCLC 18069573.

[12] Lamb, Willis E.; Scully, Marlan O. (1968). "The photoelectric effect without photons" (PDF).

[13] "Observing the quantum behavior of light in an undergraduate laboratory". *American Journal of Physics*. **72**: 1210. Bibcode:2004AmJPh..72.1210T. doi:10.1119/1.1737397.

[14] Zhang, Q (1996). "Intensity dependence of the photoelectric effect induced by a circularly polarized laser beam". *Physics Letters A*. **216** (1-5): 125–128. Bibcode:1996PhLA..216..125Z. doi:10.1016/0375-9601(96)00259-9.

[15] Donald H Menzel, "*Fundamental formulas of Physics*", volume 1, page 153; Gives the de Broglie wavelengths for composite particles such as protons and neutrons.

[16] Brian Greene, The Elegant Universe, page 104 "all matter has a wave-like character"

[17] See this Science Channel production (Season II, Episode VI "How Does The Universe Work?"), presented by Morgan Freeman, https://www.youtube.com/watch?v=W9yWv5dqSKk

[18] Bohmian Mechanics, *Stanford Encyclopedia of Philosophy*.

[19] Bell, J. S., "Speakable and Unspeakable in Quantum Mechanics", Cambridge: Cambridge University Press, 1987.

[20] Y. Couder, A. Boudaoud, S. Protière, Julien Moukhtar, E. Fort: *Walking droplets: a form of wave-particle duality at macroscopic level?*, doi:10.1051/epn/2010101, (PDF)

[21] Estermann, I.; Stern O. (1930). "Beugung von Molekularstrahlen". *Zeitschrift für Physik*. **61** (1-2): 95–125. Bibcode:1930ZPhy...61...95E. doi:10.1007/BF01340293.

[22] R. Colella, A. W. Overhauser and S. A. Werner, Observation of Gravitationally Induced Quantum Interference, *Phys. Rev. Lett.* **34**, 1472–1474 (1975).

[23] Arndt, Markus; O. Nairz; J. Voss-Andreae, C. Keller, G. van der Zouw, A. Zeilinger (14 October 1999). "Wave–particle duality of C_{60}". *Nature*. **401** (6754): 680–682. Bibcode:1999Natur.401..680A. doi:10.1038/44348. PMID 18494170.

[24] Juffmann, Thomas; et al. (25 March 2012). "Real-time single-molecule imaging of quantum interference". Nature Nanotechnology. Retrieved 27 March 2012.

[25] Quantumnanovienna. "Single molecules in a quantum interference movie". Retrieved 2012-04-21.

[26] Hackermüller, Lucia; Stefan Uttenthaler; Klaus Hornberger; Elisabeth Reiger; Björn Brezger; Anton Zeilinger; Markus Arndt (2003). "The wave nature of biomolecules and fluorofullerenes". *Phys. Rev. Lett.* **91** (9): 090408. arXiv:quant-ph/0309016⊘. Bibcode:2003PhRvL..91i0408H. doi:10.1103/PhysRevLett.91.090408. PMID 14525169.

[27] Clauser, John F.; S. Li (1994). "Talbot von Lau interefometry with cold slow potassium atoms.". *Phys. Rev. A*. **49** (4): R2213–17. Bibcode:1994PhRvA..49.2213C. doi:10.1103/PhysRevA.49.R2213. PMID 9910609.

[28] Brezger, Björn; Lucia Hackermüller; Stefan Uttenthaler; Julia Petschinka; Markus Arndt; Anton Zeilinger (2002). "Matter-wave interferometer for large molecules". *Phys. Rev. Lett.* **88** (10): 100404. arXiv:quant-ph/0202158⊘. Bibcode:2002PhRvL..88j0404B. doi:10.1103/PhysRevLett.88.100404. PMID 11909334.

[29] Hornberger, Klaus; Stefan Uttenthaler; Björn Brezger; Lucia Hackermüller; Markus Arndt; Anton Zeilinger (2003). "Observation of Collisional Decoherence in Interferometry". *Phys. Rev. Lett.* **90** (16): 160401. arXiv:quant-ph/0303093⊘. Bibcode:2003PhRvL..90p0401H. doi:10.1103/PhysRevLett.90.160401. PMID 12731960.

[30] Hackermüller, Lucia; Klaus Hornberger; Björn Brezger; Anton Zeilinger; Markus Arndt (2004). "Decoherence of matter waves by thermal emission of radiation". *Nature*. **427** (6976): 711–714. arXiv:quant-ph/0402146⊘. Bibcode:2004Natur.427..711H. doi:10.1038/nature02276. PMID 14973478.

[31] Gerlich, Stefan; et al. (2011). "Quantum interference of large organic molecules". *Nature Communications*. **2** (263). Bibcode:2011NatCo...2E.263G.

doi:10.1038/ncomms1263. PMC 3104521⊖. PMID 21468015.

[32] Eibenberger, S.; Gerlich, S.; Arndt, M.; Mayor, M.; Tüxen, J. (2013). "Matter–wave interference of particles selected from a molecular library with masses exceeding 10 000 amu". *Physical Chemistry Chemical Physics.* **15** (35): 14696–14700. doi:10.1039/c3cp51500a. PMID 23900710.

[33] Peter Gabriel Bergmann, *The Riddle of Gravitation*, Courier Dover Publications, 1993 ISBN 0-486-27378-4 online

[34] http://www.youtube.com/watch?v=W9yWv5dqSKk - You Tube video - Yves Couder Explains Wave/Particle Duality via Silicon Droplets

[35] Y. Couder, E. Fort, *Single-Particle Diffraction and Interference at a Macroscopic Scale*, PRL 97, 154101 (2006) online

[36] A. Eddi, E. Fort, F. Moisy, Y. Couder, *Unpredictable Tunneling of a Classical Wave–Particle Association*, PRL 102, 240401 (2009)

[37] Fort, E.; Eddi, A.; Boudaoud, A.; Moukhtar, J.; Couder, Y. (2010). "Path-memory induced quantization of classical orbits". *PNAS.* **107** (41): 17515–17520. doi:10.1073/pnas.1007386107.

[38] http://prl.aps.org/abstract/PRL/v108/i26/e264503 - Level Splitting at Macroscopic Scale

[39] (Buchanan pp. 29–31)

[40] Afshar S.S. et al: Paradox in Wave Particle Duality. Found. Phys. 37, 295 (2007) http://arxiv.org/abs/quant-ph/0702188 arXiv:quant-ph/0702188

[41] David Haddon. "Recovering Rational Science". *Touchstone.* Retrieved 2007-09-12.

[42] Paul Arthur Schilpp, ed, *Albert Einstein: Philosopher-Scientist*, Open Court (1949), ISBN 0-87548-133-7, p 51.

[43] See section VI(e) of Everett's thesis: *The Theory of the Universal Wave Function*, in Bryce Seligman DeWitt, R. Neill Graham, eds, *The Many-Worlds Interpretation of Quantum Mechanics*, Princeton Series in Physics, Princeton University Press (1973), ISBN 0-691-08131-X, pp 3–140.

[44] Horodecki, R. (1981). "De broglie wave and its dual wave". *Phys. Lett. A.* **87** (3): 95–97. Bibcode:1981PhLA...87...95H. doi:10.1016/0375-9601(81)90571-5.

[45] Horodecki, R. (1983). "Superluminal singular dual wave". *Lett. Novo Cimento.* **38**: 509–511.

[46] Duane, W. (1923). The transfer in quanta of radiation momentum to matter, *Proc. Natl. Acad. Sci.* **9**(5): 158–164.

[47] Landé, A. (1951). *Quantum Mechanics*, Sir Isaac Pitman and Sons, London, pp. 19–22.

[48] Heisenberg, W. (1930). *The Physical Principles of the Quantum Theory*, translated by C. Eckart and F.C. Hoyt, University of Chicago Press, Chicago, pp. 77–78.

[49] Eddington, Arthur Stanley (1928). *The Nature of the Physical World*. Cambridge, UK.: MacMillan. p. 201.

[50] Penrose, Roger (2007). *The Road to Reality: A Complete Guide to the Laws of the Universe*. Vintage. p. 521, §21.10. ISBN 978-0-679-77631-4.

[51] http://www.quantum-relativity.org/Quantum-Relativity.pdf. See Q. Zheng and T. Kobayashi, *Quantum Optics as a Relativistic Theory of Light*; Physics Essays 9 (1996) 447. Annual Report, Department of Physics, School of Science, University of Tokyo (1992) 240.

[52] Ecole Polytechnique Federale de Lausanne. "The first ever photograph of light as both a particle and wave". Retrieved 15 July 2016.

9.9 External links

- Animation, applications and research linked to the wave-particle duality and other basic quantum phenomena (Université Paris Sud)

- H. Nikolic. "Quantum mechanics: Myths and facts". arXiv:quant-ph/0609163⊖.

- Young & Geller. "College Physics".

- B. Crowell. "Light as a Particle" (Web page). Retrieved December 10, 2006.

- E.H. Carlson, *Wave–Particle Duality: Light* on Project PHYSNET

- R. Nave. "Wave–Particle Duality" (Web page). *HyperPhysics*. Georgia State University, Department of Physics and Astronomy. Retrieved December 12, 2005.

- Juffmann, Thomas; et al. (25 March 2012). "Real-time single-molecule imaging of quantum interference". *Nature Nanotechnology*. Retrieved 21 January 2014.

Chapter 10

Squeezed coherent state

In physics, a **squeezed coherent state** is any state of the quantum mechanical Hilbert space such that the uncertainty principle is saturated. That is, the product of the corresponding two operators takes on its minimum value:

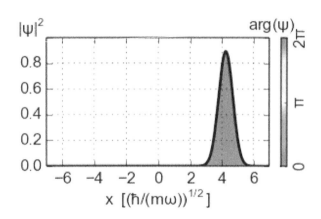

Animated position-wavefunction of an 2dB amplitude-squeezed coherent state of $\alpha=3$.

Husimi distribution of the squeezed coherent state

$$\Delta x \Delta p = \frac{\hbar}{2}$$

The simplest such state is the ground state $|0\rangle$ of the quantum harmonic oscillator. The next simple class of states that satisfies this identity are the family of coherent states $|\alpha\rangle$.

Often, the term *squeezed state* is used for any such state with $\Delta x \neq \Delta p$ in "natural oscillator units". The idea behind this is that the circle denoting a coherent state in a quadrature diagram (see below) has been "squeezed" to an ellipse of the same area. [1] [2] [3] [4] [5]

10.1 Mathematical definition

The most general wave function that satisfies the identity above is the **squeezed coherent state** (we work in units with $\hbar = 1$)

$$\psi(x) = C \exp\left(-\frac{(x-x_0)^2}{2w_0^2} + ip_0 x\right)$$

where C, x_0, w_0, p_0 are constants (a normalization constant, the center of the wavepacket, its width, and the expectation value of its momentum). The new feature relative to a coherent state is the free value of the width w_0, which is the reason why the state is called "squeezed".

The squeezed state above is an eigenstate of a linear operator

$$\hat{x} + i\hat{p}w_0^2$$

and the corresponding eigenvalue equals $x_0 + ip_0 w_0^2$. In this sense, it is a generalization of the ground state as well as the coherent state.

10.2 Operator representation of squeezed coherent states

The general form of a squeezed coherent state for a quantum harmonic oscillator is given by

$$|\alpha, \zeta\rangle = D(\alpha)S(\zeta)|0\rangle$$

where $|0\rangle$ is the vacuum state, $D(\alpha)$ is the displacement operator and $S(\zeta)$ is the squeeze operator, given by

$$D(\alpha) = \exp(\alpha \hat{a}^\dagger - \alpha^* \hat{a}) \quad \text{and} \quad S(\zeta) = \exp(\frac{1}{2}(\zeta^* \hat{a}^2 - \zeta \hat{a}^{\dagger 2}))$$

where $\hat{a}$ and $\hat{a}^\dagger$ are annihilation and creation operators, respectively. For a quantum harmonic oscillator of angular frequency ω, these operators are given by

$$\hat{a}^\dagger = \sqrt{\frac{m\omega}{2\hbar}}\left(x - \frac{ip}{m\omega}\right) \quad \text{and} \quad \hat{a} = \sqrt{\frac{m\omega}{2\hbar}}\left(x + \frac{ip}{m\omega}\right)$$

For a real ζ, the uncertainty in x and p are given by

$$(\Delta x)^2 = \frac{\hbar}{2m\omega}e^{-2r} \quad \text{and} \quad (\Delta p)^2 = \frac{m\hbar\omega}{2}e^{2r}$$

Therefore, a squeezed coherent state saturates the Heisenberg Uncertainty Principle $\Delta x \Delta p = \frac{\hbar}{2}$, with reduced uncertainty in one of its quadrature components and increased uncertainty in the other.

10.3 Examples of squeezed coherent states

Depending on at which phase the state's quantum noise is reduced, one can distinguish amplitude-squeezed and phase-squeezed states or general quadrature squeezed states. If no coherent excitation exists the state is called a squeezed vacuum. The figures below give a nice visual demonstration of the close connection between squeezed states and Heisenberg's uncertainty relation: Diminishing the quantum noise at a specific quadrature (phase) of the wave has as a direct consequence an enhancement of the noise of the complementary quadrature, that is, the field at the phase shifted by $\pi/2$.

From the top:

- Vacuum state
- Squeezed vacuum state
- Phase-squeezed state
- arbitrary squeezed state
- Amplitude-squeezed state

As can be seen at once, in contrast to the coherent state the quantum noise for a squeezed state is no longer independent of the phase of the light wave. A characteristic broadening and narrowing of the noise during one oscillation period can be observed. The wave packet of a squeezed state is defined by the square of the wave function introduced in the last paragraph. They correspond to the probability distribution of the electric field strength of the light wave. The moving wave packets display an oscillatory motion combined with the widening and narrowing of their distribution: the "breathing" of the wave packet. For an amplitude-squeezed state, the most narrow distribution of the wave packet is reached at the field maximum, resulting in an amplitude that is defined more precisely than the one of a coherent state. For a phase-squeezed state, the most narrow distribution is reached at field zero, resulting in an average phase value that is better defined than the one of a coherent state.

In phase space, quantum mechanical uncertainties can be depicted by the Wigner quasi-probability distribution. The intensity of the light wave, its coherent excitation, is given by the displacement of the Wigner distribution from the origin. A change in the phase of the squeezed quadrature results in a rotation of the distribution.

10.4 Photon number distributions and phase distributions of squeezed states

The squeezing angle, that is the phase with minimum quantum noise, has a large influence on the photon number distribution of the light wave and its phase distribution as well.

For amplitude squeezed light the photon number distribution is usually narrower than the one of a coherent state of the same amplitude resulting in sub-Poissonian light, whereas its phase distribution is wider. The opposite is true for the phase-squeezed light, which displays a large intensity (photon number) noise but a narrow phase distribution. Nevertheless, the statistics of amplitude squeezed light was not observed directly with photon number resolving detector due to experimental difficulty.[7]

For the squeezed vacuum state the photon number distribution displays odd-even-oscillations. This can be explained by the mathematical form of the squeezing operator, that

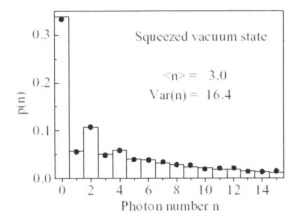

Figure 6: Reconstructed and theoretical photon number distributions for a squeezed-vacuum state. A pure squeezed vacuum state would have no contribution from odd-photon-number states. The non-zero contribution in the above figure is because the detected state is not a pure state - losses in the setup convert the pure squeezed vacuum into a mixed state.[6] (source: link 1)

resembles the operator for two-photon generation and annihilation processes. Photons in a squeezed vacuum state are more likely to appear in pairs.

10.5 Classification of squeezed states

10.5.1 Based on the number of modes

Squeezed states of light are broadly classified into single-mode squeezed states and two-mode squeezed states,[8] depending on the number of modes of the electromagnetic field involved in the process. Recent studies have looked into multimode squeezed states showing quantum correlations among more than two modes as well.

Single-mode squeezed states

Single-mode squeezed states, as the name suggests, consists of a single mode of the electromagnetic field whose one quadrature has fluctuations below the shot noise level, and the orthogonal quadrature has excess noise. Specifically, a single-mode squeezed *vacuum* (SMSV) state can be mathematically represented as,

$$|SMSV\rangle = S(\zeta)|0\rangle$$

where the squeezing operator S is the same as introduced in the section on operator representations above. In the photon number basis, this can be expanded as,

$$|SMSV\rangle = \frac{1}{\sqrt{\cosh r}} \sum_{n=0}^{\infty} (-\tanh r)^n \frac{\sqrt{(2n)!}}{2^n n!} |2n\rangle$$

which explicitly shows that the pure SMSV consists entirely of even-photon Fock state superpositions. Single mode squeezed states are typically generated by degenerate parametric oscillation in an optical parametric oscillator,[9] or using four-wave mixing.[5]

Two-mode squeezed states

Two-mode squeezing involves two modes of the electromagnetic field which exhibit quantum noise reduction below the shot noise level in a linear combination of the quadratures of the two fields. For example, the field produced by a nondegenerate parametric oscillator above threshold shows squeezing in the amplitude difference quadrature. The first experimental demonstration of two-mode squeezing in optics was by Heidmann et al..[10] More recently, two-mode squeezing was generated on-chip using a four-wave mixing OPO above threshold.[11] Two-mode squeezing is often seen as a precursor to continuous-variable entanglement, and hence a demonstration of the Einstein-Podolsky-Rosen paradox in its original formulation in terms of continuous position and momentum observables.[12][13] A two-mode squeezed vacuum (TMSV)state can be mathematically represented as,

$$|TMSV\rangle = S_2(\zeta)|0\rangle = \exp(\zeta^* \hat{a}\hat{b} - \zeta \hat{a}^\dagger \hat{b}^\dagger)|0\rangle$$

and in the photon number basis as,

$$|TMSV\rangle = \frac{1}{\cosh r} \sum_{n=0}^{\infty} (\tanh r)^n |nn\rangle$$

If the individual modes of a TMSV are considered separately (i.e., $|nn\rangle = |n\rangle|n\rangle$, say by tracing over or absorbing the other mode), then the remaining mode is left in a thermal state.

10.5.2 Based on the presence of a mean field

Squeezed states of light can be divided into squeezed vacuum and bright squeezed light, depending on the absence or presence of a non-zero mean field (also called a carrier), respectively. Interestingly, an Optical Parametric Oscillator operated below threshold produces squeezed vacuum, whereas the same OPO operated above threshold produces bright squeezed light. Bright squeezed light can be advantageous for certain quantum information processing applications as it obviates the need of sending local oscillator to provide a phase reference, whereas squeezed vacuum is considered more suitable for quantum enhanced sensing applications. The AdLIGO and GEO600 gravitational wave detectors use squeezed vacuum to achieve enhanced sensitivity beyond the standard quantum limit.[14][15]

10.6 Experimental realizations of squeezed coherent states

There has been a whole variety of successful demonstrations of squeezed states. The first demonstrations were experiments with light fields using lasers and non-linear optics (see optical parametric oscillator). This is achieved by a simple process of four-wave mixing with a $\chi^{(3)}$ crystal; similarly traveling wave phase-sensitive amplifiers generate spatially multimode quadrature-squeezed states of light when the $\chi^{(2)}$ crystal is pumped in absence of any signal. Sub-Poissonian current sources driving semiconductor laser diodes have led to amplitude squeezed light.[16]

Squeezed states have also been realized via motional states of an ion in a trap, phonon states in crystal lattices, and spin states in neutral atom ensembles.[17][18] Much progress has been made on the creation and observation of spin squeezed states in ensembles of neutral atoms and ions, which can be used to enhancement measurements of time, accelerations, and fields, and the current state of the art for measurement enhancement is 20 dB.[19][19][20][21][22] Generation of spin squeezed states have been demonstrated using both coherent evolution of a coherent spin state and projective, coherence-preserving measurements. Even macroscopic oscillators were driven into classical motional states that were very similar to squeezed coherent states. Current state of the art in noise suppression, for laser radiation using squeezed light, amounts to 12.7 dB.[23]

10.7 Applications

Squeezed states of the light field can be used to enhance precision measurements. For example, phase-squeezed light can improve the phase read out of interferometric measurements (see for example gravitational waves). Amplitude-squeezed light can improve the readout of very weak spectroscopic signals.

Spin squeezed states of atoms can be used to improve the precision of atom clocks.[24][25] This is an important problem in atomic clocks and other sensors that use small ensembles of cold atoms where the quantum projection noise represents a fundamental limitation to the precision of the sensor.[26]

Various squeezed coherent states, generalized to the case of many degrees of freedom, are used in various calculations in quantum field theory, for example Unruh effect and Hawking radiation, and generally, particle production in curved backgrounds and Bogoliubov transformations.

Recently, the use of squeezed states for quantum information processing in the continuous variables (CV) regime has been increasing rapidly.[27] Continuous variable quantum optics uses squeezing of light as an essential resource to realize CV protocols for quantum communication, unconditional quantum teleportation and one-way quantum computing.[28][29] This is in contrast to quantum information processing with single photons or photon pairs as qubits. CV quantum information processing relies heavily on the fact that squeezing is intimately related to quantum entanglement, as the quadratures of a squeezed state exhibit sub-shot-noise quantum correlations.

10.8 See also

- Negative energy
- Nonclassical light
- Optical Phase Space
- Quantum optics
- Squeeze operator
- Squeezed vacuum

10.9 References

[1] Loudon, Rodney, *The Quantum Theory of Light* (Oxford University Press, 2000), [ISBN 0-19-850177-3]

[2] D. F. Walls and G.J. Milburn, *Quantum Optics*, Springer Berlin 1994

[3] C W Gardiner and Peter Zoller, "Quantum Noise", 3rd ed, Springer Berlin 2004

[4] D. Walls, *Squeezed states of light*, Nature 306, 141 (1983)

[5] R. E. Slusher et al., *Observation of squeezed states generated by four wave mixing in an optical cavity*, Phys. Rev. Lett. 55 (22), 2409 (1985)

[6] G. Breitenbach, S. Schiller, and J. Mlynek, "Measurement of the quantum states of squeezed light", Nature, 387, 471 (1997)

[7] Entanglement evaluation with Fisher information - http://arxiv.org/pdf/quant-ph/0612099

[8] A. I. Lvovsky, "Squeezed light," http://arxiv.org/abs/1401.4118

[9] L.-A. Wu, M. Xiao, and H. J. Kimble, "Squeezed states of light from an optical parametric oscillator," J. Opt. Soc. Am. B 4, 1465 (1987).

[10] A. Heidmann, R. Horowicz, S. Reynaud, E. Giacobino, C. Fabre, and G. Camy, "Observation of Quantum Noise Reduction on Twin Laser Beams," Physical Review Letters 59, 2555 (1987).

[11] A. Dutt, K. Luke, S. Manipatruni, A. L. Gaeta, P. Nussenzveig, and M. Lipson, "On-Chip Optical Squeezing," Physical Review Applied 3, 044005 (2015). http://arxiv.org/abs/1309.6371

[12] Z. Y. Ou, S. F. Pereira, H. J. Kimble, and K. C. Peng, "Realization of the Einstein-Podolsky-Rosen paradox for continuous variables," Phys. Rev. Lett. 68, 3663 (1992).

[13] A. S. Villar, L. S. Cruz, K. N. Cassemiro, M. Martinelli, and P. Nussenzveig, "Generation of Bright Two-Color Continuous Variable Entanglement," Phys. Rev. Lett. 95, 243603 (2005).

[14] H. Grote, K. Danzmann, K. L. Dooley, R. Schnabel, J. Slutsky, and H. Vahlbruch, "First Long-Term Application of Squeezed States of Light in a Gravitational-Wave Observatory," Phys. Rev. Lett. 110, 181101 (2013) http://arxiv.org/abs/1302.2188

[15] The LIGO Scientific Collaboration, "A gravitational wave observatory operating beyond the quantum shot-noise limit," Nature Physics 7, 962 (2011).

[16] Machida, S.; Yamamoto, Y.; Itaya, Y. (9 March 1987). "Observation of amplitude squeezing in a constant-current driven semiconductor laser". *Physical Review Letters*. **58** (10): 1000–1003. Bibcode:1987PhRvL..58.1000M. doi:10.1103/PhysRevLett.58.1000.

[17] O. V. Misochko, J. Hu, K. G. Nakamura, "Controlling phonon squeezing and correlation via one- and two-phonon interference," http://arxiv.org/abs/1011.2001

[18] Ma, Jian; Wang, Xiaoguang; Sun, C.P.; Nori, Franco (December 2011). "Quantum spin squeezing". *Physics Reports*. **509** (2-3): 89–165. arXiv:1011.2978. Bibcode:2011PhR...509...89M. doi:10.1016/j.physrep.2011.08.003.

[19] Hosten, Onur; Engelsen, Nils J.; Krishnakumar, Rajiv; Kasevich, Mark A. (11 January 2016). "Measurement noise 100 times lower than the quantum-projection limit using entangled atoms". *Nature*. **529**. doi:10.1038/nature16176.

[20] Cox, Kevin C.; Greve, Graham P.; Weiner, Joshua M.; Thompson, James K. (4 March 2016). "Deterministic Squeezed States with Collective Measurements and Feedback". *Physical Review Letters*. **116** (9). doi:10.1103/PhysRevLett.116.093602.

[21] Bohnet, J. G.; Cox, K. C.; Norcia, M. A.; Weiner, J. M.; Chen, Z.; Thompson, J. K. (13 July 2014). "Reduced spin measurement back-action for a phase sensitivity ten times beyond the standard quantum limit". *Nature Photonics*. **8** (9): 731–736. arXiv:1310.3177. Bibcode:2014NaPho...8..731B. doi:10.1038/nphoton.2014.151.

[22] Lücke, Bernd; Peise, Jan; Vitagliano, Giuseppe; Arlt, Jan; Santos, Luis; Tóth, Géza; Klempt, Carsten (17 April 2014). "Detecting Multiparticle Entanglement of Dicke States". *Physical Review Letters*. **112** (15). arXiv:1403.4542. Bibcode:2014PhRvL.112o5304L. doi:10.1103/PhysRevLett.112.155304.

[23] Eberle, Tobias; Steinlechner, Sebastian; Bauchrowitz, Jöran; Händchen, Vitus; Vahlbruch, Henning; Mehmet, Moritz; Müller-Ebhardt, Helge; Schnabel, Roman (22 June 2010). "Quantum Enhancement of the Zero-Area Sagnac Interferometer Topology for Gravitational Wave Detection". *Physical Review Letters*. **104** (25). arXiv:1007.0574. Bibcode:2010PhRvL.104y1102E. doi:10.1103/PhysRevLett.104.251102.

[24] Leroux, Ian D.; Schleier-Smith, Monika H.; Vuletić, Vladan (25 June 2010). "Orientation-Dependent Entanglement Lifetime in a Squeezed Atomic Clock". *Physical Review Letters*. **104** (25): 250801. arXiv:1004.1725. Bibcode:2010PhRvL.104y0801L. doi:10.1103/PhysRevLett.104.250801.

[25] Louchet-Chauvet, Anne; Appel, Jürgen; Renema, Jelmer J; Oblak, Daniel; Kjaergaard, Niels; Polzik, Eugene S (28 June 2010). "Entanglement-assisted atomic clock beyond the projection noise limit". *New Journal of Physics*. **12** (6): 065032. arXiv:0912.3895. Bibcode:2010NJPh...12f5032L. doi:10.1088/1367-2630/12/6/065032.

[26] Kitagawa, Masahiro; Ueda, Masahito (1 June 1993). "Squeezed spin states". *Physical Review A*. **47** (6): 5138–5143. Bibcode:1993PhRvA..47.5138K. doi:10.1103/PhysRevA.47.5138.

[27] Braunstein, Samuel L.; van Loock, Peter (29 June 2005). "Quantum information with continuous variables". *Reviews of Modern Physics*. **77** (2): 513–577. arXiv:quant-ph/0410100. Bibcode:2005RvMP...77..513B. doi:10.1103/RevModPhys.77.513.

[28] Furusawa, A. (23 October 1998). "Unconditional Quantum Teleportation". *Science*. **282** (5389): 706–709. Bibcode:1998Sci...282..706F. doi:10.1126/science.282.5389.706.

[29] Menicucci, Nicolas C.; Flammia, Steven T.; Pfister, Olivier (22 September 2008). "One-Way Quantum Computing in the Optical Frequency Comb". *Physical Review Letters*. **101** (13): 13501. arXiv:0804.4468. Bibcode:2008PhRvL.101m0501M. doi:10.1103/PhysRevLett.101.130501.

10.10 External links

- Tutorial about quantum optics of the light field
- www.squeezed-light.de

Chapter 11

Uncertainty principle

For other uses, see Uncertainty principle (disambiguation).

In quantum mechanics, the **uncertainty principle**, also known as **Heisenberg's uncertainty principle**, is any of a variety of mathematical inequalities[1] asserting a fundamental limit to the precision with which certain pairs of physical properties of a particle, known as complementary variables, such as position x and momentum p, can be known.

Introduced first in 1927, by the German physicist Werner Heisenberg, it states that the more precisely the position of some particle is determined, the less precisely its momentum can be known, and vice versa.[2] The formal inequality relating the standard deviation of position σx and the standard deviation of momentum σp was derived by Earle Hesse Kennard[3] later that year and by Hermann Weyl[4] in 1928:

($\hbar$ is the reduced Planck constant, $h/2\pi$).

Historically, the uncertainty principle has been confused[5][6] with a somewhat similar effect in physics, called the observer effect, which notes that measurements of certain systems cannot be made without affecting the systems. Heisenberg offered such an observer effect at the quantum level (see below) as a physical "explanation" of quantum uncertainty.[7] It has since become clear, however, that the uncertainty principle is inherent in the properties of all wave-like systems,[8] and that it arises in quantum mechanics simply due to the matter wave nature of all quantum objects. Thus, *the uncertainty principle actually states a fundamental property of quantum systems, and is not a statement about the observational success of current technology*.[9] It must be emphasized that *measurement* does not mean only a process in which a physicist-observer takes part, but rather any interaction between classical and quantum objects regardless of any observer.[10] (N.B. on *precision*: If δx and δp are the precisions of position and momentum obtained in an *individual* measurement and σ_x, σ_p their standard deviations in an *ensemble* of individual measurements on similarly prepared systems, then "*There are, in principle, no restrictions on the precisions of individual measurements δx and δp, but the standard deviations will always satisfy $\sigma_x \sigma_p \geq \hbar/2$*".[11])

Since the uncertainty principle is such a basic result in quantum mechanics, typical experiments in quantum mechanics routinely observe aspects of it. Certain experiments, however, may deliberately test a particular form of the uncertainty principle as part of their main research program. These include, for example, tests of number–phase uncertainty relations in superconducting[12] or quantum optics[13] systems. Applications dependent on the uncertainty principle for their operation include extremely low noise technology such as that required in gravitational-wave interferometers.[14]

11.1 Introduction

Main article: Introduction to quantum mechanics

The uncertainty principle is not readily apparent on the macroscopic[15] scales of everyday experience. So it is helpful to demonstrate how it applies to more easily understood physical situations. Two alternative frameworks for quantum physics offer different explanations for the uncertainty principle. The wave mechanics picture of the uncertainty principle is more visually intuitive, but the more abstract matrix mechanics picture formulates it in a way that generalizes more easily.

Mathematically, in wave mechanics, the uncertainty relation between position and momentum arises because the expressions of the wavefunction in the two corresponding orthonormal bases in Hilbert space are Fourier transforms of one another (i.e., position and momentum are conjugate variables). A nonzero function and its Fourier transform cannot both be sharply localized. A similar tradeoff between the variances of Fourier conjugates arises in all systems underlain by Fourier analysis, for example in sound

11.1. INTRODUCTION

89

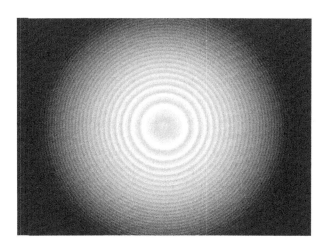

Click to see animation. The evolution of an initially very localized gaussian wave function of a free particle in two-dimensional space, with colour and intensity indicating phase and amplitude. The spreading of the wave function in all directions shows that the initial momentum has a spread of values, unmodified in time; while the spread in position increases in time: as a result, the uncertainty $\Delta x\, \Delta p$ increases in time.

The superposition of several plane waves to form a wave packet. This wave packet becomes increasingly localized with the addition of many waves. The Fourier transform is a mathematical operation that separates a wave packet into its individual plane waves. Note that the waves shown here are real for illustrative purposes only, whereas in quantum mechanics the wave function is generally complex.

waves: A pure tone is a sharp spike at a single frequency, while its Fourier transform gives the shape of the sound wave in the time domain, which is a completely delocalized sine wave. In quantum mechanics, the two key points are that the position of the particle takes the form of a matter wave, and momentum is its Fourier conjugate, assured by the de Broglie relation $p = \hbar k$, where k is the wavenumber.

In matrix mechanics, the mathematical formulation of quantum mechanics, any pair of non-commuting self-adjoint operators representing observables are subject to similar uncertainty limits. An eigenstate of an observable represents the state of the wavefunction for a certain measurement value (the eigenvalue). For example, if a measurement of an observable A is performed, then the system is in a particular eigenstate Ψ of that observable. However, the particular eigenstate of the observable A need not be an eigenstate of another observable B: If so, then it does not have a unique associated measurement for it, as the system is not in an eigenstate of that observable.[16]

11.1.1 Wave mechanics interpretation

(Ref [10])

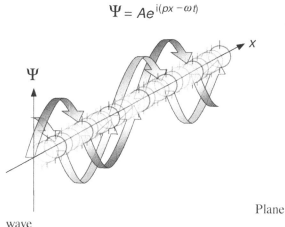

Plane wave

Wave packet

Propagation of de Broglie waves in 1d—real part of the complex amplitude is blue, imaginary part is green. The probability (shown as the colour opacity) of finding the particle at a given point x is spread out like a waveform, there is no definite position of the particle. As the amplitude increases above zero the curvature reverses sign, so the amplitude begins to decrease again, and vice versa—the result is an alternating amplitude: a wave.

Main article: Wave packet
Main article: Schrödinger equation

According to the de Broglie hypothesis, every object in the

universe is a wave, a situation which gives rise to this phenomenon. The position of the particle is described by a wave function $\Psi(x,t)$. The time-independent wave function of a single-moded plane wave of wavenumber k_0 or momentum p_0 is

$$\psi(x) \propto e^{ik_0 x} = e^{ip_0 x/\hbar}.$$

The Born rule states that this should be interpreted as a probability density amplitude function in the sense that the probability of finding the particle between a and b is

$$P[a \leq X \leq b] = \int_a^b |\psi(x)|^2 \, dx.$$

In the case of the single-moded plane wave, $|\psi(x)|^2$ is a uniform distribution. In other words, the particle position is extremely uncertain in the sense that it could be essentially anywhere along the wave packet. Consider a wave function that is a sum of many waves, however, we may write this as

$$\psi(x) \propto \sum_n A_n e^{ip_n x/\hbar},$$

where An represents the relative contribution of the mode pn to the overall total. The figures to the right show how with the addition of many plane waves, the wave packet can become more localized. We may take this a step further to the continuum limit, where the wave function is an integral over all possible modes

$$\psi(x) = \frac{1}{\sqrt{2\pi\hbar}} \int_{-\infty}^{\infty} \phi(p) \cdot e^{ipx/\hbar} \, dp,$$

with $\phi(p)$ representing the amplitude of these modes and is called the wave function in momentum space. In mathematical terms, we say that $\phi(p)$ is the *Fourier transform* of $\psi(x)$ and that x and p are conjugate variables. Adding together all of these plane waves comes at a cost, namely the momentum has become less precise, having become a mixture of waves of many different momenta.

One way to quantify the precision of the position and momentum is the standard deviation σ. Since $|\psi(x)|^2$ is a probability density function for position, we calculate its standard deviation.

The precision of the position is improved, i.e. reduced σ_x, by using many plane waves, thereby weakening the precision of the momentum, i.e. increased σ_p. Another way of stating this is that σ_x and σ_p have an inverse relationship or are at least bounded from below. This is the uncertainty principle, the exact limit of which is the Kennard bound. Click the *show* button below to see a semi-formal derivation of the Kennard inequality using wave mechanics.

11.1.2 Matrix mechanics interpretation

(Ref [10])

Main article: Matrix mechanics

In matrix mechanics, observables such as position and momentum are represented by self-adjoint operators. When considering pairs of observables, an important quantity is the *commutator*. For a pair of operators $\hat{A}$ and $\hat{B}$, one defines their commutator as

$$[\hat{A}, \hat{B}] = \hat{A}\hat{B} - \hat{B}\hat{A}.$$

In the case of position and momentum, the commutator is the canonical commutation relation

$$[\hat{x}, \hat{p}] = i\hbar.$$

The physical meaning of the non-commutativity can be understood by considering the effect of the commutator on position and momentum eigenstates. Let $|\psi\rangle$ be a right eigenstate of position with a constant eigenvalue x_0. By definition, this means that $\hat{x}|\psi\rangle = x_0|\psi\rangle$. Applying the commutator to $|\psi\rangle$ yields

$$[\hat{x}, \hat{p}]|\psi\rangle = (\hat{x}\hat{p} - \hat{p}\hat{x})|\psi\rangle = (\hat{x} - x_0\hat{I}) \cdot \hat{p}|\psi\rangle = i\hbar|\psi\rangle,$$

where $\hat{I}$ is the identity operator.

Suppose, for the sake of proof by contradiction, that $|\psi\rangle$ is also a right eigenstate of momentum, with constant eigenvalue p_0. If this were true, then one could write

$$(\hat{x} - x_0\hat{I}) \cdot \hat{p}|\psi\rangle = (\hat{x} - x_0\hat{I}) \cdot p_0|\psi\rangle = (x_0\hat{I} - x_0\hat{I}) \cdot p_0|\psi\rangle = 0.$$

On the other hand, the above canonical commutation relation requires that

$$[\hat{x}, \hat{p}]|\psi\rangle = i\hbar|\psi\rangle \neq 0.$$

This implies that no quantum state can simultaneously be both a position and a momentum eigenstate.

When a state is measured, it is projected onto an eigenstate in the basis of the relevant observable. For example, if a particle's position is measured, then the state amounts to a position eigenstate. This means that the state is *not* a momentum eigenstate, however, but rather it can be represented as a sum of multiple momentum basis eigenstates.

In other words, the momentum must be less precise. This precision may be quantified by the standard deviations,

$$\sigma_x = \sqrt{\langle \hat{x}^2 \rangle - \langle \hat{x} \rangle^2}$$

$$\sigma_p = \sqrt{\langle \hat{p}^2 \rangle - \langle \hat{p} \rangle^2}.$$

As in the wave mechanics interpretation above, one sees a tradeoff between the respective precisions of the two, quantified by the uncertainty principle.

11.2 Robertson–Schrödinger uncertainty relations

The most common general form of the uncertainty principle is the *Robertson uncertainty relation*.[17]

For an arbitrary Hermitian operator $\hat{\mathcal{O}}$ we can associate a standard deviation

$$\sigma_{\mathcal{O}} = \sqrt{\langle \hat{\mathcal{O}}^2 \rangle - \langle \hat{\mathcal{O}} \rangle^2},$$

where the brackets $\langle \mathcal{O} \rangle$ indicate an expectation value. For a pair of operators $\hat{A}$ and $\hat{B}$, we may define their *commutator* as

$$[\hat{A}, \hat{B}] = \hat{A}\hat{B} - \hat{B}\hat{A},$$

In this notation, the Robertson uncertainty relation is given by

$$\sigma_A \sigma_B \geq \left| \frac{1}{2i} \langle [\hat{A}, \hat{B}] \rangle \right| = \frac{1}{2} \left| \langle [\hat{A}, \hat{B}] \rangle \right|,$$

The Robertson uncertainty relation immediately follows from a slightly stronger inequality, the *Schrödinger uncertainty relation*,[18]

where we have introduced the *anticommutator*,

$$\{\hat{A}, \hat{B}\} = \hat{A}\hat{B} + \hat{B}\hat{A}.$$

Since the Robertson and Schrödinger relations are for general operators, the relations can be applied to any two observables to obtain specific uncertainty relations. A few of the most common relations found in the literature are given below.

- For position and linear momentum, the canonical commutation relation $[\hat{x}, \hat{p}] = i\hbar$ implies the Kennard inequality from above:

$$\sigma_x \sigma_p \geq \frac{\hbar}{2}$$

- For two orthogonal components of the total angular momentum operator of an object:

$$\sigma_{J_i} \sigma_{J_j} \geq \frac{\hbar}{2} |\langle J_k \rangle|,$$

  where i, j, k are distinct and Ji denotes angular momentum along the xi axis. This relation implies that unless all three components vanish together, only a single component of a system's angular momentum can be defined with arbitrary precision, normally the component parallel to an external (magnetic or electric) field. Moreover, for $[J_x, J_y] = i\hbar\epsilon_{xyz}J_z$, a choice $\hat{A} = J_x$, $\hat{B} = J_y$, in angular momentum multiplets, $\psi = |j, m\rangle$, bounds the Casimir invariant (angular momentum squared, $\langle J_x^2 + J_y^2 + J_z^2 \rangle$) from below and thus yields useful constraints such as $j(j+1) \geq m(m+1)$, and hence $j \geq m$, among others.

- In non-relativistic mechanics, time is privileged as an independent variable. Nevertheless, in 1945, L. I. Mandelshtam and I. E. Tamm derived a non-relativistic **time–energy uncertainty relation**, as follows.[26][27] For a quantum system in a non-stationary state ψ and an observable B represented by a self-adjoint operator $\hat{B}$, the following formula holds:

$$\sigma_E \frac{\sigma_B}{\left| \frac{d\langle \hat{B} \rangle}{dt} \right|} \geq \frac{\hbar}{2}.$$

  where σE is the standard deviation of the energy operator (Hamiltonian) in the state ψ, σB stands for the standard deviation of B. Although the second factor in the left-hand side has dimension of time, it is different from the time parameter that enters the Schrödinger equation. It is a *lifetime* of the state ψ with respect to the observable *B*: In other words, this is the *time interval* (*Δt*) after which the expectation value $\langle \hat{B} \rangle$ changes appreciably.

An informal, heuristic meaning of the principle is the following: A state that only exists for a short time cannot have a definite energy. To have a definite energy, the frequency of the state must be defined accurately, and this requires the state to hang around for many cycles, the reciprocal of the required accuracy. For example, in spectroscopy, excited states have a finite lifetime. By the time–energy uncertainty principle, they do not have a definite energy, and, each time they decay, the energy they release is slightly different. The average energy of the outgoing photon has a peak at the theoretical energy of the state, but the distribution has a finite width called the *natural linewidth*. Fast-decaying states have a broad linewidth, while slow decaying states have a narrow linewidth.[28]

The same linewidth effect also makes it difficult to specify the rest mass of unstable, fast-decaying particles in particle physics. The faster the particle decays (the shorter its lifetime), the less certain is its mass (the larger the particle's width).

- For the number of electrons in a superconductor and the phase of its Ginzburg–Landau order parameter[29][30]

$$\Delta N \Delta \phi \geq 1.$$

11.3 Examples

(Refs [10][19])

11.3.1 Quantum harmonic oscillator stationary states

Main articles: Quantum harmonic oscillator and Stationary state

Consider a one-dimensional quantum harmonic oscillator (QHO). It is possible to express the position and momentum operators in terms of the creation and annihilation operators:

$$\hat{x} = \sqrt{\frac{\hbar}{2m\omega}}(a + a^\dagger)$$

$$\hat{p} = i\sqrt{\frac{m\omega\hbar}{2}}(a^\dagger - a)$$

Using the standard rules for creation and annihilation operators on the eigenstates of the QHO,

$$a^\dagger |n\rangle = \sqrt{n+1}|n+1\rangle$$
$$a|n\rangle = \sqrt{n}|n-1\rangle$$

the variances may be computed directly,

$$\sigma_x^2 = \frac{\hbar}{m\omega}\left(n + \frac{1}{2}\right)$$

$$\sigma_p^2 = \hbar m\omega \left(n + \frac{1}{2}\right).$$

The product of these standard deviations is then

$$\sigma_x \sigma_p = \hbar\left(n + \frac{1}{2}\right) \geq \frac{\hbar}{2}.$$

In particular, the above Kennard bound[3] is saturated for the ground state *n*=0, for which the probability density is just the normal distribution.

11.3.2 Quantum harmonic oscillator with Gaussian initial condition

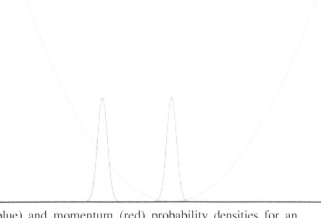

(blue) and momentum (red) probability densities for an initially Gaussian distribution. From top to bottom, the animations show the cases Ω=ω, Ω=2ω, and Ω=ω/2. Note the tradeoff between the widths of the distributions.

In a quantum harmonic oscillator of characteristic angular frequency ω, place a state that is offset from the bottom of the potential by some displacement x_0 as

$$\psi(x) = \left(\frac{m\Omega}{\pi\hbar}\right)^{1/4} \exp\left(-\frac{m\Omega(x-x_0)^2}{2\hbar}\right)$$

where Ω describes the width of the initial state but need not be the same as ω. Through integration over the propagator, we can solve for the full time-dependent solution. After many cancelations, the probability densities reduce to

$$|\Psi(x,t)|^2 \sim \mathcal{N}\left(x_0\cos(\omega t), \frac{\hbar}{2m\Omega}\left(\cos^2(\omega t) + \frac{\Omega^2}{\omega^2}\sin^2(\omega t)\right)\right)$$

$$|\Phi(p,t)|^2 \sim \mathcal{N}\left(-mx_0\omega\sin(\omega t), \frac{\hbar m\Omega}{2}\left(\cos^2(\omega t) + \frac{\omega^2}{\Omega^2}\sin^2(\omega t)\right)\right)$$

where we have used the notation $\mathcal{N}(\mu, \sigma^2)$ to denote a normal distribution of mean μ and variance σ^2. Copying the variances above and applying trigonometric identities, we can write the product of the standard deviations as

$$\sigma_x \sigma_p = \frac{\hbar}{2}\sqrt{\left(\cos^2(\omega t) + \frac{\Omega^2}{\omega^2}\sin^2(\omega t)\right)\left(\cos^2(\omega t) + \frac{\omega^2}{\Omega^2}\sin^2(\omega t)\right)}$$

$$= \frac{\hbar}{4}\sqrt{3 + \frac{1}{2}\left(\frac{\Omega^2}{\omega^2} + \frac{\omega^2}{\Omega^2}\right) - \left(\frac{1}{2}\left(\frac{\Omega^2}{\omega^2} + \frac{\omega^2}{\Omega^2}\right) - 1\right)\cos(4\omega t)}$$

From the relations

$$\frac{\Omega^2}{\omega^2} + \frac{\omega^2}{\Omega^2} \geq 2, \quad |\cos(4\omega t)| \leq 1$$

we can conclude

$$\sigma_x \sigma_p \geq \frac{\hbar}{4}\sqrt{3 + \frac{1}{2}\left(\frac{\Omega^2}{\omega^2} + \frac{\omega^2}{\Omega^2}\right) - \left(\frac{1}{2}\left(\frac{\Omega^2}{\omega^2} + \frac{\omega^2}{\Omega^2}\right) - 1\right)} = \frac{\hbar}{2}$$

11.3.3 Coherent states

Main article: Coherent state

A coherent state is a right eigenstate of the annihilation operator,

$$\hat{a}|\alpha\rangle = \alpha|\alpha\rangle$$

which may be represented in terms of Fock states as

$$|\alpha\rangle = e^{-\frac{|\alpha|^2}{2}} \sum_{n=0}^{\infty} \frac{\alpha^n}{\sqrt{n!}}|n\rangle$$

In the picture where the coherent state is a massive particle in a QHO, the position and momentum operators may be expressed in terms of the annihilation operators in the same formulas above and used to calculate the variances,

$$\sigma_x^2 = \frac{\hbar}{2m\omega}$$

$$\sigma_p^2 = \frac{\hbar m\omega}{2}$$

Therefore, every coherent state saturates the Kennard bound

$$\sigma_x \sigma_p = \sqrt{\frac{\hbar}{2m\omega}}\sqrt{\frac{\hbar m\omega}{2}} = \frac{\hbar}{2}$$

with position and momentum each contributing an amount $\sqrt{\hbar/2}$ in a "balanced" way. Moreover, every squeezed coherent state also saturates the Kennard bound although the individual contributions of position and momentum need not be balanced in general.

11.3.4 Particle in a box

Consider a particle in a one-dimensional box of length L. The eigenfunctions in position and momentum space are

$$\psi_n(x,t) = \begin{cases} A\sin(k_n x)e^{-i\omega_n t}, & 0 < x < L, \\ 0, & \text{otherwise}, \end{cases}$$

and

$$\phi_n(p,t) = \sqrt{\frac{\pi L}{\hbar}}\frac{n\left(1-(-1)^n e^{-ikL}\right)e^{-i\omega_n t}}{\pi^2 n^2 - k^2 L^2}$$

where $\omega_n = \frac{\pi^2 \hbar n^2}{8L^2 m}$ and we have used the de Broglie relation $p = \hbar k$. The variances of x and p can be calculated explicitly:

$$\sigma_x^2 = \frac{L^2}{12}\left(1 - \frac{6}{n^2\pi^2}\right)$$

$$\sigma_p^2 = \left(\frac{\hbar n\pi}{L}\right)^2$$

The product of the standard deviations is therefore

$$\sigma_x \sigma_p = \frac{\hbar}{2}\sqrt{\frac{n^2\pi^2}{3} - 2}.$$

For all $n = 1, 2, 3\ldots$, the quantity $\sqrt{\frac{n^2\pi^2}{3} - 2}$ is greater than 1, so the uncertainty principle is never violated. For numerical concreteness, the smallest value occurs when $n = 1$, in which case

$$\sigma_x \sigma_p = \frac{\hbar}{2}\sqrt{\frac{\pi^2}{3} - 2} \approx 0.568\hbar > \frac{\hbar}{2}$$

11.3.5 Constant momentum

Main article: Wave packet

Assume a particle initially has a momentum space wave

Position space probability density of an initially Gaussian state moving at minimally uncertain, constant momentum in free space

function described by a normal distribution around some constant momentum p_0 according to

$$\phi(p) = \left(\frac{x_0}{\hbar\sqrt{\pi}}\right)^{1/2} \cdot \exp\left(\frac{-x_0^2(p-p_0)^2}{2\hbar^2}\right),$$

where we have introduced a reference scale $x_0 = \sqrt{\hbar/m\omega_0}$, with $\omega_0 > 0$ describing the width of the distribution—cf. nondimensionalization. If the state is allowed to evolve in free space, then the time-dependent momentum and position space wave functions are

$$\Phi(p,t) = \left(\frac{x_0}{\hbar\sqrt{\pi}}\right)^{1/2} \cdot \exp\left(\frac{-x_0^2(p-p_0)^2}{2\hbar^2} - \frac{ip^2 t}{2m\hbar}\right),$$

$$\Psi(x,t) = \left(\frac{1}{x_0\sqrt{\pi}}\right)^{1/2} \cdot \frac{e^{-x_0^2 p_0^2/2\hbar^2}}{\sqrt{1+i\omega_0 t}} \cdot \exp\left(-\frac{(x - ix_0^2 p_0/\hbar)^2}{2x_0^2(1+i\omega_0 t)}\right).$$

Since $\langle p(t)\rangle = p_0$ and $\sigma_p(t) = \hbar/x_0\sqrt{2}$, this can be interpreted as a particle moving along with constant momentum at arbitrarily high precision. On the other hand, the standard deviation of the position is

$$\sigma_x = \frac{x_0}{\sqrt{2}}\sqrt{1 + \omega_0^2 t^2}$$

such that the uncertainty product can only increase with time as

$$\sigma_x(t)\sigma_p(t) = \frac{\hbar}{2}\sqrt{1 + \omega_0^2 t^2}$$

11.4 Additional uncertainty relations

11.4.1 Mixed states

The Robertson–Schrödinger uncertainty relation may be generalized in a straightforward way to describe mixed states.[31]

$$\sigma_A^2 \sigma_B^2 \geq \left(\frac{1}{2}\text{tr}(\rho\{A,B\}) - \text{tr}(\rho A)\text{tr}(\rho B)\right)^2 + \left(\frac{1}{2i}\text{tr}(\rho[A,B])\right)^2$$

11.4.2 Phase space

In the phase space formulation of quantum mechanics, the Robertson–Schrödinger relation follows from a positivity condition on a real star-square function. Given a Wigner function $W(x,p)$ with star product $\star$ and a function f, the following is generally true:[32]

$$\langle f^* \star f\rangle = \int (f^* \star f) W(x,p)\, dx dp \geq 0.$$

Choosing $f = a + bx + cp$, we arrive at

$$\langle f^* \star f\rangle = \begin{bmatrix} a^* & b^* & c^* \end{bmatrix} \begin{bmatrix} 1 & \langle x\rangle & \langle p\rangle \\ \langle x\rangle & \langle x \star x\rangle & \langle x \star p\rangle \\ \langle p\rangle & \langle p \star x\rangle & \langle p \star p\rangle \end{bmatrix} \begin{bmatrix} a \\ b \\ c \end{bmatrix} \geq 0.$$

11.4. ADDITIONAL UNCERTAINTY RELATIONS

Since this positivity condition is true for *all a*, *b*, and *c*, it follows that all the eigenvalues of the matrix are positive. The positive eigenvalues then imply a corresponding positivity condition on the determinant:

$$\det \begin{bmatrix} 1 & \langle x \rangle & \langle p \rangle \\ \langle x \rangle & \langle x \star x \rangle & \langle x \star p \rangle \\ \langle p \rangle & \langle p \star x \rangle & \langle p \star p \rangle \end{bmatrix} = \det \begin{bmatrix} 1 & \langle x \rangle & \langle p \rangle \\ \langle x \rangle & \langle x^2 \rangle & \langle xp + \frac{i\hbar}{2} \rangle \\ \langle p \rangle & \langle xp - \frac{i\hbar}{2} \rangle & \langle p^2 \rangle \end{bmatrix} \geq 0,$$

or, explicitly, after algebraic manipulation,

$$\sigma_x^2 \sigma_p^2 = (\langle x^2 \rangle - \langle x \rangle^2)(\langle p^2 \rangle - \langle p \rangle^2) \geq (\langle xp \rangle - \langle x \rangle \langle p \rangle)^2 + \frac{\hbar^2}{4}.$$

11.4.3 Systematic and statistical errors

The inequalities above focus on the *statistical imprecision* of observables as quantified by the standard deviation σ. Heisenberg's original version, however, was dealing with the *systematic error*, a disturbance of the quantum system produced by the measuring apparatus, i.e., an observer effect.

If we let ϵ_A represent the error (i.e., inaccuracy) of a measurement of an observable A and η_B the disturbance produced on a subsequent measurement of the conjugate variable B by the former measurement of A, then the inequality proposed by Ozawa[6] — encompassing both systematic and statistical errors — holds:

Heisenberg uncertainty principle, as originally described in the 1927 formulation, mentions only the first term of Ozawa inequality, regarding the *systematic error*. Using the notation above to describe the *error/disturbance* effect of *sequential measurements* (first A, then B), it could be written as

The formal derivation of Heisenberg relation is possible but far from intuitive. It was *not* proposed by Heisenberg, but formulated in a mathematically consistent way only in recent years.[33][34] Also, it must be stressed that the Heisenberg formulation is not taking into account the intrinsic statistical errors σ_A and σ_B. There is increasing experimental evidence[8][35][36][37] that the total quantum uncertainty cannot be described by the Heisenberg term alone, but requires the presence of all the three terms of the Ozawa inequality.

Using the same formalism,[1] it is also possible to introduce the other kind of physical situation, often confused with the previous one, namely the case of *simultaneous measurements* (A and B at the same time):

The two simultaneous measurements on A and B are necessarily[38] *unsharp* or *weak*.

It is also possible to derive an uncertainty relation that, as the Ozawa's one, combines both the statistical and systematic error components, but keeps a form very close to the Heisenberg original inequality. By adding Robertson[1] and Ozawa relations we obtain

$$\epsilon_A \eta_B + \epsilon_A \sigma_B + \sigma_A \eta_B + \sigma_A \sigma_B \geq \left| \langle [\hat{A}, \hat{B}] \rangle \right|.$$

The four terms can be written as:

$$(\epsilon_A + \sigma_A)(\eta_B + \sigma_B) \geq \left| \langle [\hat{A}, \hat{B}] \rangle \right|.$$

Defining:

$$\bar{\epsilon}_A \equiv (\epsilon_A + \sigma_A)$$

as the *inaccuracy* in the measured values of the variable A and

$$\bar{\eta}_B \equiv (\eta_B + \sigma_B)$$

as the *resulting fluctuation* in the conjugate variable B, Fujikawa[39] established an uncertainty relation similar to the Heisenberg original one, but valid both for *systematic and statistical errors*:

11.4.4 Quantum entropic uncertainty principle

For many distributions, the standard deviation is not a particularly natural way of quantifying the structure. For example, uncertainty relations in which one of the observables is an angle has little physical meaning for fluctuations larger than one period.[24][40][41][42] Other examples include highly bimodal distributions, or unimodal distributions with divergent variance.

A solution that overcomes these issues is an uncertainty based on entropic uncertainty instead of the product of variances. While formulating the many-worlds interpretation of quantum mechanics in 1957, Hugh Everett III conjectured a stronger extension of the uncertainty principle based on entropic certainty.[43] This conjecture, also studied by Hirschman[44] and proven in 1975 by Beckner[45] and by Iwo Bialynicki-Birula and Jerzy Mycielski[46] is that, for two normalized, dimensionless Fourier transform pairs $f(a)$ and $g(b)$ where

$$f(a) = \int_{-\infty}^{\infty} g(b)\, e^{2\pi i a b}\, dx \text{ and } g(b) = \int_{-\infty}^{\infty} f(a)\, e^{-2\pi i a b}\, dx$$

the Shannon information entropies

$$H_a = \int_{-\infty}^{\infty} f(a) \log(f(a))\, dx,$$

and

$$H_b = \int_{-\infty}^{\infty} g(b) \log(g(b))\, dy$$

are subject to the following constraint,

where the logarithms may be in any base.

The probability distribution functions associated with the position wave function $\psi(x)$ and the momentum wave function $\varphi(x)$ have dimensions of inverse length and momentum respectively, but the entropies may be rendered dimensionless by

$$H_x = -\int |\psi(x)|^2 \ln(x_0 |\psi(x)|^2)\, dx = -\left\langle \ln(x_0 |\psi(x)|^2) \right\rangle$$

$$H_p = -\int |\phi(p)|^2 \ln(p_0 |\phi(p)|^2)\, dp = -\left\langle \ln(p_0 |\phi(p)|^2) \right\rangle$$

where x_0 and p_0 are some arbitrarily chosen length and momentum respectively, which render the arguments of the logarithms dimensionless. Note that the entropies will be functions of these chosen parameters. Due to the Fourier transform relation between the position wave function $\psi(x)$ and the momentum wavefuction $\varphi(p)$, the above constraint can be written for the corresponding entropies as

where h is Planck's constant.

Depending on one's choice of the $x_0\, p_0$ product, the expression may be written in many ways. If $x_0\, p_0$ is chosen to be h, then

$$H_x + H_p \geq \log\left(\frac{e}{2}\right)$$

If, instead, $x_0\, p_0$ is chosen to be $\hbar$, then

$$H_x + H_p \geq \log(e\,\pi)$$

If x_0 and p_0 are chosen to be unity in whatever system of units are being used, then

$$H_x + H_p \geq \log\left(\frac{e\,h}{2}\right)$$

where h is interpreted as a dimensionless number equal to the value of Planck's constant in the chosen system of units.

The quantum entropic uncertainty principle is more restrictive than the Heisenberg uncertainty principle. From the inverse logarithmic Sobolev inequalities[47]

$$H_x \leq \frac{1}{2} \log(2e\pi\sigma_x^2/x_0^2),$$

$$H_p \leq \frac{1}{2} \log(2e\pi\sigma_p^2/p_0^2),$$

(equivalently, from the fact that normal distributions maximize the entropy of all such with a given variance), it readily follows that this entropic uncertainty principle is *stronger than the one based on standard deviations*, because

$$\sigma_x \sigma_p \geq \frac{\hbar}{2} \exp\left(H_x + H_p - \log\left(\frac{e\,h}{2\, x_0\, p_0}\right)\right) \geq \frac{\hbar}{2}.$$

In other words, the Heisenberg uncertainty principle, is a consequence of the quantum entropic uncertainty principle, but not vice versa. A few remarks on these inequalities.

First, the choice of base e is a matter of popular convention in physics. The logarithm can alternatively be in any base, provided that it be consistent on both sides of the inequality. Second, recall the Shannon entropy has been used, *not* the quantum von Neumann entropy. Finally, the normal distribution saturates the inequality, and it is the only distribution with this property, because it is the maximum entropy probability distribution among those with fixed variance (cf. here for proof).

11.5 Harmonic analysis

Main article: Fourier transform § Uncertainty principle

In the context of harmonic analysis, a branch of mathematics, the uncertainty principle implies that one cannot at the same time localize the value of a function and its Fourier transform. To wit, the following inequality holds,

$$\left(\int_{-\infty}^{\infty} x^2 |f(x)|^2 \, dx\right) \left(\int_{-\infty}^{\infty} \xi^2 |\hat{f}(\xi)|^2 \, d\xi\right) \geq \frac{\|f\|_2^4}{16\pi^2}.$$

Further mathematical uncertainty inequalities, including the above entropic uncertainty, hold between a function f and its Fourier transform ƒ:[48][49][50]

$$H_x + H_\xi \geq \log(e/2)$$

11.5.1 Signal processing

In the context of signal processing, and in particular time–frequency analysis, uncertainty principles are referred to as the **Gabor limit**, after Dennis Gabor, or sometimes the *Heisenberg–Gabor limit*. The basic result, which follows from "Benedicks's theorem", below, is that a function cannot be both time limited and band limited (a function and its Fourier transform cannot both have bounded domain)—see bandlimited versus timelimited.

Stated alternatively, "One cannot simultaneously sharply localize a signal (function f) in both the time domain and frequency domain (ƒ, its Fourier transform)".

When applied to filters, the result implies that one cannot achieve high temporal resolution and frequency resolution at the same time; a concrete example are the resolution issues of the short-time Fourier transform—if one uses a wide window, one achieves good frequency resolution at the cost of temporal resolution, while a narrow window has the opposite trade-off.

Alternate theorems give more precise quantitative results, and, in time–frequency analysis, rather than interpreting the (1-dimensional) time and frequency domains separately, one instead interprets the limit as a lower limit on the support of a function in the (2-dimensional) time–frequency plane. In practice, the Gabor limit limits the *simultaneous* time–frequency resolution one can achieve without interference; it is possible to achieve higher resolution, but at the cost of different components of the signal interfering with each other.

11.5.2 Benedicks's theorem

Amrein-Berthier[51] and Benedicks's theorem[52] intuitively says that the set of points where f is non-zero and the set of points where ƒ is nonzero cannot both be small.

Specifically, it is impossible for a function f in $L^2(\mathbf{R})$ and its Fourier transform ƒ to both be supported on sets of finite Lebesgue measure. A more quantitative version is[53][54]

$$\|f\|_{L^2(\mathbf{R}^d)} \leq Ce^{C|S||\Sigma|}\left(\|f\|_{L^2(S^c)} + \|\hat{f}\|_{L^2(\Sigma^c)}\right).$$

One expects that the factor $Ce^{C|S||\Sigma|}$ may be replaced by $Ce^{C(|S||\Sigma|)^{1/d}}$, which is only known if either S or Σ is convex.

11.5.3 Hardy's uncertainty principle

The mathematician G. H. Hardy formulated the following uncertainty principle:[55] it is not possible for f and ƒ to both be "very rapidly decreasing." Specifically, if f in $L^2(\mathbf{R})$ is such that

$$|f(x)| \leq C(1+|x|)^N e^{-a\pi x^2}$$

and

$$|\hat{f}(\xi)| \leq C(1+|\xi|)^N e^{-b\pi \xi^2} \quad (C>0, N \text{ an integer}),$$

then, if $ab > 1$, $f = 0$, while if $ab = 1$, then there is a polynomial P of degree ≤ N such that

$$f(x) = P(x)e^{-a\pi x^2}.$$

This was later improved as follows: if $f \in L^2(\mathbf{R}^d)$ is such that

$$\int_{\mathbf{R}^d}\int_{\mathbf{R}^d}|f(x)||\hat{f}(\xi)|\frac{e^{\pi|\langle x,\xi\rangle|}}{(1+|x|+|\xi|)^N}\,dx\,d\xi<+\infty\,,$$

then

$$f(x)=P(x)e^{-\pi\langle Ax,x\rangle}\,,$$

where P is a polynomial of degree $(N-d)/2$ and A is a real $d{\times}d$ positive definite matrix.

This result was stated in Beurling's complete works without proof and proved in Hörmander[56] (the case $d=1, N=0$) and Bonami, Demange, and Jaming[57] for the general case. Note that Hörmander–Beurling's version implies the case $ab>1$ in Hardy's Theorem while the version by Bonami–Demange–Jaming covers the full strength of Hardy's Theorem. A different proof of Beurling's theorem based on Liouville's theorem appeared in ref.[58]

A full description of the case $ab<1$ as well as the following extension to Schwartz class distributions appears in ref.[59]

Theorem. If a tempered distribution $f\in\mathcal{S}'(\mathbb{R}^d)$ is such that

$$e^{\pi|x|^2}f\in\mathcal{S}'(\mathbb{R}^d)$$

and

$$e^{\pi|\xi|^2}\hat{f}\in\mathcal{S}'(\mathbb{R}^d)\,,$$

then

$$f(x)=P(x)e^{-\pi\langle Ax,x\rangle}\,,$$

for some convenient polynomial P and real positive definite matrix A of type $d\times d$.

11.6 History

Werner Heisenberg formulated the Uncertainty Principle at Niels Bohr's institute in Copenhagen, while working on the mathematical foundations of quantum mechanics.[60]

In 1925, following pioneering work with Hendrik Kramers, Heisenberg developed matrix mechanics, which replaced the ad hoc old quantum theory with modern quantum mechanics. The central premise was that the classical concept

Werner Heisenberg and Niels Bohr

of motion does not fit at the quantum level, as electrons in an atom do not travel on sharply defined orbits. Rather, their motion is smeared out in a strange way: the Fourier transform of its time dependence only involves those frequencies that could be observed in the quantum jumps of their radiation.

Heisenberg's paper did not admit any unobservable quantities like the exact position of the electron in an orbit at any time; he only allowed the theorist to talk about the Fourier components of the motion. Since the Fourier components were not defined at the classical frequencies, they could not be used to construct an exact trajectory, so that the formalism could not answer certain overly precise questions about where the electron was or how fast it was going.

In March 1926, working in Bohr's institute, Heisenberg realized that the non-commutativity implies the uncertainty principle. This implication provided a clear physical interpretation for the non-commutativity, and it laid the foundation for what became known as the Copenhagen interpretation of quantum mechanics. Heisenberg showed that the commutation relation implies an uncertainty, or in Bohr's language a complementarity.[61] Any two variables that do not commute cannot be measured simultaneously—the more precisely one is known, the less precisely the other can be known. Heisenberg wrote:

> It can be expressed in its simplest form as follows: One can never know with perfect accuracy both of those two important factors which determine the movement of one of the smallest particles—its position and its velocity. It is impossible to determine accurately *both* the position and the direction and speed of a particle *at the same instant*.[62]

In his celebrated 1927 paper, "Über den anschaulichen Inhalt der quantentheoretischen Kinematik und Mechanik" ("On the Perceptual Content of Quantum Theoretical Kinematics and Mechanics"), Heisenberg established this expression as the minimum amount of unavoidable momentum disturbance caused by any position measurement,[2] but he did not give a precise definition for the uncertainties Δx and Δp. Instead, he gave some plausible estimates in each case separately. In his Chicago lecture[63] he refined his principle:

Kennard[3] in 1927 first proved the modern inequality:

where $\hbar = h/2\pi$, and σx, σp are the standard deviations of position and momentum. Heisenberg only proved relation (**2**) for the special case of Gaussian states.[63]

11.6.1 Terminology and translation

Throughout the main body of his original 1927 paper, written in German, Heisenberg used the word, "Ungenauigkeit" ("indeterminacy"),[2] to describe the basic theoretical principle. Only in the endnote did he switch to the word, "Unsicherheit" ("uncertainty"). When the English-language version of Heisenberg's textbook, *The Physical Principles of the Quantum Theory*, was published in 1930, however, the translation "uncertainty" was used, and it became the more commonly used term in the English language thereafter.[64]

11.6.2 Heisenberg's microscope

Main article: Heisenberg's microscope

The principle is quite counter-intuitive, so the early students of quantum theory had to be reassured that naive measurements to violate it were bound always to be unworkable. One way in which Heisenberg originally illustrated the intrinsic impossibility of violating the uncertainty principle is by using an imaginary microscope as a measuring device.[63]

He imagines an experimenter trying to measure the position and momentum of an electron by shooting a photon at it.

> Problem 1 – If the photon has a short wavelength, and therefore, a large momentum, the position can be measured accurately. But the photon scatters in a random direction, transferring a large

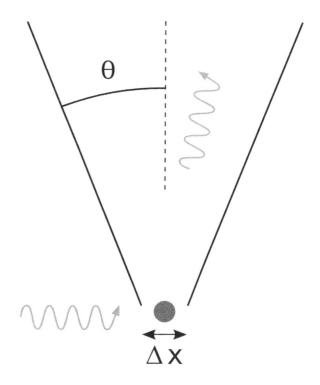

Heisenberg's gamma-ray microscope for locating an electron (shown in blue). The incoming gamma ray (shown in green) is scattered by the electron up into the microscope's aperture angle θ. The scattered gamma-ray is shown in red. Classical optics shows that the electron position can be resolved only up to an uncertainty Δx that depends on θ and the wavelength λ of the incoming light.

and uncertain amount of momentum to the electron. If the photon has a long wavelength and low momentum, the collision does not disturb the electron's momentum very much, but the scattering will reveal its position only vaguely.

> Problem 2 – If a large aperture is used for the microscope, the electron's location can be well resolved (see Rayleigh criterion); but by the principle of conservation of momentum, the transverse momentum of the incoming photon affects the electrons beamline momentum and hence, the new momentum of the electron resolves poorly. If a small aperture is used, the accuracy of both resolutions is the other way around.

The combination of these trade-offs imply that no matter what photon wavelength and aperture size are used, the product of the uncertainty in measured position and measured momentum is greater than or equal to a lower limit, which is (up to a small numerical factor) equal to Planck's constant.[65] Heisenberg did not care to formulate the uncertainty principle as an exact limit (which is elaborated below), and preferred to use it instead, as a heuristic quantitative statement, correct up to small numerical factors, which makes the radically new noncommutativity of quantum mechanics inevitable.

11.7 Critical reactions

Main article: Bohr–Einstein debates

The Copenhagen interpretation of quantum mechanics and Heisenberg's Uncertainty Principle were, in fact, seen as twin targets by detractors who believed in an underlying determinism and realism. According to the Copenhagen interpretation of quantum mechanics, there is no fundamental reality that the quantum state describes, just a prescription for calculating experimental results. There is no way to say what the state of a system fundamentally is, only what the result of observations might be.

Albert Einstein believed that randomness is a reflection of our ignorance of some fundamental property of reality, while Niels Bohr believed that the probability distributions are fundamental and irreducible, and depend on which measurements we choose to perform. Einstein and Bohr debated the uncertainty principle for many years. Some experiments within the first decade of the twenty-first century have cast doubt on observer effect aspects of the uncertainty principle.[66][67]

11.7.1 Einstein's slit

The first of Einstein's thought experiments challenging the uncertainty principle went as follows:

> Consider a particle passing through a slit of width d. The slit introduces an uncertainty in momentum of approximately h/d because the particle passes through the wall. But let us determine the momentum of the particle by measuring the recoil of the wall. In doing so, we find the momentum of the particle to arbitrary accuracy by conservation of momentum.

Bohr's response was that the wall is quantum mechanical as well, and that to measure the recoil to accuracy Δp, the momentum of the wall must be known to this accuracy before the particle passes through. This introduces an uncertainty in the position of the wall and therefore the position of the slit equal to $h/\Delta p$, and if the wall's momentum is known precisely enough to measure the recoil, the slit's position is uncertain enough to disallow a position measurement.

A similar analysis with particles diffracting through multiple slits is given by Richard Feynman.[68]

11.7.2 Einstein's box

Bohr was present when Einstein proposed the thought experiment which has become known as Einstein's box. Einstein argued that "Heisenberg's uncertainty equation implied that the uncertainty in time was related to the uncertainty in energy, the product of the two being related to Planck's constant."[69] Consider, he said, an ideal box, lined with mirrors so that it can contain light indefinitely. The box could be weighed before a clockwork mechanism opened an ideal shutter at a chosen instant to allow one single photon to escape. "We now know, explained Einstein, precisely the time at which the photon left the box."[70] "Now, weigh the box again. The change of mass tells the energy of the emitted light. In this manner, said Einstein, one could measure the energy emitted and the time it was released with any desired precision, in contradiction to the uncertainty principle."[69]

Bohr spent a sleepless night considering this argument, and eventually realized that it was flawed. He pointed out that if the box were to be weighed, say by a spring and a pointer on a scale, "since the box must move vertically with a change in its weight, there will be uncertainty in its vertical velocity and therefore an uncertainty in its height above the table. ... Furthermore, the uncertainty about the elevation above the earth's surface will result in an uncertainty in the rate of the clock,"[71] because of Einstein's own theory of gravity's effect on time. "Through this chain of uncertainties, Bohr showed that Einstein's light box experiment could not simultaneously measure exactly both the energy of the photon and the time of its escape."[72]

11.7.3 EPR paradox for entangled particles

Bohr was compelled to modify his understanding of the uncertainty principle after another thought experiment by Einstein. In 1935, Einstein, Podolsky and Rosen (see EPR paradox) published an analysis of widely separated entangled particles. Measuring one particle, Einstein realized, would alter the probability distribution of the other, yet here the other particle could not possibly be disturbed. This example led Bohr to revise his understanding of the principle, concluding that the uncertainty was not caused by a direct interaction.[73]

But Einstein came to much more far-reaching conclusions from the same thought experiment. He believed the "natural basic assumption" that a complete description of reality, would have to predict the results of experiments from "locally changing deterministic quantities", and therefore, would have to include more information than the maximum possible allowed by the uncertainty principle.

In 1964, John Bell showed that this assumption can be falsified, since it would imply a certain inequality between the probabilities of different experiments. Experimental results confirm the predictions of quantum mechanics, ruling out Einstein's basic assumption that led him to the suggestion of his *hidden variables*. Ironically this fact is one of the best pieces of evidence supporting Karl Popper's philosophy of invalidation of a theory by falsification-experiments. That is to say, here Einstein's "basic assumption" became falsified by experiments based on Bell's inequalities. For the objections of Karl Popper to the Heisenberg inequality itself, see below.

While it is possible to assume that quantum mechanical predictions are due to nonlocal, hidden variables, and in fact David Bohm invented such a formulation, this resolution is not satisfactory to the vast majority of physicists. The question of whether a random outcome is predetermined by a nonlocal theory can be philosophical, and it can be potentially intractable. If the hidden variables are not constrained, they could just be a list of random digits that are used to produce the measurement outcomes. To make it sensible, the assumption of nonlocal hidden variables is sometimes augmented by a second assumption— that the size of the observable universe puts a limit on the computations that these variables can do. A nonlocal theory of this sort predicts that a quantum computer would encounter fundamental obstacles when attempting to factor numbers of approximately 10,000 digits or more; a potentially achievable task in quantum mechanics.[74]

11.7.4 Popper's criticism

Main article: Popper's experiment

Karl Popper approached the problem of indeterminacy as a logician and metaphysical realist.[75] He disagreed with the application of the uncertainty relations to individual particles rather than to ensembles of identically prepared particles, referring to them as "statistical scatter relations".[75][76] In this statistical interpretation, a *particular* measurement may be made to arbitrary precision without invalidating the quantum theory. This directly contrasts with the Copenhagen interpretation of quantum mechanics, which is non-deterministic but lacks local hidden variables.

In 1934, Popper published *Zur Kritik der Ungenauigkeitsrelationen* (*Critique of the Uncertainty Relations*) in *Naturwissenschaften*,[77] and in the same year *Logik der Forschung* (translated and updated by the author as *The Logic of Scientific Discovery* in 1959), outlining his arguments for the statistical interpretation. In 1982, he further developed his theory in *Quantum theory and the schism in Physics*, writing:

> [Heisenberg's] formulae are, beyond all doubt, derivable *statistical formulae* of the quantum theory. But they have been *habitually misinterpreted* by those quantum theorists who said that these formulae can be interpreted as determining some upper limit to the *precision of our measurements*.[original emphasis][78]

Popper proposed an experiment to falsify the uncertainty relations, although he later withdrew his initial version after discussions with Weizsäcker, Heisenberg, and Einstein; this experiment may have influenced the formulation of the EPR experiment.[75][79]

11.7.5 Many-worlds uncertainty

Main article: Many-worlds interpretation

The many-worlds interpretation originally outlined by Hugh Everett III in 1957 is partly meant to reconcile the differences between Einstein's and Bohr's views by replacing Bohr's wave function collapse with an ensemble of deterministic and independent universes whose *distribution* is governed by wave functions and the Schrödinger equation. Thus, uncertainty in the many-worlds interpretation follows from each observer within any universe having no knowledge of what goes on in the other universes.

11.7.6 Free will

Some scientists including Arthur Compton[80] and Martin Heisenberg[81] have suggested that the uncertainty principle, or at least the general probabilistic nature of quantum mechanics, could be evidence for the two-stage model of free will. One critique, however, is that apart from the basic role of quantum mechanics as a foundation for chemistry, nontrivial biological mechanisms requiring quantum mechanics are unlikely, due to the rapid decoherence time of quantum systems at room temperature.[82] The standard view, however, is that this decoherence is overcome by both screening and decoherence-free subspaces found in biological cells.[83]

11.8 See also

- Afshar experiment
- Canonical commutation relation
- Correspondence principle
- Correspondence rules

- Gromov's non-squeezing theorem
- Discrete Fourier transform#Uncertainty principle
- Heisenbug
- Introduction to quantum mechanics
- Operationalization
- Observer effect (information technology)
- Observer effect (physics)
- Quantum indeterminacy
- Quantum non-equilibrium
- Quantum tunnelling
- *The Part and The Whole* (book)
- Weak measurement

11.9 Notes

[1] Sen, D. (2014). "The uncertainty relations in quantum mechanics" (PDF). *Current Science*. **107** (2): 203–218.

[2] Heisenberg, W. (1927), "Über den anschaulichen Inhalt der quantentheoretischen Kinematik und Mechanik", *Zeitschrift für Physik* (in German), **43** (3–4): 172–198, Bibcode:1927ZPhy...43..172H, doi:10.1007/BF01397280.. Annotated pre-publication proof sheet of Über den anschaulichen Inhalt der quantentheoretischen Kinematik und Mechanik, March 21, 1927.

[3] Kennard, E. H. (1927), "Zur Quantenmechanik einfacher Bewegungstypen", *Zeitschrift für Physik* (in German), **44** (4–5): 326, Bibcode:1927ZPhy...44..326K, doi:10.1007/BF01391200.

[4] Weyl, H. (1928), *Gruppentheorie und Quantenmechanik*, Leipzig: Hirzel

[5] Furuta, Aya (2012), "One Thing Is Certain: Heisenberg's Uncertainty Principle Is Not Dead", *Scientific American*

[6] Ozawa, Masanao (2003), "Universally valid reformulation of the Heisenberg uncertainty principle on noise and disturbance in measurement", *Physical Review A*, **67** (4): 42105, arXiv:quant-ph/0207121, Bibcode:2003PhRvA..67d2105O, doi:10.1103/PhysRevA.67.042105

[7] Werner Heisenberg, *The Physical Principles of the Quantum Theory*, p. 20

[8] Rozema, L. A.; Darabi, A.; Mahler, D. H.; Hayat, A.; Soudagar, Y.; Steinberg, A. M. (2012). "Violation of Heisenberg's Measurement-Disturbance Relationship by Weak Measurements". *Physical Review Letters*. **109** (10). arXiv:1208.0034v2. Bibcode:2012PhRvL.109j0404R. doi:10.1103/PhysRevLett.109.100404.

[9] Indian Institute of Technology Madras, Professor V. Balakrishnan, Lecture 1 – Introduction to Quantum Physics; Heisenberg's uncertainty principle, National Programme of Technology Enhanced Learning on YouTube

[10] L.D. Landau, E.M. Lifshitz (1977). *Quantum Mechanics: Non-Relativistic Theory*. Vol. 3 (3rd ed.). Pergamon Press. ISBN 978-0-08-020940-1. Online copy.

[11] Section 3.2 of Ballentine, Leslie E. (1970), "The Statistical Interpretation of Quantum Mechanics", *Reviews of Modern Physics*, **42** (4): 358–381, doi:10.1103/RevModPhys.42.358. This fact is experimentally well-known for example in quantum optics (see e.g. chap. 2 and Fig. 2.1 Leonhardt, Ulf (1997), *Measuring the Quantum State of Light*, Cambridge: Cambridge University Press, ISBN 0 521 49730 2

[12] Elion, W. J.; M. Matters, U. Geigenmüller & J. E. Mooij; Geigenmüller, U.; Mooij, J. E. (1994), "Direct demonstration of Heisenberg's uncertainty principle in a superconductor", *Nature*, **371** (6498): 594–595, Bibcode:1994Natur.371..594E, doi:10.1038/371594a0

[13] Smithey, D. T.; M. Beck, J. Cooper, M. G. Raymer; Cooper, J.; Raymer, M. G. (1993), "Measurement of number–phase uncertainty relations of optical fields", *Phys. Rev. A*, **48** (4): 3159–3167, Bibcode:1993PhRvA..48.3159S, doi:10.1103/PhysRevA.48.3159, PMID 9909968

[14] Caves, Carlton (1981), "Quantum-mechanical noise in an interferometer", *Phys. Rev. D*, **23** (8): 1693–1708, Bibcode:1981PhRvD..23.1693C, doi:10.1103/PhysRevD.23.1693

[15] Jaeger, Gregg (September 2014). "What in the (quantum) world is macroscopic?". *American Journal of Physics*. **82** (9): 896–905. Bibcode:2014AmJPh..82..896J. doi:10.1119/1.4878358.

[16] Claude Cohen-Tannoudji; Bernard Diu; Franck Laloë (1996), *Quantum mechanics*, Wiley-Interscience: Wiley, pp. 231–233, ISBN 978-0-471-56952-7

[17] Robertson, H. P. (1929), "The Uncertainty Principle", *Phys. Rev.*, **34**: 163–64, Bibcode:1929PhRv...34..163R, doi:10.1103/PhysRev.34.163

[18] Schrödinger, E. (1930), "Zum Heisenbergschen Unschärfeprinzip", *Sitzungsberichte der Preussischen Akademie der Wissenschaften, Physikalisch-mathematische Klasse*, **14**: 296-303

[19] Griffiths, David (2005), *Quantum Mechanics*, New Jersey: Pearson

11.9. NOTES

[20] Riley, K. F.; M. P. Hobson and S. J. Bence (2006), *Mathematical Methods for Physics and Engineering*, Cambridge, p. 246

[21] Davidson, E. R. (1965), "On Derivations of the Uncertainty Principle", *J. Chem. Phys.*, **42** (4): 1461, Bibcode:1965JChPh..42.1461D, doi:10.1063/1.1696139

[22] Hall, B. C. (2013), *Quantum Theory for Mathematicians*, Springer, p. 245

[23] Jackiw, Roman (1968), "Minimum Uncertainty Product, Number-Phase Uncertainty Product, and Coherent States", *J. Math. Phys.*, **9** (3): 339, Bibcode:1968JMP.....9..339J, doi:10.1063/1.1664585

[24] Carruthers, P.; Nieto, M. M. (1968), "Phase and Angle Variables in Quantum Mechanics", *Rev. Mod. Phys.*, **40** (2): 411, Bibcode:1968RvMP...40..411C, doi:10.1103/RevModPhys.40.411

[25] Hall, B. C. (2013), *Quantum Theory for Mathematicians*, Springer

[26] L. I. Mandelshtam, I. E. Tamm, *The uncertainty relation between energy and time in nonrelativistic quantum mechanics*, 1945

[27] Hilgevoord, Jan (1996). "The uncertainty principle for energy and time." *American Journal of Physics* **64.12**, 1451-1456, ; Hilgevoord, Jan (1998). "The uncertainty principle for energy and time. II." *American Journal of Physics* **66.5** 396-402.

[28] The broad linewidth of fast decaying states makes it difficult to accurately measure the energy of the state, and researchers have even used detuned microwave cavities to slow down the decay rate, to get sharper peaks. Gabrielse, Gerald; H. Dehmelt (1985), "Observation of Inhibited Spontaneous Emission", *Physical Review Letters*, **55** (1): 67–70, Bibcode:1985PhRvL..55...67G, doi:10.1103/PhysRevLett.55.67, PMID 10031682

[29] Likharev, K.K.; A.B. Zorin (1985), "Theory of Bloch-Wave Oscillations in Small Josephson Junctions", *J. Low Temp. Phys.*, **59** (3/4): 347–382, Bibcode:1985JLTP...59..347L, doi:10.1007/BF00683782

[30] Anderson, P.W. (1964), "Special Effects in Superconductivity", in Caianiello, E.R., *Lectures on the Many-Body Problem, Vol. 2*, New York: Academic Press

[31] Steiger, Nathan. "Quantum Uncertainty and Conservation Law Restrictions on Gate Fidelity". Brigham Young University. Retrieved 19 June 2011.

[32] Curtright, T.; Zachos, C. (2001). "Negative Probability and Uncertainty Relations". *Modern Physics Letters A*. **16** (37): 2381–2385. arXiv:hep-th/0105226. Bibcode:2001MPLA...16.2381C. doi:10.1142/S021773230100576X.

[33] Busch, P.; Lahti, P.; Werner, R. F. (2013). "Proof of Heisenberg's Error-Disturbance Relation". *Physical Review Letters*. **111** (16). arXiv:1306.1565. Bibcode:2013PhRvL.111p0405B. doi:10.1103/PhysRevLett.111.160405.

[34] Busch, P.; Lahti, P.; Werner, R. F. (2014). "Heisenberg uncertainty for qubit measurements". *Physical Review A*. **89**. arXiv:1311.0837. Bibcode:2014PhRvA..89a2129B. doi:10.1103/PhysRevA.89.012129.

[35] Erhart, J.; Sponar, S.; Sulyok, G.; Badurek, G.; Ozawa, M.; Hasegawa, Y. (2012). "Experimental demonstration of a universally valid error-disturbance uncertainty relation in spin measurements". *Nature Physics*. **8** (3): 185–189. arXiv:1201.1833. Bibcode:2012NatPh...8..185E. doi:10.1038/nphys2194.

[36] Baek, S.-Y.; Kaneda, F.; Ozawa, M.; Edamatsu, K. (2013). "Experimental violation and reformulation of the Heisenberg's error-disturbance uncertainty relation". *Scientific Reports*. **3**: 2221. Bibcode:2013NatSR...3E2221B. doi:10.1038/srep02221.

[37] Ringbauer, M.; Biggerstaff, D.N.; Broome, M.A.; Fedrizzi, A.; Branciard, C.; White, A.G. (2014). "Experimental Joint Quantum Measurements with Minimum Uncertainty". *Physical Review Letters*. **112**: 020401. arXiv:1308.5688. Bibcode:2014PhRvL.112b0401R. doi:10.1103/PhysRevLett.112.020401.

[38] Björk, G.; Söderholm, J.; Trifonov, A.; Tsegaye, T.; Karlsson, A. (1999). "Complementarity and the uncertainty relations". *Physical Review*. **A60**: 1878. arXiv:quant-ph/9904069. Bibcode:1999PhRvA..60.1874B. doi:10.1103/PhysRevA.60.1874.

[39] Fujikawa, Kazuo (2012). "Universally valid Heisenberg uncertainty relation". *Physical Review A*. **85** (6). arXiv:1205.1360. Bibcode:2012PhRvA..85f2117F. doi:10.1103/PhysRevA.85.062117.

[40] Judge, D. (1964), "On the uncertainty relation for angle variables", *Il Nuovo Cimento*, **31** (2): 332–340, doi:10.1007/BF02733639

[41] Bouten, M.; Maene, N.; Van Leuven, P. (1965), "On an uncertainty relation for angle variables", *Il Nuovo Cimento*, **37** (3): 1119–1125, doi:10.1007/BF02773197

[42] Louisell, W. H. (1963), "Amplitude and phase uncertainty relations", *Physics Letters*, **7** (1): 60–61, Bibcode:1963PhL......7...60L, doi:10.1016/0031-9163(63)90442-6

[43] DeWitt, B. S.; Graham, N. (1973), *The Many-Worlds Interpretation of Quantum Mechanics*, Princeton: Princeton University Press, pp. 52–53, ISBN 0-691-08126-3

[44] Hirschman, I. I., Jr. (1957), "A note on entropy", *American Journal of Mathematics*, **79** (1): 152–156, doi:10.2307/2372390, JSTOR 2372390.

[45] Beckner, W. (1975), "Inequalities in Fourier analysis", *Annals of Mathematics*, **102** (6): 159–182, doi:10.2307/1970980, JSTOR 1970980.

[46] Bialynicki-Birula, I.; Mycielski, J. (1975), "Uncertainty Relations for Information Entropy in Wave Mechanics", *Communications in Mathematical Physics*, **44** (2): 129, Bibcode:1975CMaPh..44..129B, doi:10.1007/BF01608825

[47] Chafaï, D. (2003), *Gaussian maximum of entropy and reversed log-Sobolev inequality*, arXiv:math/0102227, doi:10.1007/978-3-540-36107-7_5, ISBN 978-3-540-00072-3

[48] Havin, V.; Jöricke, B. (1994), *The Uncertainty Principle in Harmonic Analysis*, Springer-Verlag

[49] Folland, Gerald; Sitaram, Alladi (May 1997), "The Uncertainty Principle: A Mathematical Survey", *Journal of Fourier Analysis and Applications*, **3** (3): 207–238, doi:10.1007/BF02649110, MR 98f:42006

[50] Sitaram, A (2001), "Uncertainty principle, mathematical", in Hazewinkel, Michiel, *Encyclopedia of Mathematics*, Springer, ISBN 978-1-55608-010-4

[51] Amrein, W.O.; Berthier, A.M. (1977), "On support properties of L^p-functions and their Fourier transforms", *Journal of Functional Analysis*, **24** (3): 258–267, doi:10.1016/0022-1236(77)90056-8.

[52] Benedicks, M. (1985), "On Fourier transforms of functions supported on sets of finite Lebesgue measure", *J. Math. Anal. Appl.*, **106** (1): 180–183, doi:10.1016/0022-247X(85)90140-4

[53] Nazarov, F. (1994), "Local estimates for exponential polynomials and their applications to inequalities of the uncertainty principle type,", *St. Petersburg Math. J.*, **5**: 663–717

[54] Jaming, Ph. (2007), "Nazarov's uncertainty principles in higher dimension", *J. Approx. Theory*, **149** (1): 30–41, doi:10.1016/j.jat.2007.04.005

[55] Hardy, G.H. (1933), "A theorem concerning Fourier transforms", *Journal of the London Mathematical Society*, **8** (3): 227–231, doi:10.1112/jlms/s1-8.3.227

[56] Hörmander, L. (1991), "A uniqueness theorem of Beurling for Fourier transform pairs", *Ark. Mat.*, **29**: 231–240, Bibcode:1991ArM....29..237H, doi:10.1007/BF02384339

[57] Bonami, A.; Demange, B.; Jaming, Ph. (2003), "Hermite functions and uncertainty principles for the Fourier and the windowed Fourier transforms", *Rev. Mat. Iberoamericana*, **19**: 23–55., arXiv:math/0102111, Bibcode:2001math......2111B, doi:10.4171/RMI/337

[58] Hedenmalm, H. (2012), "Heisenberg's uncertainty principle in the sense of Beurling", *J. Anal. Math.*, **118** (2): 691–702, doi:10.1007/s11854-012-0048-9

[59] Demange, Bruno (2009), *Uncertainty Principles Associated to Non-degenerate Quadratic Forms*, Société Mathématique de France, ISBN 978-2-85629-297-6

[60] American Physical Society online exhibit on the Uncertainty Principle

[61] Bohr, Niels; Noll, Waldemar (1958), "Atomic Physics and Human Knowledge", *American Journal of Physics*, New York: Wiley, **26** (8): 38, Bibcode:1958AmJPh..26..596B, doi:10.1119/1.1934707

[62] Heisenberg, W., *Die Physik der Atomkerne*, Taylor & Francis, 1952, p. 30.

[63] Heisenberg, W. (1930), *Physikalische Prinzipien der Quantentheorie* (in German), Leipzig: Hirzel English translation *The Physical Principles of Quantum Theory*. Chicago: University of Chicago Press, 1930.

[64] Cassidy, David; Saperstein, Alvin M. (2009), "Beyond Uncertainty: Heisenberg, Quantum Physics, and the Bomb", *Physics Today*, New York: Bellevue Literary Press, **63**: 185, Bibcode:2010PhT....63a..49C, doi:10.1063/1.3293416

[65] Tipler, Paul A.; Llewellyn, Ralph A. (1999), "5–5", *Modern Physics* (3rd ed.), W. H. Freeman and Co., ISBN 1-57259-164-1

[66] R&D Magazine & University of Toronto, September 10, 2012 Scientists cast doubt on the uncertainty principle retrieved Sept 10, 2012

[67] "U of T scientists cast doubt on the uncertainty principle". *utoronto.ca*. Retrieved 22 June 2016.

[68] Feynman lectures on Physics, vol 3, 2–2

[69] Gamow, G., *The great physicists from Galileo to Einstein*, Courier Dover, 1988, p.260.

[70] Kumar, M., *Quantum: Einstein, Bohr and the Great Debate About the Nature of Reality*, Icon, 2009, p. 282.

[71] Gamow, G., *The great physicists from Galileo to Einstein*, Courier Dover, 1988, p. 260–261.

[72] Kumar, M., *Quantum: Einstein, Bohr and the Great Debate About the Nature of Reality*, Icon, 2009, p. 287.

[73] Isaacson, Walter (2007), *Einstein: His Life and Universe*, New York: Simon & Schuster, p. 452, ISBN 978-0-7432-6473-0

[74] Gerardus 't Hooft has at times advocated this point of view.

[75] Popper, Karl (1959), *The Logic of Scientific Discovery*, Hutchinson & Co.

[76] Jarvie, Ian Charles; Milford, Karl; Miller, David W (2006), *Karl Popper: a centenary assessment*, **3**, Ashgate Publishing, ISBN 978-0-7546-5712-5

[77] Popper, Karl; Carl Friedrich von Weizsäcker (1934), "Zur Kritik der Ungenauigkeitsrelationen (Critique of the Uncertainty Relations)", *Naturwissenschaften*, **22** (48): 807–808, Bibcode:1934NW.....22..807P, doi:10.1007/BF01496543.

[78] Popper, K. *Quantum theory and the schism in Physics*, Unwin Hyman Ltd, 1982, pp. 53–54.

[79] Mehra, Jagdish; Rechenberg, Helmut (2001), *The Historical Development of Quantum Theory*, Springer, ISBN 978-0-387-95086-0

[80] Compton, A. H. (1931). "The Uncertainty Principle and Free Will". *Science*. **74** (1911): 172. Bibcode:1931Sci....74..172C. doi:10.1126/science.74.1911.172. PMID 17808216.

[81] Heisenberg, M. (2009). "Is free will an illusion?". *Nature*. **459** (7244): 164. Bibcode:2009Natur.459..164H. doi:10.1038/459164a.

[82] Davies, P. C. W. (2004). "Does quantum mechanics play a non-trivial role in life?". *Biosystems*. **78** (1–3): 69–79. doi:10.1016/j.biosystems.2004.07.001. PMID 15555759.

[83] Davies, P. C. W. (2004). "Does quantum mechanics play a non-trivial role in life?". *Biosystems*. **78** (1–3): 69–79. doi:10.1016/j.biosystems.2004.07.001. PMID 15555759.

11.10 External links

- Hazewinkel, Michiel, ed. (2001), "Uncertainty principle", *Encyclopedia of Mathematics*, Springer, ISBN 978-1-55608-010-4

- Matter as a Wave – a chapter from an online textbook

- Quantum mechanics: Myths and facts

- Stanford Encyclopedia of Philosophy entry

- Fourier Transforms and Uncertainty at MathPages

- aip.org: Quantum mechanics 1925–1927 – The uncertainty principle

- Eric Weisstein's World of Physics – Uncertainty principle

- John Baez on the time–energy uncertainty relation

- The certainty principle

- Common Interpretation of Heisenberg's Uncertainty Principle Is Proved False

Chapter 12

De Broglie–Bohm theory

The **de Broglie–Bohm theory**, also known as the **pilot-wave theory**, **Bohmian mechanics**, the **Bohm or Bohm's interpretation**, and the **causal interpretation**, is an interpretation of quantum theory. In addition to a wavefunction on the space of all possible configurations, it also postulates an actual configuration that exists even when unobserved. The evolution over time of the configuration (that is, of the positions of all particles or the configuration of all fields) is defined by the wave function via a guiding equation. The evolution of the wave function over time is given by Schrödinger's equation. The theory is named after Louis de Broglie (1892–1987), and David Bohm (1917–1992).

The theory is deterministic[1] and explicitly nonlocal: the velocity of any one particle depends on the value of the guiding equation, which depends on the configuration of the system given by its wavefunction; the latter depends on the boundary conditions of the system, which in principle may be the entire universe.

The theory results in a measurement formalism, analogous to thermodynamics for classical mechanics, that yields the standard quantum formalism generally associated with the Copenhagen interpretation. The theory's explicit non-locality resolves the "measurement problem", which is conventionally delegated to the topic of interpretations of quantum mechanics in the Copenhagen interpretation. The Born rule in Broglie–Bohm theory is not a basic law. Rather, in this theory the link between the probability density and the wave function has the status of a hypothesis, called the quantum equilibrium hypothesis, which is additional to the basic principles governing the wave function.

The theory was historically developed by de Broglie in the 1920s, who in 1927 was persuaded to abandon it in favour of the then-mainstream Copenhagen interpretation. David Bohm, dissatisfied with the prevailing orthodoxy, rediscovered de Broglie's pilot wave theory in 1952. Bohm's suggestions were not widely received then, partly due to reasons unrelated to their content, connected to Bohm's youthful communist affiliations.[2] De Broglie–Bohm theory was widely deemed unacceptable by mainstream theorists, mostly because of its explicit non-locality. Bell's theorem (1964) was inspired by Bell's discovery of the work of David Bohm and his subsequent wondering if the obvious nonlocality of the theory could be eliminated. Since the 1990s, there has been renewed interest in formulating extensions to de Broglie–Bohm theory, attempting to reconcile it with special relativity and quantum field theory, besides other features such as spin or curved spatial geometries.[3]

The *Stanford Encyclopedia of Philosophy* article on Quantum decoherence (Guido Bacciagaluppi, 2012) groups "approaches to quantum mechanics" into five groups, of which "pilot-wave theories" are one (the others being the Copenhagen interpretation, objective collapse theories, many-world interpretations and modal interpretations).

There are several equivalent mathematical formulations of the theory and it is known by a number of different names. The de Broglie wave has a macroscopic analogy termed Faraday wave.[4]

12.1 Overview

De Broglie–Bohm theory is based on the following postulates:

- There is a configuration q of the universe, described by coordinates q^k, which is an element of the configuration space Q. The configuration space is different for different versions of pilot wave theory. For example, this may be the space of positions $\mathbf{Q}_k$ of N particles, or, in case of field theory, the space of field configurations $\phi(x)$. The configuration evolves (for spin=0) according to the guiding equation

$$m_k \frac{dq^k}{dt}(t) = \hbar \nabla_k \operatorname{Im} \ln\psi(q,t) =$$

$$\hbar \operatorname{Im}\left(\frac{\nabla_k \psi}{\psi}\right)(q,t) = \frac{m_k \mathbf{j}_k}{\psi^* \psi}$$

$$= \operatorname{Re}\left(\frac{\hat{\mathbf{P}}_k \Psi}{\Psi}\right)$$

Where **j** is the probability current or probability flux and **P̂** is the momentum operator. Here, $\psi(q,t)$ is the standard complex-valued wavefunction known from quantum theory, which evolves according to Schrödinger's equation

$$i\hbar \frac{\partial}{\partial t}\psi(q,t) = -\sum_{i=1}^{N} \frac{\hbar^2}{2m_i}\nabla_i^2 \psi(q,t) + V(q)\psi(q,t)$$

This already completes the specification of the theory for any quantum theory with Hamilton operator of type $H = \sum \frac{1}{2m_i}\hat{p}_i^2 + V(\hat{q})$.

- The configuration is distributed according to $|\psi(q,t)|^2$ at some moment of time t, and this consequently holds for all times. Such a state is named quantum equilibrium. With quantum equilibrium, this theory agrees with the results of standard quantum mechanics.

Notably, even if this latter relation is frequently presented as an axiom of the theory, in Bohm's original papers of 1952 it was presented as derivable from statistical-mechanical arguments. This argument was further supported by the work of Bohm in 1953 and was substantiated by Vigier and Bohm's paper of 1954 in which they introduced stochastic *fluid fluctuations* that drive a process of asymptotic relaxation from quantum non-equilibrium to quantum equilibrium ($\rho \to |\psi|^2$).[5]

12.1.1 Double-slit experiment

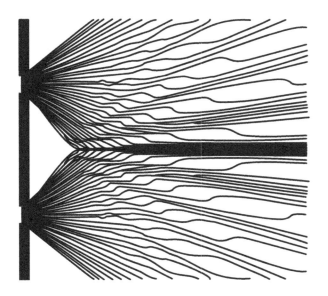

The Bohmian trajectories for an electron going through the two-slit experiment. A similar pattern was also extrapolated from weak measurements of single photons.[6]

The double-slit experiment is an illustration of wave-particle duality. In it, a beam of particles (such as electrons) travels through a barrier that has two slits. If one puts a detector screen on the side beyond the barrier, the pattern of detected particles shows interference fringes characteristic of waves arriving at the screen from two sources (the two slits); however, the interference pattern is made up of individual dots corresponding to particles that had arrived on the screen. The system seems to exhibit the behaviour of both waves (interference patterns) and particles (dots on the screen).

If we modify this experiment so that one slit is closed, no interference pattern is observed. Thus, the state of both slits affects the final results. We can also arrange to have a minimally invasive detector at one of the slits to detect which slit the particle went through. When we do that, the interference pattern disappears.

The Copenhagen interpretation states that the particles are not localised in space until they are detected, so that, if there is not any detector on the slits, there is no information about which slit the particle has passed through. If one slit has a detector on it, then the wavefunction collapses due to that detection.

In de Broglie–Bohm theory, the wavefunction is defined at both slits, but each particle has a well-defined trajectory that passes through exactly one of the slits. The final position of the particle on the detector screen and the slit through which the particle passes is determined by the initial position of the particle. Such initial position is not knowable or controllable by the experimenter, so there is an appearance of randomness in the pattern of detection. In Bohm's 1952 papers he used the wavefunction to construct a quantum potential that, when included in Newton's equations, gave the trajectories of the particles streaming through the two slits. In effect the wave function interferes with itself and guides the particles via the quantum potential in such a way that the particles avoid the regions in which the interference is destructive and are attracted to the regions in which the interference is constructive, resulting in the interference pattern on the detector screen.

To explain the behavior when the particle is detected to go through one slit, one needs to appreciate the role of the conditional wavefunction and how it results in the collapse of the wavefunction; this is explained below. The basic idea is that the environment registering the detection effectively separates the two wave packets in configuration space.

12.2 The theory

12.2.1 The ontology

The ontology of de Broglie-Bohm theory consists of a configuration $q(t) \in Q$ of the universe and a pilot wave $\psi(q,t) \in \mathbb{C}$. The configuration space Q can be chosen differently, as in classical mechanics and standard quantum mechanics.

Thus, the ontology of pilot wave theory contains as the trajectory $q(t) \in Q$ we know from classical mechanics, as the wave function $\psi(q,t) \in \mathbb{C}$ of quantum theory. So, at every moment of time there exists not only a wave function, but also a well-defined configuration of the whole universe (i.e., the system as defined by the boundary conditions used in solving the Schrödinger equation). The correspondence to our experiences is made by the identification of the configuration of our brain with some part of the configuration of the whole universe $q(t) \in Q$, as in classical mechanics.

While the ontology of classical mechanics is part of the ontology of de Broglie–Bohm theory, the dynamics are very different. In classical mechanics, the accelerations of the particles are imparted directly by forces, which exist in physical three-dimensional space. In de Broglie–Bohm theory, the velocities of the particles are given by the wavefunction, which exists in a 3N-dimensional configuration space, where N corresponds to the number of particles in the system;[7] Bohm hypothesized that each particle has a "complex and subtle inner structure" that provides the capacity to react to the information provided by the wavefunction via the quantum potential.[8] Also, unlike in classical mechanics, physical properties (e.g., mass, charge) are spread out over the wavefunction in de Broglie-Bohm theory, not localized at the position of the particle.[9][10]

The wavefunction itself, and not the particles, determines the dynamical evolution of the system: the particles do not act back onto the wave function. As Bohm and Hiley worded it, "the Schrödinger equation for the quantum field does not have sources, nor does it have any other way by which the field could be directly affected by the condition of the particles [...] the quantum theory can be understood completely in terms of the assumption that the quantum field has no sources or other forms of dependence on the particles."[11] P. Holland considers this lack of reciprocal action of particles and wave function to be one "[a]mong the many nonclassical properties exhibited by this theory".[12] It should be noted however that Holland has later called this a merely *apparent* lack of back reaction, due to the incompleteness of the description.[13]

In what follows below, we will give the setup for one particle moving in $\mathbb{R}^3$ followed by the setup for N particles moving in 3 dimensions. In the first instance, configuration space and real space are the same while in the second, real space is still $\mathbb{R}^3$, but configuration space becomes $\mathbb{R}^{3N}$. While the particle positions themselves are in real space, the velocity field and wavefunction are on configuration space, which is how particles are entangled with each other in this theory.

Extensions to this theory include spin and more complicated configuration spaces.

We use variations of $\mathbf{Q}$ for particle positions while ψ represents the complex-valued wavefunction on configuration space.

12.2.2 Guiding equation

For a spinless single particle moving in $\mathbb{R}^3$, the particle's velocity is given

$$\frac{d\mathbf{Q}}{dt}(t) = \frac{\hbar}{m} \operatorname{Im}\left(\frac{\nabla \psi}{\psi}\right)(\mathbf{Q}, t)$$

For many particles, we label them as $\mathbf{Q}_k$ for the k th particle and their velocities are given by

$$\frac{d\mathbf{Q}_k}{dt}(t) = \frac{\hbar}{m_k} \operatorname{Im}\left(\frac{\nabla_k \psi}{\psi}\right)(\mathbf{Q}_1, \mathbf{Q}_2, \ldots, \mathbf{Q}_N, t)$$

The main fact to notice is that this velocity field depends on the actual positions of all of the N particles in the universe. As explained below, in most experimental situations, the influence of all of those particles can be encapsulated into an effective wavefunction for a subsystem of the universe.

12.2.3 Schrödinger's equation

The one particle Schrödinger equation governs the time evolution of a complex-valued wavefunction on $\mathbb{R}^3$. The equation represents a quantized version of the total energy of a classical system evolving under a real-valued potential function V on $\mathbb{R}^3$:

$$i\hbar \frac{\partial}{\partial t}\psi = -\frac{\hbar^2}{2m}\nabla^2 \psi + V\psi$$

For many particles, the equation is the same except that ψ and V are now on configuration space, $\mathbb{R}^{3N}$.

$$i\hbar \frac{\partial}{\partial t}\psi = -\sum_{k=1}^{N} \frac{\hbar^2}{2m_k}\nabla_k^2 \psi + V\psi$$

This is the same wavefunction of conventional quantum mechanics.

12.2.4 Relation to the Born Rule

In Bohm's original papers [Bohm 1952], he discusses how de Broglie–Bohm theory results in the usual measurement results of quantum mechanics. The main idea is that this is true if the positions of the particles satisfy the statistical distribution given by $|\psi|^2$. And that distribution is guaranteed to be true for all time by the guiding equation if the initial distribution of the particles satisfies $|\psi|^2$.

For a given experiment, we can postulate this as being true and verify experimentally that it does indeed hold true, as it does. But, as argued in Dürr et al.,[14] one needs to argue that this distribution for subsystems is typical. They argue that $|\psi|^2$ by virtue of its equivariance under the dynamical evolution of the system, is the appropriate measure of typicality for initial conditions of the positions of the particles. They then prove that the vast majority of possible initial configurations will give rise to statistics obeying the Born rule (i.e., $|\psi|^2$) for measurement outcomes. In summary, in a universe governed by the de Broglie–Bohm dynamics, Born rule behavior is typical.

The situation is thus analogous to the situation in classical statistical physics. A low entropy initial condition will, with overwhelmingly high probability, evolve into a higher entropy state: behavior consistent with the second law of thermodynamics is typical. There are, of course, anomalous initial conditions that would give rise to violations of the second law. However, in the absence of some very detailed evidence supporting the actual realization of one of those special initial conditions, it would be quite unreasonable to expect anything but the actually observed uniform increase of entropy. Similarly, in the de Broglie–Bohm theory, there are anomalous initial conditions that would produce measurement statistics in violation of the Born rule (i.e., in conflict with the predictions of standard quantum theory). But the typicality theorem shows that, in the absence of some specific reason to believe that one of those special initial conditions was in fact realized, the Born rule behavior is what one should expect.

It is in that qualified sense that the Born rule is, for the de Broglie–Bohm theory, a theorem rather than (as in ordinary quantum theory) an additional postulate.

It can also be shown that a distribution of particles that is *not* distributed according to the Born rule (that is, a distribution 'out of quantum equilibrium') and evolving under the de Broglie-Bohm dynamics is overwhelmingly likely to evolve dynamically into a state distributed as $|\psi|^2$. See, for example Ref. .[15] A video of the electron density in a 2D box evolving under this process is available here.

12.2.5 The conditional wave function of a subsystem

In the formulation of the De Broglie–Bohm theory, there is only a wave function for the entire universe (which always evolves by the Schrödinger equation). It should however be noted that the "universe" is simply the system limited by the same boundary conditions used to solve the Schrödinger equation. However, once the theory is formulated, it is convenient to introduce a notion of wave function also for subsystems of the universe. Let us write the wave function of the universe as $\psi(t, q^I, q^{II})$, where q^I denotes the configuration variables associated to some subsystem (I) of the universe and q^{II} denotes the remaining configuration variables. Denote, respectively, by $Q^I(t)$ and by $Q^{II}(t)$ the actual configuration of subsystem (I) and of the rest of the universe. For simplicity, we consider here only the spinless case. The *conditional wave function* of subsystem (I) is defined by:

$$\psi^I(t, q^I) = \psi(t, q^I, Q^{II}(t)).$$

It follows immediately from the fact that $Q(t) = (Q^I(t), Q^{II}(t))$ satisfies the guiding equation that also the configuration $Q^I(t)$ satisfies a guiding equation identical to the one presented in the formulation of the theory, with the universal wave function ψ replaced with the conditional wave function ψ^I. Also, the fact that $Q(t)$ is random with probability density given by the square modulus of $\psi(t, \cdot)$ implies that the conditional probability density of $Q^I(t)$ given $Q^{II}(t)$ is given by the square modulus of the (normalized) conditional wave function $\psi^I(t, \cdot)$ (in the terminology of Dürr et al.[16] this fact is called the *fundamental conditional probability formula*).

Unlike the universal wave function, the conditional wave function of a subsystem does not always evolve by the Schrödinger equation, but in many situations it does. For instance, if the universal wave function factors as:

$$\psi(t, q^I, q^{II}) = \psi^I(t, q^I)\psi^{II}(t, q^{II})$$

then the conditional wave function of subsystem (I) is (up to an irrelevant scalar factor) equal to ψ^I (this is what Standard Quantum Theory would regard as the wave function of subsystem (I)). If, in addition, the Hamiltonian does not contain an interaction term between subsystems (I) and (II) then ψ^I does satisfy a Schrödinger equation. More generally, assume that the universal wave function ψ can be written in the form:

$$\psi(t, q^I, q^{II}) = \psi^I(t, q^I)\psi^{II}(t, q^{II}) + \phi(t, q^I, q^{II}),$$

where ϕ solves Schrödinger equation and $\phi(t, q^I, Q^{II}(t)) =$ 0 for all t and q^I. Then, again, the conditional wave function of subsystem (I) is (up to an irrelevant scalar factor) equal to ψ^I and if the Hamiltonian does not contain an interaction term between subsystems (I) and (II), ψ^I satisfies a Schrödinger equation.

The fact that the conditional wave function of a subsystem does not always evolve by the Schrödinger equation is related to the fact that the usual collapse rule of Standard Quantum Theory emerges from the Bohmian formalism when one considers conditional wave functions of subsystems.

12.3 Extensions

12.3.1 Relativity

Pilot wave theory is explicitly nonlocal, which is in ostensible conflict with special relativity. Various extensions of "Bohm-like" mechanics exist that attempt to resolve this problem. Bohm himself in 1953 presented an extension of the theory satisfying the Dirac equation for a single particle. However, this was not extensible to the many-particles case because it used an absolute time.[17] A renewed interest in constructing Lorentz-invariant extensions of Bohmian theory arose in the 1990s; see Bohm and Hiley: The Undivided Universe, and , , and references therein. Another approach is given in the work of Dürr et al.[18] in which they use Bohm-Dirac models and a Lorentz-invariant foliation of space-time.

Thus, Dürr et al. (1999) showed that it is possible to formally restore Lorentz invariance for the Bohm-Dirac theory by introducing additional structure. This approach still requires a foliation of space-time. While this is in conflict with the standard interpretation of relativity, the preferred foliation, if unobservable, does not lead to any empirical conflicts with relativity. In 2013, Dürr et al. suggested that the required foliation could be covariantly determined by the wave function.[19]

The relation between nonlocality and preferred foliation can be better understood as follows. In de Broglie–Bohm theory, nonlocality manifests as the fact that the velocity and acceleration of one particle depends on the instantaneous positions of all other particles. On the other hand, in the theory of relativity the concept of instantaneousness does not have an invariant meaning. Thus, to define particle trajectories, one needs an additional rule that defines which space-time points should be considered instantaneous. The simplest way to achieve this is to introduce a preferred foliation of space-time by hand, such that each hypersurface of the foliation defines a hypersurface of equal time.

Initially, it had been considered impossible to set out a description of photon trajectories in the de Broglie–Bohm theory in view of the difficulties of describing bosons relativistically.[20] In 1996, Partha Ghose had presented a relativistic quantum mechanical description of spin-0 and spin-1 bosons starting from the Duffin–Kemmer–Petiau equation, setting out Bohmian trajectories for massive bosons and for massless bosons (and therefore photons).[20] In 2001, Jean-Pierre Vigier emphasized the importance of deriving a well-defined description of light in terms of particle trajectories in the framework of either the Bohmian mechanics or the Nelson stochastic mechanics.[21] The same year, Ghose worked out Bohmian photon trajectories for specific cases.[22] Subsequent weak measurement experiments yielded trajectories that coincide with the predicted trajectories.[23][24]

Chris Dewdney and G. Horton have proposed a relativistically covariant, wave-functional formulation of Bohm's quantum field theory[25][26] and has extended it to a form that allows the inclusion of gravity.[27]

Nikolić has proposed a Lorentz-covariant formulation of the Bohmian interpretation of many-particle wave functions.[28] He has developed a generalized relativistic-invariant probabilistic interpretation of quantum theory,[29][30][31] in which $|\psi|^2$ is no longer a probability density in space, but a probability density in space-time. He uses this generalized probabilistic interpretation to formulate a relativistic-covariant version of de Broglie–Bohm theory without introducing a preferred foliation of space-time. His work also covers the extension of the Bohmian interpretation to a quantization of fields and strings.[32]

See also: Quantum potential § Relativistic and field-theoretic extensions

Roderick I. Sutherland at the University in Sydney has a Lagrangian formalism for the pilot wave and its beables. I draws on Yakir Aharonov's retrocasual weak measurements to explain many-particle entanglement in a special relativistic way without the need for configuration space. The basic idea was already published by Costa de Beauregard in the 1950s and is also used by John Cramer in his transactional interpretation sans the beables that exist between the von Neumann strong projection operator measurements. Sutherland's Lagrangian includes two-way action-reaction between pilot wave and beables. Therefore, it is a post-quantum non-statistical theory with final boundary conditions that violate the no-signal theorems of quantum theory. Just as special relativity is a limiting case of general relativity when the spacetime curvature vanishes, so, too is statistical no-entanglement signaling quantum theory with the Born rule a limiting case of the post-quantum action-reaction Lagrangian when the reaction is set to zero and the final boundary condition is integrated out.[33]

12.3.2 Spin

To incorporate spin, the wavefunction becomes complex-vector valued. The value space is called spin space; for a spin-½ particle, spin space can be taken to be $\mathbb{C}^2$. The guiding equation is modified by taking inner products in spin space to reduce the complex vectors to complex numbers. The Schrödinger equation is modified by adding a Pauli spin term.

$$\frac{d\mathbf{Q}_k}{dt}(t) = \frac{\hbar}{m_k} Im\left(\frac{(\psi, D_k\psi)}{(\psi,\psi)}\right)(\mathbf{Q}_1, \mathbf{Q}_2, \ldots, \mathbf{Q}_N, t)$$

$$i\hbar\frac{\partial}{\partial t}\psi = \left(-\sum_{k=1}^{N}\frac{\hbar^2}{2m_k}D_k^2 + V - \sum_{k=1}^{N}\mu_k \mathbf{S}_k/S_k \cdot \mathbf{B}(\mathbf{q}_k)\right)\psi$$

where μ_k is the magnetic moment of the k th particle, $\mathbf{S}_k$ is the appropriate spin operator acting in the k th particle's spin space, S_k is spin of the particle ($S_k = 1/2$ for electron),

$$D_k = \nabla_k - \frac{ie_k}{}$$

mechanics in so far as the latter has predictions. However, while standard quantum mechanics is limited to discussing the results of 'measurements', de Broglie–Bohm theory is a theory that governs the dynamics of a system without the intervention of outside observers (p. 117 in Bell[38]).

The basis for agreement with standard quantum mechanics is that the particles are distributed according to $|\psi|^2$. This is a statement of observer ignorance, but it can be proven[14] that for a universe governed by this theory, this will typically be the case. There is apparent collapse of the wave function governing subsystems of the universe, but there is no collapse of the universal wavefunction.

12.4.1 Measuring spin and polarization

According to ordinary quantum theory, it is not possible to measure the spin or polarization of a particle directly; instead, the component in one direction is measured; the outcome from a single particle may be 1, meaning that the particle is aligned with the measuring apparatus, or −1, meaning that it is aligned the opposite way. For an ensemble of particles, if we expect the particles to be aligned, the results are all 1. If we expect them to be aligned oppositely, the results are all −1. For other alignments, we expect some results to be 1 and some to be −1 with a probability that depends on the expected alignment. For a full explanation of this, see the Stern-Gerlach Experiment.

In de Broglie–Bohm theory, the results of a spin experiment cannot be analyzed without some knowledge of the experimental setup. It is possible[39] to modify the setup so that the trajectory of the particle is unaffected, but that the particle with one setup registers as spin up while in the other setup it registers as spin down. Thus, for the de Broglie–Bohm theory, the particle's spin is not an intrinsic property of the particle—instead spin is, so to speak, in the wave function of the particle in relation to the particular device being used to measure the spin. This is an illustration of what is sometimes referred to as contextuality, and is related to naive realism about operators.[40]

12.4.2 Measurements, the quantum formalism, and observer independence

De Broglie–Bohm theory gives the same results as quantum mechanics. It treats the wavefunction as a fundamental object in the theory as the wavefunction describes how the particles move. This means that no experiment can distinguish between the two theories. This section outlines the ideas as to how the standard quantum formalism arises out of quantum mechanics. References include Bohm's original 1952 paper and Dürr et al.[14]

Collapse of the wavefunction

De Broglie–Bohm theory is a theory that applies primarily to the whole universe. That is, there is a single wavefunction governing the motion of all of the particles in the universe according to the guiding equation. Theoretically, the motion of one particle depends on the positions of all of the other particles in the universe. In some situations, such as in experimental systems, we can represent the system itself in terms of a de Broglie–Bohm theory in which the wavefunction of the system is obtained by conditioning on the environment of the system. Thus, the system can be analyzed with Schrödinger's equation and the guiding equation, with an initial $|\psi|^2$ distribution for the particles in the system (see the section on the conditional wave function of a subsystem for details).

It requires a special setup for the conditional wavefunction of a system to obey a quantum evolution. When a system interacts with its environment, such as through a measurement, the conditional wavefunction of the system evolves in a different way. The evolution of the universal wavefunction can become such that the wavefunction of the system appears to be in a superposition of distinct states. But if the environment has recorded the results of the experiment, then using the actual Bohmian configuration of the environment to condition on, the conditional wavefunction collapses to just one alternative, the one corresponding with the measurement results.

Collapse of the universal wavefunction never occurs in de Broglie–Bohm theory. Its entire evolution is governed by Schrödinger's equation and the particles' evolutions are governed by the guiding equation. Collapse only occurs in a phenomenological way for systems that seem to follow their own Schrödinger's equation. As this is an effective description of the system, it is a matter of choice as to what to define the experimental system to include and this will affect when "collapse" occurs.

Operators as observables

In the standard quantum formalism, measuring observables is generally thought of as measuring operators on the Hilbert space. For example, measuring position is considered to be a measurement of the position operator. This relationship between physical measurements and Hilbert space operators is, for standard quantum mechanics, an additional axiom of the theory. The de Broglie–Bohm theory, by contrast, requires no such measurement axioms (and measurement as such is not a dynamically distinct or special sub-category of physical processes in the theory). In particular, the usual operators-as-observables formalism is, for de Broglie–Bohm theory, a theorem.[41] A major point of the analysis is that many of the measurements of the observables do not correspond to properties of the particles; they are (as in the case of spin discussed above) measurements of the wavefunction.

In the history of de Broglie–Bohm theory, the proponents have often had to deal with claims that this theory is impossible. Such arguments are generally based on inappropriate analysis of operators as observables. If one believes that spin measurements are indeed measuring the spin of a particle that existed prior to the measurement, then one does reach contradictions. De Broglie–Bohm theory deals with this by noting that spin is not a feature of the particle, but rather that of the wavefunction. As such, it only has a definite outcome once the experimental apparatus is chosen. Once that is taken into account, the impossibility theorems become irrelevant.

There have also been claims that experiments reject the Bohm trajectories in favor of the standard QM lines. But as shown in and , such experiments cited above only disprove a misinterpretation of the de Broglie–Bohm theory, not the theory itself.

There are also objections to this theory based on what it says about particular situations usually involving eigenstates of an operator. For example, the ground state of hydrogen is a real wavefunction. According to the guiding equation, this means that the electron is at rest when in this state. Nevertheless, it is distributed according to $|\psi|^2$ and no contradiction to experimental results is possible to detect.

Operators as observables leads many to believe that many operators are equivalent. De Broglie–Bohm theory, from this perspective, chooses the position observable as a favored observable rather than, say, the momentum observable. Again, the link to the position observable is a consequence of the dynamics. The motivation for de Broglie–Bohm theory is to describe a system of particles. This implies that the goal of the theory is to describe the positions of those particles at all times. Other observables do not have this compelling ontological status. Having definite positions explains having definite results such as flashes on a detector screen. Other observables would not lead to that conclusion, but there need not be any problem in defining a mathematical theory for other observables; see Hyman et al.[42] for an exploration of the fact that a probability density and probability current can be defined for any set of commuting operators.

Hidden variables

De Broglie–Bohm theory is often referred to as a "hidden variable" theory. Bohm used this description in his original papers on the subject, writing, "From the point of view of the usual interpretation, these additional elements or parameters [permitting a detailed causal and continuous description of all processes] could be called 'hidden' variables." Bohm and Hiley later stated that they found Bohm's choice of the term "hidden variables" to be too restrictive. In particular, they argued that a particle is not actually hidden but rather "is what is most directly manifested in an observation [though] its properties cannot be observed with arbitrary precision (within the limits set by uncertainty principle)".[43] However, others nevertheless treat the term "hidden variable" as a suitable description.[44]

Generalized particle trajectories can be extrapolated from numerous weak measurements on an ensemble of equally prepared systems, and such trajectories coincide with the de Broglie–Bohm trajectories. In particular, an experiment with two entangled photons, in which a set of Bohmian trajectories for one of the photons was determined using weak measurements and postselection, can be understood in terms of a nonlocal connection between that photon's trajectory and the other photon's polarisation.[45][46][47] However, not only the De Broglie-Bohm interpretation, but also many other interpretation of quantum mechanics that do not include such trajectories are consistent with such experimental evidence.

12.4.3 Heisenberg's uncertainty principle

The Heisenberg uncertainty principle states that when two complementary measurements are made, there is a limit to the product of their accuracy. As an example, if one measures the position with an accuracy of Δx, and the momentum with an accuracy of Δp, then $\Delta x \Delta p \gtrsim h$. If we make further measurements in order to get more information, we disturb the system and change the trajectory into a new one depending on the measurement setup; therefore, the measurement results are still subject to Heisenberg's uncertainty relation.

In de Broglie–Bohm theory, there is always a matter of fact about the position and momentum of a particle. Each particle has a well-defined trajectory, as well as a wave function. Observers have limited knowledge as to what this trajectory is (and thus of the position and momentum). It is the lack of knowledge of the particle's trajectory that accounts for the uncertainty relation. What one can know about a particle at any given time is described by the wavefunction. Since the uncertainty relation can be derived from the wavefunction in other interpretations of quantum mechanics, it can be likewise derived (in the epistemic sense mentioned above), on the de Broglie–Bohm theory.

To put the statement differently, the particles' positions are only known statistically. As in classical mechanics, successive observations of the particles' positions refine the experimenter's knowledge of the particles' initial conditions. Thus, with succeeding observations, the initial conditions become more and more restricted. This formalism is consistent with the normal use of the Schrödinger equation.

For the derivation of the uncertainty relation, see Heisenberg uncertainty principle, noting that it describes it from the viewpoint of the Copenhagen interpretation.

12.4.4 Quantum entanglement, Einstein-Podolsky-Rosen paradox, Bell's theorem, and nonlocality

De Broglie–Bohm theory highlighted the issue of nonlocality: it inspired John Stewart Bell to prove his now-famous theorem,[48] which in turn led to the Bell test experiments.

In the Einstein–Podolsky–Rosen paradox, the authors describe a thought-experiment one could perform on a pair of particles that have interacted, the results of which they interpreted as indicating that quantum mechanics is an incomplete theory.[49]

Decades later John Bell proved Bell's theorem (see p. 14 in Bell[38]), in which he showed that, if they are to agree with the empirical predictions of quantum mechanics, all such "hidden-variable" completions of quantum mechanics must either be nonlocal (as the Bohm interpretation is) or give up the assumption that experiments produce unique results (see counterfactual definiteness and many-worlds interpretation). In particular, Bell proved that any local theory with unique results must make empirical predictions satisfying a statistical constraint called "Bell's inequality".

Alain Aspect performed a series of Bell test experiments that test Bell's inequality using an EPR-type setup. Aspect's results show experimentally that Bell's inequality is in fact violated—meaning that the relevant quantum mechanical predictions are correct. In these Bell test experiments, entangled pairs of particles are created; the particles are separated, traveling to remote measuring apparatus. The orientation of the measuring apparatus can be changed while the particles are in flight, demonstrating the apparent nonlocality of the effect.

The de Broglie–Bohm theory makes the same (empirically correct) predictions for the Bell test experiments as ordinary quantum mechanics. It is able to do this because it is manifestly nonlocal. It is often criticized or rejected based on this; Bell's attitude was: "It is a merit of the de Broglie–Bohm version to bring this [nonlocality] out so explicitly that it cannot be ignored."[50]

The de Broglie–Bohm theory describes the physics in the Bell test experiments as follows: to understand the evolution of the particles, we need to set up a wave equation for both particles; the orientation of the apparatus affects the wavefunction. The particles in the experiment follow the guidance of the wavefunction. It is the wavefunction that carries the faster-than-light effect of changing the orientation of the apparatus. An analysis of exactly what kind of nonlocality is present and how it is compatible with relativity can be found in Maudlin.[51] Note that in Bell's work, and in more detail in Maudlin's work, it is shown that the nonlocality does not allow for signaling at speeds faster than light.

12.4.5 Classical limit

Bohm's formulation of de Broglie–Bohm theory in terms of a classical-looking version has the merits that the emergence of classical behavior seems to follow immediately for any situation in which the quantum potential is negligible, as noted by Bohm in 1952. Modern methods of decoherence are relevant to an analysis of this limit. See Allori et al.[52] for steps towards a rigorous analysis.

12.4.6 Quantum trajectory method

Work by Robert E. Wyatt in the early 2000s attempted to use the Bohm "particles" as an adaptive mesh that follows the actual trajectory of a quantum state in time and space. In the "quantum trajectory" method, one samples the quantum wavefunction with a mesh of quadrature points. One then evolves the quadrature points in time according to the Bohm equations of motion. At each time-step, one then re-synthesizes the wavefunction from the points, recomputes the quantum forces, and continues the calculation. (QuickTime movies of this for H+H$_2$ reactive scattering can be found on the Wyatt group web-site at UT Austin.) This approach has been adapted, extended, and used by a number of researchers in the Chemical Physics community as a way to compute semi-classical and quasi-classical molecular dynamics. A recent (2007) issue of the Journal of Physical Chemistry A was dedicated to Prof. Wyatt and his work on "Computational Bohmian Dynamics".

Eric R. Bittner's group at the University of Houston has advanced a statistical variant of this approach that uses Bayesian sampling technique to sample the quantum density and compute the quantum potential on a structureless mesh of points. This technique was recently used to estimate quantum effects in the heat-capacity of small clusters Ne$_n$ for n~100.

There remain difficulties using the Bohmian approach, mostly associated with the formation of singularities in the quantum potential due to nodes in the quantum wavefunction. In general, nodes forming due to interference effects lead to the case where $R^{-1}\nabla^2 R \to \infty$. This results in an infinite force on the sample particles forcing them to move away from the node and often crossing the path of other sample points (which violates single-valuedness). Various schemes have been developed to overcome this; however, no general solution has yet emerged.

These methods, as does Bohm's Hamilton-Jacobi formulation, do not apply to situations in which the full dynamics of spin need to be taken into account.

12.4.7 Occam's razor criticism

Both Hugh Everett III and Bohm treated the wavefunction as a physically real field. Everett's many-worlds interpretation is an attempt to demonstrate that the wavefunction alone is sufficient to account for all our observations. When we see the particle detectors flash or hear the click of a Geiger counter then Everett's theory interprets this as our *wavefunction* responding to changes in the detector's *wavefunction*, which is responding in turn to the passage of another *wavefunction* (which we think of as a "particle", but is actually just another wave-packet).[53] No particle (in the Bohm sense of having a defined position and velocity) exists, according to that theory. For this reason Everett sometimes referred to his own many-worlds approach as the "pure wave theory". Talking of Bohm's 1952 approach, Everett says:

In the Everettian view, then, the Bohm particles are superfluous entities, similar to, and equally as unnecessary as, for example, the luminiferous ether, which was found to be unnecessary in special relativity. This argument of Everett's is sometimes called the "redundancy argument", since the superfluous particles are redundant in the sense of Occam's razor.[55]

Many authors have expressed critical views of the de Broglie-Bohm theory, by comparing it to Everett's many worlds approach. Many (but not all) proponents of the de Broglie-Bohm theory (such as Bohm and Bell) interpret the universal wave function as physically real. According to some supporters of Everett's theory, if the (never collapsing) wave function is taken to be physically real, then it is natural to interpret the theory as having the same many worlds as Everett's theory. In the Everettian view the role of the Bohm particle is to act as a "pointer", tagging, or selecting, just one branch of the universal wavefunction (the assumption that this branch indicates which *wave packet* determines the observed result of a given experiment is called the "result assumption"[53]); the other branches are designated "empty" and implicitly assumed by Bohm to be devoid of conscious observers.[53] H. Dieter Zeh comments on these "empty" branches:

David Deutsch has expressed the same point more "acerbically":[53]

According to Brown & Wallace[53] the de Broglie-Bohm particles play no role in the solution of the measurement problem. These authors claim[53] that the "result assumption" (see above) is inconsistent with the view that there is no measurement problem in the predictable outcome (i.e. single-outcome) case. These authors also claim[53] that a standard tacit assumption of the de Broglie-Bohm theory (that an observer becomes aware of configurations of particles of ordinary objects by means of correlations between such configurations and the configuration of the particles in the observer's brain) is unreasonable. This conclusion has been challenged by Valentini[58] who argues that the entirety of such objections arises from a failure to interpret de Broglie-Bohm theory on its own terms.

According to Peter R. Holland, in a wider Hamiltonian framework, theories can be formulated in which particles *do* act back on the wave function.[59]

12.4.8 Non-equivalence

In 2016, Pisin Chen and Hagen Kleinert argued that the Copenhagen interpretation and the De Broglie–Bohm theory yield different results for the ratio of peak intensities in the double-slit experiment. They concluded that they are thus not mathematically equivalent.[60]

12.5 Derivations

De Broglie–Bohm theory has been derived many times and in many ways. Below are six derivations all of which are very different and lead to different ways of understanding and extending this theory.

- Schrödinger's equation can be derived by using Einstein's light quanta hypothesis: $E = \hbar\omega$ and de Broglie's hypothesis: $\mathbf{p} = \hbar\mathbf{k}$.

  The guiding equation can be derived in a similar fashion. We assume a plane wave: $\psi(\mathbf{x}, t) = Ae^{i(\mathbf{k}\cdot\mathbf{x}-\omega t)}$. Notice that $i\mathbf{k} = \nabla\psi/\psi$. Assuming that $\mathbf{p} = m\mathbf{v}$ for the particle's actual velocity, we have that $\mathbf{v} = \frac{\hbar}{m} Im\left(\frac{\nabla\psi}{\psi}\right)$. Thus, we have the guiding equation.

  Notice that this derivation does not use Schrödinger's equation.

- Preserving the density under the time evolution is another method of derivation. This is the method that Bell cites. It is this method that generalizes to many possible alternative theories. The starting point is the continuity equation $-\frac{\partial \rho}{\partial t} = \nabla \cdot (\rho v^\psi)$ for the density $\rho = |\psi|^2$. This equation describes a probability flow along a current. We take the velocity field associated with this current as the velocity field whose integral curves yield the motion of the particle.

- A method applicable for particles without spin is to do a polar decomposition of the wavefunction and transform Schrödinger's equation into two coupled equations: the continuity equation from above and the Hamilton–Jacobi equation. This is the method used by Bohm in 1952. The decomposition and equations are as follows:

Decomposition: $\psi(\mathbf{x}, t) = R(\mathbf{x}, t)e^{iS(\mathbf{x},t)/\hbar}$. Note $R^2(\mathbf{x}, t)$ corresponds to the probability density $\rho(\mathbf{x}, t) = |\psi(\mathbf{x}, t)|^2$.

$$-\frac{\partial \rho(\mathbf{x},t)}{\partial t} = \nabla \cdot \left(\rho(\mathbf{x},t)\frac{\nabla S(\mathbf{x},t)}{m}\right)$$

$$\frac{\partial S(\mathbf{x},t)}{\partial t} = -\left[\frac{1}{2m}(\nabla S(\mathbf{x},t))^2 + V - \frac{\hbar^2}{2m}\frac{\nabla^2 R(\mathbf{x},t)}{R(\mathbf{x},t)}\right]$$

The Hamilton–Jacobi equation is the equation derived from a Newtonian system with potential $V - \frac{\hbar^2}{2m}\frac{\nabla^2 R}{R}$ and velocity field $\frac{\nabla S}{m}$. The potential V is the classical potential that appears in Schrödinger's equation and the other term involving R is the quantum potential, terminology introduced by Bohm.

This leads to viewing the quantum theory as particles moving under the classical force modified by a quantum force. However, unlike standard Newtonian mechanics, the initial velocity field is already specified by $\frac{\nabla S}{m}$, which is a symptom of this being a first-order theory, not a second-order theory.

- A fourth derivation was given by Dürr et al.[14] In their derivation, they derive the velocity field by demanding the appropriate transformation properties given by the various symmetries that Schrödinger's equation satisfies, once the wavefunction is suitably transformed. The guiding equation is what emerges from that analysis.

- A fifth derivation, given by Dürr et al.[34] is appropriate for generalization to quantum field theory and the Dirac equation. The idea is that a velocity field can also be understood as a first order differential operator acting on functions. Thus, if we know how it acts on functions, we know what it is. Then given the Hamiltonian operator H, the equation to satisfy for all functions f (with associated multiplication operator $\hat{f}$) is

$$(v(f))(q) = \text{Re}\frac{(\psi, \frac{i}{\hbar}[H,\hat{f}]\psi)}{(\psi,\psi)}(q)$$ where (v, w) is the local Hermitian inner product on the value space of the wavefunction.

This formulation allows for stochastic theories such as the creation and annihilation of particles.

- A further derivation has been given by Peter R. Holland, on which he bases the entire work presented in his quantum physics textbook *The Quantum Theory of Motion*, a main reference book on the de Broglie–Bohm theory. It is based on three basic postulates and an additional fourth postulate that links the wave function to measurement probabilities:[61]

1. A physical system consists in a spatiotemporally propagating wave and a point particle guided by it;

2. The wave is described mathematically by a solution ψ to Schrödinger's wave equation;

3. The particle motion is described by a solution to $\mathbf{x}(t) = [\nabla S(\mathbf{x}(t), t))]/m$ in dependence on initial condition $\mathbf{x}(t = 0)$, with S the phase of ψ.

The fourth postulate is subsidiary yet consistent with the first three:

4. The probability $\rho(\mathbf{x}(t))$ to find the particle in the differential volume $d^3 x$ at time t equals $|\psi(\mathbf{x}(t))|^2$.

12.6 History

De Broglie–Bohm theory has a history of different formulations and names. In this section, each stage is given a name and a main reference.

12.6.1 Pilot-wave theory

Louis de Broglie presented his pilot wave theory at the 1927 Solvay Conference,[62] after close collaboration with Schrödinger, who developed his wave equation for de Broglie's theory. At the end of the presentation, Wolfgang Pauli pointed out that it was not compatible with a semi-classical technique Fermi had previously adopted in the case of inelastic scattering. Contrary to a popular legend, de Broglie actually gave the correct rebuttal that the particular technique could not be generalized for Pauli's purpose, although the audience might have been lost in the technical details and de Broglie's mild manner left the impression that Pauli's objection was valid. He was eventually persuaded to abandon this theory nonetheless because he was "discouraged by criticisms which [it] roused".[63] De Broglie's theory already applies to multiple spin-less particles, but lacks an adequate theory of measurement as no one understood quantum decoherence at the time. An analysis of de Broglie's presentation is given in Bacciagaluppi et al.[64][65] Also, in 1932 John von Neumann published a paper,[66] that was widely (and erroneously, as shown by Jeffrey Bub[67]) believed to prove that all hidden-variable theories are impossible. This sealed the fate of de Broglie's theory for the next two decades.

In 1926, Erwin Madelung had developed a hydrodynamic version of Schrödinger's equation, which is incorrectly considered as a basis for the density current derivation of the de Broglie–Bohm theory.[68] The Madelung equations, being quantum Euler equations (fluid dynamics), differ philosophically from the de Broglie–Bohm mechanics[69] and are the basis of the stochastic interpretation of quantum mechanics.

Peter R. Holland has pointed out that, earlier in 1927, Einstein had actually submitted a preprint with a similar proposal but, not convinced, had withdrawn it before publication.[70] According to Holland, failure to appreciate key points of the de Broglie–Bohm theory has led to confusion, the key point being "that the trajectories of a many-body quantum system are correlated not because the particles exert a direct force on one another (*à la* Coulomb) but because all are acted upon by an entity – mathematically described by the wavefunction or functions of it – that lies beyond them".[71] This entity is the quantum potential.

After publishing a popular textbook on Quantum Mechanics that adhered entirely to the Copenhagen orthodoxy, Bohm was persuaded by Einstein to take a critical look at von Neumann's theorem. The result was 'A Suggested Interpretation of the Quantum Theory in Terms of "Hidden Variables" I and II' [Bohm 1952]. It was an independent origination of the pilot wave theory, and extended it to incorporate a consistent theory of measurement, and to address a criticism of Pauli that de Broglie did not properly respond to; it is taken to be deterministic (though Bohm hinted in the original papers that there should be disturbances to this, in the way Brownian motion disturbs Newtonian mechanics). This stage is known as the *de Broglie–Bohm Theory* in Bell's work [Bell 1987] and is the basis for 'The Quantum Theory of Motion' [Holland 1993].

This stage applies to multiple particles, and is deterministic.

The de Broglie–Bohm theory is an example of a hidden variables theory. Bohm originally hoped that hidden variables could provide a local, causal, objective description that would resolve or eliminate many of the paradoxes of quantum mechanics, such as Schrödinger's cat, the measurement problem and the collapse of the wavefunction. However, Bell's theorem complicates this hope, as it demonstrates that there can be no local hidden variable theory that is compatible with the predictions of quantum mechanics. The Bohmian interpretation is causal but not local.

Bohm's paper was largely ignored or panned by other physicists. Albert Einstein, who had suggested that Bohm search for a realist alternative to the prevailing Copenhagen approach, did not consider Bohm's interpretation to be a satisfactory answer to the quantum nonlocality question, calling it "too cheap",[72] while Werner Heisenberg considered it a "superfluous 'ideological superstructure' ".[73] Wolfgang Pauli, who had been unconvinced by de Broglie in 1927, conceded to Bohm as follows:

> I just received your long letter of 20th November, and I also have studied more thoroughly the details of your paper. I do not see any longer the possibility of any logical contradiction as long as your results agree completely with those of the usual wave mechanics and as long as no means is given to measure the values of your hidden parameters both in the measuring apparatus and in the observe [sic] system. As far as the whole matter stands now, your 'extra wave-mechanical predictions' are still a check, which cannot be cashed.[74]

He subsequently described Bohm's theory as "artificial metaphysics".[75]

According to physicist Max Dresden, when Bohm's theory was presented at the Institute for Advanced Study in Princeton, many of the objections were ad hominem, focusing on Bohm's sympathy with communists as exemplified by his refusal to give testimony to the House Un-American Activities Committee.[76]

In 1979, Chris Philippidis, Chris Dewdney and Basil Hiley were the first to perform numeric computations on the basis of the quantum potential to deduce ensembles of particle trajectories.[77][78] Their work renewed the interests of physicists in the Bohm interpretation of quantum physics.[79]

Eventually John Bell began to defend the theory. In "Speakable and Unspeakable in Quantum Mechanics" [Bell 1987], several of the papers refer to hidden variables theories (which include Bohm's).

The trajectories of the Bohm model that would result for particular experimental arrangements were termed "surreal" by some.[80][81] Still in 2016, mathematical physicist

Sheldon Goldstein said about Bohm's theory: "There was a time when you couldn't even talk about it because it was heretical. It probably still is the kiss of death for a physics career to be actually working on Bohm, but maybe that's changing."[47]

12.6.2 Bohmian mechanics

This term is used to describe the same theory, but with an emphasis on the notion of current flow, which is determined on the basis of the quantum equilibrium hypothesis that the probability follows the Born rule. The term "Bohmian mechanics" is also often used to include most of the further extensions past the spin-less version of Bohm. While de Broglie–Bohm theory has Lagrangians and Hamilton-Jacobi equations as a primary focus and backdrop, with the icon of the quantum potential, Bohmian mechanics considers the continuity equation as primary and has the guiding equation as its icon. They are mathematically equivalent in so far as the Hamilton-Jacobi formulation applies, i.e., spin-less particles. The papers of Dürr et al. popularized the term.

All of non-relativistic quantum mechanics can be fully accounted for in this theory.

12.6.3 Causal interpretation and ontological interpretation

Bohm developed his original ideas, calling them the *Causal Interpretation*. Later he felt that *causal* sounded too much like *deterministic* and preferred to call his theory the *Ontological Interpretation*. The main reference is 'The Undivided Universe' [Bohm, Hiley 1993].

This stage covers work by Bohm and in collaboration with Jean-Pierre Vigier and Basil Hiley. Bohm is clear that this theory is non-deterministic (the work with Hiley includes a stochastic theory). As such, this theory is not, strictly speaking, a formulation of the de Broglie–Bohm theory. However, it deserves mention here because the term "Bohm Interpretation" is ambiguous between this theory and the de Broglie–Bohm theory.

An in-depth analysis of possible interpretations of Bohm's model of 1952 was given in 1996 by philosopher of science Arthur Fine.[82] R,v MA

12.7 Experiments

Researchers performed the ESSW experiment.[83] They found that the photon trajectories aren't surrealistic after all but more precisely, that the paths may seem surrealistic, but only if one fails to take into account the nonlocality inherent in Bohm's theory.[84][85][86]

12.8 See also

- David Bohm
- Faraday wave
- Interpretation of quantum mechanics
- Madelung equations
- Local hidden variable theory
- Quantum mechanics
- Pilot wave

12.9 Notes

[1] Bohm, David (1952). "A Suggested Interpretation of the Quantum Theory in Terms of 'Hidden Variables' I". *Physical Review*. **85** (2): 166–179. Bibcode:1952PhRv...85..166B. doi:10.1103/PhysRev.85.166. ("In contrast to the usual interpretation, this alternative interpretation permits us to conceive of each individual system as being in a precisely definable state, whose changes with time are determined by definite laws, analogous to (but not identical with) the classical equations of motion. Quantum-mechanical probabilities are regarded (like their counterparts in classical statistical mechanics) as only a practical necessity and not as an inherent lack of complete determination in the properties of matter at the quantum level.")

[2] F. David Peat, *Infinite Potential: The Life and Times of David Bohm* (1997), p. 133. James T. Cushing, *Quantum Mechanics: Historical Contingency and the Copenhagen Hegemony* (1994) discusses "the hegemony of the Copenhagen interpretation of quantum mechanics" over theories like Bohmian mechanics as an example of how the acceptance of scientific theories may be guided by social aspects.

[3] David Bohm and Basil J. Hiley, *The Undivided Universe - An Ontological Interpretation of Quantum Theory* appreared after Bohm's death, in 1993; reviewed by Sheldon Goldstein in *Physics Today* (1994). J. Cushing, A. Fine, S. Goldstein (eds.), *Bohmian Mechanics and Quantum Theory - An Appraisal* (1996).

[4] John W. M. Bush: "Quantum mechanics writ large"

[5] Publications of D. Bohm in 1952 and 1953 and of J.-P. Vigier in 1954 as cited in Antony Valentini; Hans Westman (8 January 2005). "Dynamical origin of quantum probabilities". *Proc. R. Soc. A*. **461** (2053): 253–272. arXiv:quant-ph/0403034. Bibcode:2005RSPSA.461..253V. doi:10.1098/rspa.2004.1394. p. 254

[6] "Observing the Average Trajectories of Single Photons in a Two-Slit Interferometer"

[7] David Bohm (1957). *Causality and Chance in Modern Physics*. Routledge & Kegan Paul and D. Van Nostrand. ISBN 0-8122-1002-6., p. 117.

[8] D. Bohm and B. Hiley: *The undivided universe: An ontological interpretation of quantum theory*, p. 37.

[9] H. R. Brown, C. Dewdney and G. Horton: "Bohm particles and their detection in the light of neutron interferometry", *Foundations of Physics*, 1995, Volume 25, Number 2, pp. 329–347.

[10] J. Anandan, "The Quantum Measurement Problem and the Possible Role of the Gravitational Field", *Foundations of Physics*, March 1999, Volume 29, Issue 3, pp. 333–348.

[11] D. Bohm and B. Hiley: *The undivided universe: An ontological interpretation of quantum theory*, p. 24

[12] Peter R. Holland: *The Quantum Theory of Motion: An Account of the De Broglie-Bohm Causal Interpretation of Quantum Mechanics*, Cambridge University Press, Cambridge (first published 25 June 1993), ISBN 0-521-35404-8 hardback, ISBN 0-521-48543-6 paperback, transferred to digital printing 2004, Chapter I. section (7) "There is no reciprocal action of the particle on the wave", p. 26

[13] • P. Holland: "Hamiltonian theory of wave and particle in quantum mechanics II: Hamilton-Jacobi theory and particle back-reaction", *Nuovo Cimento B* 116, 2001, pp. 1143–1172, full text preprint p. 31)

[14] Dürr, D.; Goldstein, S.; Zanghì, N. (1992). "Quantum Equilibrium and the Origin of Absolute Uncertainty". *Journal of Statistical Physics*. **67** (5–6): 843–907. arXiv:quant-ph/0308039. Bibcode:1992JSP....67..843D. doi:10.1007/BF01049004.

[15] Towler, M. D.; Russell, N. J.; Valentini A., pbs., "Timescales for dynamical relaxation to the Born rule" quant-ph/11031589

[16] "Quantum Equilibrium and the Origin of Absolute Uncertainty", D. Dürr, S. Goldstein and N. Zanghì, *Journal of Statistical Physics* 1992; 67, 843–907

[17] Oliver Passon, "What you always wanted to know about Bohmian mechanics but were afraid to ask", Invited talk at the spring meeting of the Deutsche Physikalische Gesellschaft, Dortmund, 2006, arXiv:quant-ph/0611032, p. 13.

[18] Dürr, D.; Goldstein, S.; Münch-Berndl, K.; Zanghì, N. (1999). "Hypersurface Bohm-Dirac Models". *Physical Review A*. **60** (4): 2729–2736. arXiv:quant-ph/9801070. Bibcode:1999PhRvA..60.2729D. doi:10.1103/physreva.60.2729.

[19] Dürr, Detlef; Goldstein, Sheldon; Norsen, Travis; Struyve, Ward; Zanghì, Nino (2013). "Can Bohmian mechanics be made relativistic?". *Proceedings of the Royal Society A: Mathematical, Physical and Engineering Sciences*. **470** (2162): 20130699. doi:10.1098/rspa.2013.0699.

[20] Ghose, Partha (1996). "Relativistic quantum mechanics of spin-0 and spin-1 bosons" (PDF). *Foundations of Physics*. **26** (11): 1441–1455. Bibcode:1996FoPh...26.1441G. doi:10.1007/BF02272366.

[21] Cufaro Petroni, Nicola; Vigier, Jean-Pierre (2001). "Remarks on Observed Superluminal Light Propagation". *Foundations of Physics Letters*. **14** (4): 395–400. doi:10.1023/A:1012321402475., therein: section *3. Conclusions*, page 399

[22] Ghose, Partha; Majumdar, A. S.; Guhab, S.; Sau, J. (2001). "Bohmian trajectories for photons" (PDF). *Physics Letters A*. **290** (5–6): 205–213. arXiv:quant-ph/0102071. Bibcode:2001PhLA..290..205G. doi:10.1016/s0375-9601(01)00677-6.

[23] Sacha Kocsis, Sylvain Ravets, Boris Braverman, Krister Shalm, Aephraim M. Steinberg: "Observing the trajectories of a single photon using weak measurement" 19th Australian Institute of Physics (AIP) Congress, 2010

[24] Kocsis, Sacha; Braverman, Boris; Ravets, Sylvain; Stevens, Martin J.; Mirin, Richard P.; Shalm, L. Krister; Steinberg, Aephraim M. (2011). "Observing the Average Trajectories of Single Photons in a Two-Slit Interferometer". *Science*. **332** (6034): 1170–1173. Bibcode:2011Sci...332.1170K. doi:10.1126/science.1202218. PMID 21636767.

[25] Dewdney, Chris; Horton, George (2002). "Relativistically invariant extension of the de Broglie Bohm theory of quantum mechanics". *Journal of Physics A: Mathematical and General*. **35** (47): 10117–10127. arXiv:quant-ph/0202104. Bibcode:2002JPhA...3510117D. doi:10.1088/0305-4470/35/47/311.

[26] Dewdney, Chris; Horton, George (2004). "A relativistically covariant version of Bohm's quantum field theory for the scalar field". *Journal of Physics A: Mathematical and General*. **37** (49): 11935–11943. arXiv:quant-ph/0407089. Bibcode:2004JPhA...3711935H. doi:10.1088/0305-4470/37/49/011.

[27] Dewdney, Chris; Horton, George (2010). "A relativistic hidden-variable interpretation for the massive vector field based on energy-momentum flows". *Foundations of Physics*. **40** (6): 658–678. Bibcode:2010FoPh...40..658H. doi:10.1007/s10701-010-9456-9.

[28] Nikolić, Hrvoje (2005). "Relativistic Quantum Mechanics and the Bohmian Interpretation". *Foundations of Physics Letters*. **18** (6): 549–561. arXiv:quant-ph/0406173. Bibcode:2005FoPhL..18..549N. doi:10.1007/s10702-005-1128-1.

[29] Nikolic, H (2010). "QFT as pilot-wave theory of particle creation and destruction". *International Journal of Modern Physics.* **25** (7): 1477–1505. arXiv:0904.2287. Bibcode:2010IJMPA..25.1477N. doi:10.1142/s0217751x10047889.

[30] Hrvoje Nikolić: "Time in relativistic and nonrelativistic quantum mechanics", arXiv:0811/0811.1905v2 (submitted 12 November 2008 (v1), revised 12 Jan 2009)

[31] Hrvoje Nikolić: "Making nonlocal reality compatible with relativity", arXiv:1002.3226v2 [quant-ph] (submitted on 17 February 2010, version of 31 May 2010)

[32] Hrvoje Nikolić: "Bohmian mechanics in relativistic quantum mechanics, quantum field theory and string theory", *2007 Journal of Physics*: Conf. Ser. 67 012035

[33] Roderick Sutherland: "Lagrangian Description for Particle Interpretations of Quantum Mechanics -- Entangled Many-Particle Case", arXiv:1509.02442

[34] Dürr, D., Goldstein, S., Tumulka, R., and Zanghì, N., 2004, "Bohmian Mechanics and Quantum Field Theory", *Physical Review Letters* 93: 090402:1–4.

[35] Dürr, D., Tumulka, R., and Zanghì, N., *Journal of Physics A*: Math. Gen. 38, R1–R43 (2005), quant-ph/0407116

[36] Dürr, D.; Goldstein, S.; Taylor, J.; Tumulka, R.; Zanghì, N. (2007). "Quantum Mechanics in Multiply-Connected Spaces". *Phys. A: Math. Theor.* **40** (12): 2997–3031. arXiv:quant-ph/0506173. Bibcode:2007JPhA...40.2997D. doi:10.1088/1751-8113/40/12/s08.

[37] Valentini, A (1991). "Signal-Locality, Uncertainty and the Subquantum H-Theorem. II". *Physics Letters A.* **158**: 1–8. Bibcode:1991PhLA..158....1V. doi:10.1016/0375-9601(91)90330-b.

[38] Bell, John S. (1987). *Speakable and Unspeakable in Quantum Mechanics.* Cambridge University Press. ISBN 0521334950.

[39] Albert, D. Z., 1992, Quantum Mechanics and Experience, Cambridge, MA: Harvard University Press

[40] Daumer, M.; Dürr, D.; Goldstein, S.; Zanghì, N. (1997). "Naive Realism About Operators". *Erkenntnis.* **45**: 379–397. arXiv:quant-ph/9601013. Bibcode:1996quant.ph..1013D.

[41] Dürr, D., Goldstein, S., and Zanghì, N., "Quantum Equilibrium and the Role of Operators as Observables in Quantum Theory" Journal of Statistical Physics 116, 959–1055 (2004)

[42] Hyman, Ross et al Bohmian mechanics with discrete operators, *Journal of Physics A*: Math. Gen. 37 L547–L558, 2004

[43] David Bohm, Basil Hiley: *The Undivided Universe: An Ontological Interpretation of Quantum Theory*, edition published in the Taylor & Francis e-library 2009 (first edition Routledge, 1993), ISBN 0-203-98038-7, p. 2

[44] "While the testable predictions of Bohmian mechanics are isomorphic to standard Copenhagen quantum mechanics, its underlying hidden variables have to be, in principle, unobservable. If one could observe them, one would be able to take advantage of that and signal faster than light, which – according to the special theory of relativity – leads to physical temporal paradoxes." J. Kofler and A. Zeiliinger, "Quantum Information and Randomness", *European Review* (2010), Vol. 18, No. 4, 469–480.

[45] Dylan H. Mahler, Lee Rozema, Kent Fisher, Lydia Vermeyden, Kevin J. Resch, Howard M. Wiseman, and Aephraim Steinberg: Experimental nonlocal and surreal Bohmian trajectories, Science Advances 19 February 2016, Vol. 2, no. 2, e1501466, DOI: 10.1126/science.1501466

[46] Researchers demonstrate 'quantum surrealism', phys.org, 19 February 2016

[47] Anil Ananthaswamy: Quantum weirdness may hide an orderly reality after all, newscientist.com, 19 February 2016

[48] Bell J. S. (1964). "On the Einstein Podolsky Rosen Paradox" (PDF). *Physics.* **1**: 195.

[49] Einstein; Podolsky; Rosen (1935). "Can Quantum Mechanical Description of Physical Reality Be Considered Complete?". *Phys. Rev.* **47** (10): 777–780. Bibcode:1935PhRv...47..777E. doi:10.1103/PhysRev.47.777.

[50] Bell, page 115

[51] Maudlin, T. (1994). *Quantum Non-Locality and Relativity: Metaphysical Intimations of Modern Physics.* Cambridge, Mass.: Blackwell. ISBN 0631186093.

[52] Allori, V.; Dürr, D.; Goldstein, S.; Zanghì, N. (2002). "Seven Steps Towards the Classical World". *Journal of Optics B.* **4** (4): 482–488. arXiv:quant-ph/0112005. Bibcode:2002JOptB...4S.482A. doi:10.1088/1464-4266/4/4/344.

[53] Brown, Harvey R; Wallace, David (2005). "Solving the measurement problem: de Broglie-Bohm loses out to Everett" (PDF). *Foundations of Physics.* **35** (4): 517–540. arXiv:quant-ph/0403094. Bibcode:2005FoPh...35..517B. doi:10.1007/s10701-004-2009-3. Abstract: "The quantum theory of de Broglie and Bohm solves the measurement problem, but the hypothetical corpuscles play no role in the argument. The solution finds a more natural home in the Everett interpretation."

[54] See section VI of Everett's thesis:*Theory of the Universal Wavefunction*, pp. 3–140 of Bryce Seligman DeWitt, R. Neill Graham, eds, *The Many-Worlds Interpretation of Quantum Mechanics*, Princeton Series in Physics, Princeton University Press (1973), ISBN 0-691-08131-X

[55] Craig Callender: "The Redundancy Argument Against Bohmian Mechanics"

[56] Daniel Dennett (2000). *With a little help from my friends*. In D. Ross, A. Brook, and D. Thompson (Eds.), *Dennett's Philosophy: a comprehensive assessment*. MIT Press/Bradford, ISBN 0-262-68117-X.

[57] David Deutsch, Comment on Lockwood. *British Journal for the Philosophy of Science* 47, 222228, 1996

[58] Valentini A., "De Broglie-Bohm pilot wave theory: many worlds in denial?" *Many Worlds? Everett, Quantum Theory, and Reality*, eds. S. Saunders et al. (Oxford University Press, 2010), pp. 476–509

[59] P. Holland, "Hamiltonian Theory of Wave and Particle in Quantum Mechanics I, II", *Nuovo Cimento B* 116, 1043, 1143 (2001) online

[60] Pisin Chen, Hagen Kleinert (2016). "Deficiencies of Bohm Trajectories in View of Basic Quantum Principles" (PDF). *Electronic Journal of Theoretical Physics (EJTP)*. **13** (35). pp. 1–12.

[61] Peter R. Holland: *The quantum theory of motion*, Cambridge University Press, 1993 (re-printed 2000, transferred to digital printing 2004), ISBN 0-521-48543-6, p. 66 ff.

[62] Solvay Conference, 1928, Electrons et Photons: Rapports et Descussions du Cinquieme Conseil de Physique tenu a Bruxelles du 24 au 29 October 1927 sous les auspices de l'Institut International Physique Solvay

[63] Louis be Broglie, in the foreword to David Bohm's *Causality and Chance in Modern Physics* (1957). p. x.

[64] Bacciagaluppi, G., and Valentini, A., "Quantum Theory at the Crossroads": Reconsidering the 1927 Solvay Conference

[65] See the brief summary by Towler, M., "Pilot wave theory, Bohmian metaphysics, and the foundations of quantum mechanics"

[66] von Neumann, J. 1932 *Mathematische Grundlagen der Quantenmechanik*

[67] Bub, Jeffrey (2010). "Von Neumann's 'No Hidden Variables' Proof: A Re-Appraisal". *Foundations of Physics*. **40** (9–10): 1333–1340. arXiv:1006.0499. Bibcode:2010FoPh...40.1333B. doi:10.1007/s10701-010-9480-9.

[68] Madelung, E. (1927). "Quantentheorie in hydrodynamischer Form". *Z. Phys.* **40** (3–4): 322–326. Bibcode:1927ZPhy...40..322M. doi:10.1007/BF01400372.

[69] Tsekov, Roumen (2012). "Bohmian Mechanics versus Madelung Quantum Hydrodynamics". doi:10.13140/RG.2.1.3663.8245 (inactive 2016-07-13).

[70] Holland, Peter (2004). "What's wrong with Einstein's 1927 hidden-variable interpretation of quantum mechanics?". *Foundations of Physics*. **35** (2): 177–196. arXiv:quant-ph/0401017. Bibcode:2005FoPh...35..177H. doi:10.1007/s10701-004-1940-7. p. 1

[71] Holland, Peter (2004). "What's wrong with Einstein's 1927 hidden-variable interpretation of quantum mechanics?". *Foundations of Physics*. **35** (2): 177–196. arXiv:quant-ph/0401017. Bibcode:2005FoPh...35..177H. doi:10.1007/s10701-004-1940-7. p. 14

[72] (Letter of 12 May 1952 from Einstein to Max Born, in *The Born–Einstein Letters*, Macmillan, 1971, p. 192.

[73] Werner Heisenberg, *Physics and Philosophy* (1958), p. 133.

[74] Pauli to Bohm, 3 December 1951, in Wolfgang Pauli, *Scientific Correspondence*, Vol IV – Part I, [ed. by Karl von Meyenn], (Berlin, 1996), pp. 436–441.

[75] Pauli, W. (1953). "Remarques sur le probleme des parametres caches dans la mecanique quantique et sur la theorie de l'onde pilote". In A. George (Ed.), *Louis de Broglie—physicien et penseur* (pp. 33–42). Paris: Editions Albin Michel.

[76] F. David Peat, *Infinite Potential: The Life and Times of David Bohm* (1997), p. 133.

[77] Statement on that they were in fact the first in: B. J. Hiley: *Nonlocality in microsystems*, in: Joseph S. King, Karl H. Pribram (eds.): *Scale in Conscious Experience: Is the Brain Too Important to be Left to Specialists to Study?*, Psychology Press, 1995, pp. 318 ff., p. 319, which takes reference to: C. Philippidis, C. Dewdney and B. J. Hiley: *Quantum interference and the quantum potential*, Il Nuovo Cimento B, Volume 52, Number 1, pp. 15-28, 1979, doi:10.1007/BF02743566 (abstract)

[78] Olival Freire, Jr.: *Continuity and change: charting David Bohm's evolving ideas on quantum mechanics*, In: Décio Krause, Antonio Videira (eds.): *Brazilian Studies in the Philosophy and History of Science*, Boston Studies in the Philosophy of Science, Springer, ISBN 978-90-481-9421-6, pp.291–300, therein p. 296–297

[79] Olival Freire jr.: *A story without an ending: the quantum physics controversy 1950–1970*, Science & Education, vol. 12, pp. 573–586, 2003, p. 576

[80] B-G. Englert, M. O. Scully, G. Sussman and H. Walther, 1992, *Surrealistic Bohm Trajectories*, Z. Naturforsch. 47a, 1175-1186.

[81] B. J. Hiley, R.E Callaghan, O. Maroney: *Quantum trajectories, real, surreal or an approximation to a deeper process?* (submitted on 5 Oct 2000, version of 5 Nov 2000 - arXiv:quant-ph/0010020

[82] A. Fine: "On the interpretation of Bohmian mechanics", in: J. T. Cushing, A. Fine, S. Goldstein (Eds.): *Bohmian mechanics and quantum theory: an appraisal*, Springer, 1996, pp. 231–250

[83] https://www.researchgate.net/publication/276382824_Surrealistic_Bohm_Trajectories Surrealistic Bohm Trajectories

[84] http://advances.sciencemag.org/content/2/2/e1501466 Experimental nonlocal and surreal Bohmian trajectories

[85] https://www.newscientist.com/article/2078251-quantum-weirdness-may-hide-an-orderly-reality-after-all

[86] http://www.wired.com/2016/05/new-support-alternative-quantum-view/

12.10 References

- Albert, David Z. (May 1994). "Bohm's Alternative to Quantum Mechanics". *Scientific American*. **270** (5): 58–67. doi:10.1038/scientificamerican0594-58.

- Barbosa, G. D.; N. Pinto-Neto (2004). "A Bohmian Interpretation for Noncommutative Scalar Field Theory and Quantum Mechanics". *Physical Review D*. **69** (6): 065014. arXiv:hep-th/0304105⊙. Bibcode:2004PhRvD..69f5014B. doi:10.1103/PhysRevD.69.065014.

- Bohm, David (1952). "A Suggested Interpretation of the Quantum Theory in Terms of "Hidden Variables" I". *Physical Review*. **85** (2): 166–179. Bibcode:1952PhRv...85..166B. doi:10.1103/PhysRev.85.166. (full text)

- Bohm, David (1952). "A Suggested Interpretation of the Quantum Theory in Terms of "Hidden Variables", II". *Physical Review*. **85** (2): 180–193. Bibcode:1952PhRv...85..180B. doi:10.1103/PhysRev.85.180. (full text)

- Bohm, David (1990). "A new theory of the relationship of mind and matter" (PDF). *Philosophical Psychology*. **3** (2): 271–286. doi:10.1080/09515089008573004.

- Bohm, David; B.J. Hiley (1993). *The Undivided Universe: An ontological interpretation of quantum theory*. London: Routledge. ISBN 0-415-12185-X.

- Dürr, Detlef; Sheldon Goldstein; Roderich Tumulka; Nino Zanghì (December 2004). "Bohmian Mechanics" (PDF). *Physical Review Letters*. **93** (9): 090402. arXiv:quant-ph/0303156⊙. Bibcode:2004PhRvL..93i0402D. doi:10.1103/PhysRevLett.93.090402. ISSN 0031-9007. PMID 15447078.

- Goldstein, Sheldon (2001). "Bohmian Mechanics". *Stanford Encyclopedia of Philosophy*.

- Hall, Michael J. W. (2004). "Incompleteness of trajectory-based interpretations of quantum mechanics". *Journal of Physics A: Mathematical and General*. **37** (40): 9549–9556. arXiv:quant-ph/0406054⊙. Bibcode:2004JPhA...37.9549H. doi:10.1088/0305-4470/37/40/015. (Demonstrates incompleteness of the Bohm interpretation in the face of fractal, differentialble-nowhere wavefunctions.)

- Holland, Peter R. (1993). *The Quantum Theory of Motion : An Account of the de Broglie–Bohm Causal Interpretation of Quantum Mechanics*. Cambridge: Cambridge University Press. ISBN 0-521-48543-6.

- Nikolic, H. (2004). "Relativistic quantum mechanics and the Bohmian interpretation". *Foundations of Physics Letters*. **18** (6): 549–561. arXiv:quant-ph/0406173⊙. Bibcode:2005FoPhL..18..549N. doi:10.1007/s10702-005-1128-1.

- Passon, Oliver (2004). "Why isn't every physicist a Bohmian?". arXiv:quant-ph/0412119⊙.

- Sanz, A. S.; F. Borondo (2003). "A Bohmian view on quantum decoherence". *European Physical Journal D*. **44** (2): 319–326. arXiv:quant-ph/0310096⊙. Bibcode:2007EPJD...44..319S. doi:10.1140/epjd/e2007-00191-8.

- Sanz, A. S. (2005). "A Bohmian approach to quantum fractals". *Journal of Physics A: Mathematical and General*. **38** (26): 319. arXiv:quant-ph/0412050⊙. Bibcode:2005JPhA...38.6037S. doi:10.1088/0305-4470/38/26/013. (Describes a Bohmian resolution to the dilemma posed by non-differentiable wavefunctions.)

- Silverman, Mark P. (1993). *And Yet It Moves: Strange Systems and Subtle Questions in Physics*. Cambridge: Cambridge University Press. ISBN 0-521-44631-7.

- Streater, Ray F. (2003). "Bohmian mechanics is a 'lost cause'". Retrieved 2006-06-25.

- Valentini, Antony; Hans Westman (2004). "Dynamical Origin of Quantum Probabilities". *Proceedings of the Royal Society A: Mathematical, Physical and Engineering Sciences*. **461** (2053): 253–272. arXiv:quant-ph/0403034⊙. Bibcode:2005RSPSA.461..253V. doi:10.1098/rspa.2004.1394.

- Bohmian mechanics on arxiv.org

12.11 Further reading

- John S. Bell: *Speakable and Unspeakable in Quantum Mechanics: Collected Papers on Quantum Philosophy*, Cambridge University Press, 2004, ISBN 0-521-81862-1

- David Bohm, Basil Hiley: *The Undivided Universe: An Ontological Interpretation of Quantum Theory*, Routledge Chapman & Hall, 1993, ISBN 0-415-06588-7

- Detlef Dürr, Sheldon Goldstein, Nino Zanghì: *Quantum Physics Without Quantum Philosophy*, Springer, 2012, ISBN 978-3-642-30690-7

- Detlef Dürr, Stefan Teufel: *Bohmian Mechanics: The Physics and Mathematics of Quantum Theory*, Springer, 2009, ISBN 978-3-540-89343-1

- Peter R. Holland: *The quantum theory of motion*, Cambridge University Press, 1993 (re-printed 2000, transferred to digital printing 2004), ISBN 0-521-48543-6

12.12 External links

- "Pilot-Wave Hydrodynamics" Bush, J. W. M., *Annual Review of Fluid Mechanics*, 2015

- "Bohmian Mechanics" (*Stanford Encyclopedia of Philosophy*)

- Videos answering frequently asked questions about Bohmian Mechanics

- "Bohmian-Mechanics.net", the homepage of the international research network on Bohmian Mechanics that was started by D. Dürr, S. Goldstein and N. Zanghì.

- Workgroup Bohmian Mechanics at LMU Munich (D. Dürr)

- Bohmian Mechanics Group at University of Innsbruck (G. Grübl)

- "Pilot waves, Bohmian metaphysics, and the foundations of quantum mechanics", lecture course on de Broglie-Bohm theory by Mike Towler, Cambridge University.

- "21st-century directions in de Broglie-Bohm theory and beyond", August 2010 international conference on de Broglie-Bohm theory. Site contains slides for all the talks - the latest cutting-edge deBB research.

- "Observing the Trajectories of a Single Photon Using Weak Measurement"

- "Bohmian trajectories are no longer 'hidden variables'"

- The David Bohm Society

Chapter 13

Bose gas

An ideal **Bose gas** is a quantum-mechanical version of a classical ideal gas. It is composed of bosons, which have an integer value of spin, and obey Bose–Einstein statistics. The statistical mechanics of bosons were developed by Satyendra Nath Bose for photons, and extended to massive particles by Albert Einstein who realized that an ideal gas of bosons would form a condensate at a low enough temperature, unlike a classical ideal gas. This condensate is known as a Bose–Einstein condensate.

13.1 Thomas–Fermi approximation

See also: Thomas–Fermi model

The thermodynamics of an ideal Bose gas is best calculated using the grand partition function. The grand partition function for a Bose gas is given by:

$$\mathcal{Z}(z,\beta,V) = \prod_i \left(1 - ze^{-\beta\epsilon_i}\right)^{-g_i}$$

where each term in the product corresponds to a particular energy ϵ_i; g_i is the number of states with energy ϵ_i; z is the absolute activity (or "fugacity"), which may also be expressed in terms of the chemical potential μ by defining:

$$z(\beta,\mu) = e^{\beta\mu}$$

and β defined as:

$$\beta = \frac{1}{kT}$$

where k is Boltzmann's constant and T is the temperature. All thermodynamic quantities may be derived from the grand partition function and we will consider all thermodynamic quantities to be functions of only the three variables z, β (or T), and V. All partial derivatives are taken with respect to one of these three variables while the other two are held constant. It is more convenient to deal with the dimensionless grand potential defined as:

$$\Omega = -\ln(\mathcal{Z}) = \sum_i g_i \ln\left(1 - ze^{-\beta\epsilon_i}\right).$$

Following the procedure described in the gas in a box article, we can apply the Thomas–Fermi approximation which assumes that the average energy is large compared to the energy difference between levels so that the above sum may be replaced by an integral:

$$\Omega \approx \int_0^\infty \ln\left(1 - ze^{-\beta E}\right) dg.$$

The degeneracy dg may be expressed for many different situations by the general formula:

$$dg = \frac{1}{\Gamma(\alpha)} \frac{E^{\alpha-1}}{E_c^\alpha} dE$$

where α is a constant, E_c is a "critical energy", and Γ is the Gamma function. For example, for a massive Bose gas in a box, $\alpha=3/2$ and the critical energy is given by:

$$\frac{1}{(\beta E_c)^\alpha} = \frac{Vf}{\Lambda^3}$$

where Λ is the thermal wavelength. For a massive Bose gas in a harmonic trap we will have $\alpha=3$ and the critical energy is given by:

$$\frac{1}{(\beta E_c)^\alpha} = \frac{f}{(\hbar\omega\beta)^3}$$

where $V(r)=m\omega^2 r^2/2$ is the harmonic potential. It is seen that Ec is a function of volume only.

We can solve the equation for the grand potential by integrating the Taylor series of the integrand term by term, or by realizing that it is proportional to the Mellin transform of the $\text{Li}_1(z \exp(-\beta E))$ where $\text{Li}_s(x)$ is the polylogarithm function. The solution is:

$$\Omega \approx -\frac{\text{Li}_{\alpha+1}(z)}{(\beta E_c)^\alpha}.$$

The problem with this continuum approximation for a Bose gas is that the ground state has been effectively ignored, giving a degeneracy of zero for zero energy. This inaccuracy becomes serious when dealing with the Bose–Einstein condensate and will be dealt with in the next section.

13.2 Inclusion of the ground state

The total number of particles is found from the grand potential by

$$N = -z \frac{\partial \Omega}{\partial z} \approx \frac{\text{Li}_\alpha(z)}{(\beta E_c)^\alpha}$$

The polylogarithm term must remain real and positive, and the maximum value it can possibly have is at $z=1$ where it is equal to $\zeta(\alpha)$ where ζ is the Riemann zeta function. For a fixed N, the largest possible value that β can have is a critical value β_c where

$$N = \frac{\zeta(\alpha)}{(\beta_c E_c)^\alpha}$$

This corresponds to a critical temperature $T_c=1/k\beta_c$ below which the Thomas–Fermi approximation breaks down. The above equation can be solved for the critical temperature:

$$T_c = \left(\frac{N}{\zeta(\alpha)}\right)^{1/\alpha} \frac{E_c}{k}$$

For example, for $\alpha = 3/2$ and using the above noted value of E_c yields

$$T_c = \left(\frac{N}{V f \zeta(3/2)}\right)^{2/3} \frac{h^2}{2\pi m k}$$

Again, we are presently unable to calculate results below the critical temperature, because the particle numbers using the above equation become negative. The problem here is that the Thomas–Fermi approximation has set the degeneracy of the ground state to zero, which is wrong. There is no ground state to accept the condensate and so the equation breaks down. It turns out, however, that the above equation gives a rather accurate estimate of the number of particles in the excited states, and it is not a bad approximation to simply "tack on" a ground state term:

$$N = N_0 + \frac{\text{Li}_\alpha(z)}{(\beta E_c)^\alpha}$$

where $N0$ is the number of particles in the ground state condensate:

$$N_0 = \frac{g_0 z}{1 - z}$$

This equation can now be solved down to absolute zero in

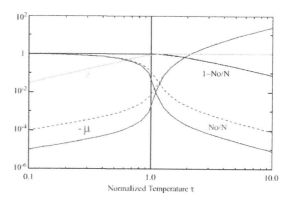

Figure 1: Various Bose gas parameters as a function of normalized temperature τ. The value of α is 3/2. Solid lines are for N=10,000, dotted lines are for N=1000. Black lines are the fraction of excited particles, blue are the fraction of condensed particles. The negative of the chemical potential μ is shown in red, and green lines are the values of z. It has been assumed that $k = \varepsilon c = 1$.

temperature. Figure 1 shows the results of the solution to this equation for $\alpha=3/2$, with $k=\varepsilon_c=1$ which corresponds to a gas of bosons in a box. The solid black line is the fraction of excited states $1-N_0/N$ for $N=10,000$ and the dotted black line is the solution for $N=1000$. The blue lines are the fraction of condensed particles N_0/N The red lines plot values of the negative of the chemical potential μ and the green lines plot the corresponding values of z. The horizontal axis is the normalized temperature τ defined by

$$\tau = \frac{T}{T_c}$$

It can be seen that each of these parameters become linear in τ^α in the limit of low temperature and, except for the chemical potential, linear in $1/\tau^\alpha$ in the limit of high temperature. As the number of particles increases, the condensed and excited fractions tend towards a discontinuity at the critical temperature.

The equation for the number of particles can be written in terms of the normalized temperature as:

$$N = \frac{g_0\, z}{1-z} + N\,\frac{\mathrm{Li}_\alpha(z)}{\zeta(\alpha)}\,\tau^\alpha$$

For a given N and τ, this equation can be solved for τ^α and then a series solution for z can be found by the method of inversion of series, either in powers of τ^α or as an asymptotic expansion in inverse powers of τ^α. From these expansions, we can find the behavior of the gas near $T=0$ and in the Maxwell–Boltzmann as T approaches infinity. In particular, we are interested in the limit as N approaches infinity, which can be easily determined from these expansions.

13.3 Thermodynamics

Adding the ground state to the equation for the particle number corresponds to adding the equivalent ground state term to the grand potential:

$$\Omega = g_0 \ln(1-z) - \frac{\mathrm{Li}_{\alpha+1}(z)}{(\beta E_c)^\alpha}$$

All thermodynamic properties may now be computed from the grand potential. The following table lists various thermodynamic quantities calculated in the limit of low temperature and high temperature, and in the limit of infinite particle number. An equal sign (=) indicates an exact result, while an approximation symbol indicates that only the first few terms of a series in τ^α is shown.

It is seen that all quantities approach the values for a classical ideal gas in the limit of large temperature. The above values can be used to calculate other thermodynamic quantities. For example, the relationship between internal energy and the product of pressure and volume is the same as that for a classical ideal gas over all temperatures:

$$U = \frac{\partial \Omega}{\partial \beta} = \alpha PV$$

A similar situation holds for the specific heat at constant volume

$$C_v = \frac{\partial U}{\partial T} = k(\alpha+1)U\beta$$

The entropy is given by:

$$TS = U + PV - G$$

Note that in the limit of high temperature, we have

$$TS = (\alpha + 1) + \ln\left(\frac{\tau^\alpha}{\zeta(\alpha)}\right)$$

which, for $\alpha=3/2$ is simply a restatement of the Sackur–Tetrode equation. In one dimension bosons with delta interaction behave as fermions, they obey Pauli exclusion principle. In one dimension Bose gas with delta interaction can be solved exactly by Bethe ansatz. The bulk free energy and thermodynamic potentials were calculated by Chen Nin Yang. In one dimensional case correlation functions also were evaluated. The In one dimension Bose gas is equivalent to quantum non-linear Schroedinger equation.

13.4 See also

- Gas in a box
- Debye model
- Bose–Einstein condensate
- Bose–Einstein condensation: a network theory approach

13.5 References

- Huang, Kerson (1967). *Statistical Mechanics*. New York: John Wiley and Sons.
- Isihara, A. (1971). *Statistical Physics*. New York: Academic Press.
- Landau, L. D.; E. M. Lifshitz (1996). *Statistical Physics, 3rd Edition Part 1*. Oxford: Butterworth-Heinemann.
- Pethick, C. J.; H. Smith (2004). *Bose–Einstein Condensation in Dilute Gases*. Cambridge: Cambridge University Press.
- Yan, Zijun (2000). "General Thermal Wavelength and its Applications" (PDF). *Eur. J. Phys.* **21** (6): 625–631. Bibcode:2000EJPh...21..625Y. doi:10.1088/0143-0807/21/6/314.

Chapter 14

Stimulated emission

Stimulated emission is the process by which an incoming photon of a specific frequency can interact with an excited atomic electron (or other excited molecular state), causing it to drop to a lower energy level. The liberated energy transfers to the electromagnetic field, creating a new photon with a phase, frequency, polarization, and direction of travel that are all identical to the photons of the incident wave. This is in contrast to spontaneous emission, which occurs at random intervals without regard to the ambient electromagnetic field.

The process is identical in form to atomic absorption in which the energy of an absorbed photon causes an identical but opposite atomic transition: from the lower level to a higher energy level. In normal media at thermal equilibrium, absorption exceeds stimulated emission because there are more electrons in the lower energy states than in the higher energy states. However, when a population inversion is present the rate of stimulated emission exceeds that of absorption, and a net optical amplification can be achieved. Such a gain medium, along with an optical resonator, is at the heart of a laser or maser. Lacking a feedback mechanism, laser amplifiers and superluminescent sources also function on the basis of stimulated emission.

Stimulated emission was a theoretical discovery by Einstein[1][2] within the framework of the old quantum theory, wherein the emission is described in terms of photons that are the quanta of the EM field. Stimulated emission can also occur in classical models, without reference to photons or quantum-mechanics.[3]

14.1 Overview

Electrons and their interactions with electromagnetic fields are important in our understanding of chemistry and physics. In the classical view, the energy of an electron orbiting an atomic nucleus is larger for orbits further from the nucleus of an atom. However, quantum mechanical effects force electrons to take on discrete positions in orbitals. Thus, electrons are found in specific energy levels of an atom, two of which are shown below:

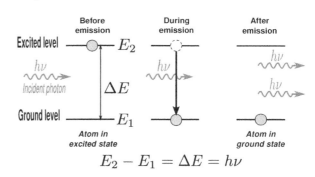

$$E_2 - E_1 = \Delta E = h\nu$$

When an electron absorbs energy either from light (photons) or heat (phonons), it receives that incident quantum of energy. But transitions are only allowed between discrete energy levels such as the two shown above. This leads to emission lines and absorption lines.

When an electron is excited from a lower to a higher energy level, it will not stay that way forever. An electron in an excited state may decay to a lower energy state which is not occupied, according to a particular time constant characterizing that transition. When such an electron decays without external influence, emitting a photon, that is called "spontaneous emission". The phase and direction associated with the photon that is emitted is random. A material with many atoms in such an excited state may thus result in radiation which has a narrow spectrum (centered around one wavelength of light), but the individual photons would have no common phase relationship and would also emanate in random directions. This is the mechanism of fluorescence and thermal emission.

An external electromagnetic field at a frequency associated with a transition can affect the quantum mechanical state of the atom without being absorbed. As the electron in the atom makes a transition between two stationary states (neither of which shows a dipole field), it enters a transition state which does have a dipole field, and which acts like a small electric dipole, and this dipole oscillates at a characteristic frequency. In response to the external electric field at this frequency, the probability of the electron's entering this transition state is greatly increased. Thus, the rate of transitions between two stationary states is increased beyond that of spontaneous emission. A transition from the higher to a lower energy state produces an additional photon with the same phase and direction as the incident photon; this is the process of **stimulated emission**.

14.2 Mathematical model

Stimulated emission can be modelled mathematically by considering an atom that may be in one of two electronic energy states, a lower level state (possibly the ground state) (1) and an *excited state* (2), with energies E_1 and E_2 respectively.

If the atom is in the excited state, it may decay into the lower state by the process of spontaneous emission, releasing the difference in energies between the two states as a photon. The photon will have frequency ν_0 and energy $h\nu_0$, given by:

$$E_2 - E_1 = h\nu_0$$

where h is Planck's constant.

Alternatively, if the excited-state atom is perturbed by an electric field of frequency ν_0, it may emit an additional photon of the same frequency and in phase, thus augmenting the external field, leaving the atom in the lower energy state. This process is known as **stimulated emission**.

In a group of such atoms, if the number of atoms in the excited state is given by N_2, the rate at which stimulated emission occurs is given by

$$\frac{\partial N_2}{\partial t} = -\frac{\partial N_1}{\partial t} = -B_{21}\,\rho(\nu)\,N_2$$

where the proportionality constant B_{21} is known as the *Einstein B coefficient* for that particular transition, and $\varrho(\nu)$ is the radiation density of the incident field at frequency ν. The rate of emission is thus proportional to the number of atoms in the excited state N_2, and to the density of incident photons.

At the same time, there will be a process of atomic absorption which *removes* energy from the field while raising electrons from the lower state to the upper state. Its rate is given by an essentially identical equation,

$$\frac{\partial N_2}{\partial t} = -\frac{\partial N_1}{\partial t} = B_{12}\,\rho(\nu)\,N_1$$

The rate of absorption is thus proportional to the number of atoms in the lower state, N_1. Einstein showed that the coefficient for this transition must be identical to that for stimulated emission:

$$B_{12} = B_{21}$$

Thus absorption and stimulated emission are reverse processes proceeding at somewhat different rates. Another way of viewing this is to look at the *net* stimulated emission or absorption viewing it as a single process. The net rate of transitions from E_2 to E_1 due to this combined process can be found by adding their respective rates, given above:

$$\frac{\partial N_1^{\text{net}}}{\partial t} = -\frac{\partial N_2^{\text{net}}}{\partial t} = B_{21}\,\rho(\nu)\,(N_2-N_1) = B_{21}\,\rho(\nu)\,\Delta N$$

Thus a net power is released into the electric field equal to the photon energy $h\nu$ times this net transition rate. In order for this to be a positive number, indicating net stimulated emission, there must be more atoms in the excited state than in the lower level: $\Delta N > 0$. Otherwise there is net absorption and the power of the wave is reduced during passage through the medium. The special condition $N_2 > N_1$ is known as a population inversion, a rather unusual condition that must be effected in the gain medium of a laser.

The notable characteristic of stimulated emission compared to everyday light sources (which depend on spontaneous emission) is that the emitted photons have the same frequency, phase, polarization, and direction of propagation as the incident photons. The photons involved are thus mutually coherent. When a population inversion ($\Delta N > 0$) is present, therefore, optical amplification of incident radiation will take place.

Although energy generated by stimulated emission is always at the exact frequency of the field which has stimulated it, the above rate equation refers only to excitation at the particular optical frequency ν_0 corresponding to the energy of the transition. At frequencies offset from ν_0 the strength of stimulated (or spontaneous) emission will be decreased according to the so-called line shape. Considering only homogeneous broadening affecting an atomic or molecular resonance, the spectral line shape function is described as a Lorentzian distribution

$$g'(\nu) = \frac{1}{\pi}\frac{(\Gamma/2)}{(\nu-\nu_0)^2 + (\Gamma/2)^2}$$

where Γ is the full width at half maximum or FWHM bandwidth.

The peak value of the Lorentzian line shape occurs at the line center, $\nu = \nu_0$. A line shape function can be normalized so that its value at ν_0 is unity; in the case of a Lorentzian we obtain

$$g(\nu) = \frac{g'(\nu)}{g'(\nu_0)} = \frac{(\Gamma/2)^2}{(\nu - \nu_0)^2 + (\Gamma/2)^2}$$

Thus stimulated emission at frequencies away from ν_0 is reduced by this factor. In practice there may also be broadening of the line shape due to inhomogeneous broadening, most notably due to the Doppler effect resulting from the distribution of velocities in a gas at a certain temperature. This has a Gaussian shape and reduces the peak strength of the line shape function. In a practical problem the full line shape function can be computed through a convolution of the individual line shape functions involved. Therefore, optical amplification will add power to an incident optical field at frequency ν at a rate given by

$$P = h\nu\, g(\nu)\, B_{21}\, \rho(\nu)\, \Delta N$$

14.3 Stimulated emission cross section

The stimulated emission cross section is

$$\sigma_{21}(\nu) = A_{21} \frac{\lambda^2}{8\pi n^2} g(\nu)$$

where

- A_{21} is the Einstein A coefficient,
- λ is the wavelength in vacuum,
- n is the refractive index of the medium (dimensionless), and
- $g(\nu)$ is the spectral line shape function.

14.4 Optical amplification

Stimulated emission can provide a physical mechanism for optical amplification. If an external source of energy stimulates more than 50% of the atoms in the ground state to transition into the excited state, then what is called a population inversion is created. When light of the appropriate frequency passes through the inverted medium, the photons are either absorbed by the atoms that remain in the ground state or the photons stimulate the excited atoms to emit additional photons of the same frequency, phase, and direction. Since more atoms are in the excited state than in the ground state then an amplification of the input intensity results.

The population inversion, in units of atoms per cubic meter, is

$$\Delta N_{21} = N_2 - \frac{g_2}{g_1} N_1$$

where g_1 and g_2 are the degeneracies of energy levels 1 and 2, respectively.

14.4.1 Small signal gain equation

The intensity (in watts per square meter) of the stimulated emission is governed by the following differential equation:

$$\frac{dI}{dz} = \sigma_{21}(\nu) \cdot \Delta N_{21} \cdot I(z)$$

as long as the intensity $I(z)$ is small enough so that it does not have a significant effect on the magnitude of the population inversion. Grouping the first two factors together, this equation simplifies as

$$\frac{dI}{dz} = \gamma_0(\nu) \cdot I(z)$$

where

$$\gamma_0(\nu) = \sigma_{21}(\nu) \cdot \Delta N_{21}$$

is the *small-signal gain coefficient* (in units of radians per meter). We can solve the differential equation using separation of variables:

$$\frac{dI}{I(z)} = \gamma_0(\nu) \cdot dz$$

Integrating, we find:

$$\ln\left(\frac{I(z)}{I_{in}}\right) = \gamma_0(\nu) \cdot z$$

or

$$I(z) = I_{in} e^{\gamma_0(\nu) z}$$

where

$$I_{in} = I(z=0)$$

14.4.2 Saturation intensity

The saturation intensity I_S is defined as the input intensity at which the gain of the optical amplifier drops to exactly half of the small-signal gain. We can compute the saturation intensity as

$$I_S = \frac{h\nu}{\sigma(\nu) \cdot \tau_S}$$

where

- h is Planck's constant, and
- τ_S is the saturation time constant, which depends on the spontaneous emission lifetimes of the various transitions between the energy levels related to the amplification.
- ν is the frequency in Hz

14.4.3 General gain equation

The general form of the gain equation, which applies regardless of the input intensity, derives from the general differential equation for the intensity I as a function of position z in the gain medium:

$$\frac{dI}{dz} = \frac{\gamma_0(\nu)}{1 + \bar{g}(\nu)\frac{I(z)}{I_S}} \cdot I(z)$$

where I_S is saturation intensity. To solve, we first rearrange the equation in order to separate the variables, intensity I and position z:

$$\frac{dI}{I(z)}\left[1 + \bar{g}(\nu)\frac{I(z)}{I_S}\right] = \gamma_0(\nu) \cdot dz$$

Integrating both sides, we obtain

$$\ln\left(\frac{I(z)}{I_{in}}\right) + \bar{g}(\nu)\frac{I(z) - I_{in}}{I_S} = \gamma_0(\nu) \cdot z$$

or

$$\ln\left(\frac{I(z)}{I_{in}}\right) + \bar{g}(\nu)\frac{I_{in}}{I_S}\left(\frac{I(z)}{I_{in}} - 1\right) = \gamma_0(\nu) \cdot z$$

The gain G of the amplifier is defined as the optical intensity I at position z divided by the input intensity:

$$G = G(z) = \frac{I(z)}{I_{in}}$$

Substituting this definition into the prior equation, we find the **general gain equation**:

$$\ln(G) + \bar{g}(\nu)\frac{I_{in}}{I_S}(G - 1) = \gamma_0(\nu) \cdot z$$

14.4.4 Small signal approximation

In the special case where the input signal is small compared to the saturation intensity, in other words,

$$I_{in} \ll I_S$$

then the general gain equation gives the small signal gain as

$$\ln(G) = \ln(G_0) = \gamma_0(\nu) \cdot z$$

or

$$G = G_0 = e^{\gamma_0(\nu)z}$$

which is identical to the small signal gain equation (see above).

14.4.5 Large signal asymptotic behavior

For large input signals, where

$$I_{in} \gg I_S$$

the gain approaches unity

$$G \to 1$$

and the general gain equation approaches a linear asymptote:

$$I(z) = I_{in} + \frac{\gamma_0(\nu) \cdot z}{\bar{g}(\nu)}I_S$$

14.5 References

[1] Einstein, A (1916). "Strahlungs-emission und - absorption nach der Quantentheorie". *Verhandlungen der Deutschen Physikalischen Gesellschaft.* **18**: 318–323. Bibcode:1916DPhyG..18..318E.

[2] Einstein, A (1917). "Zur Quantentheorie der Strahlung". *Physikalische Zeitschrift.* **18**: 121–128. Bibcode:1917PhyZ...18..121E.

[3] Fain, B.; Milonni, P. W. (1987). "Classical stimulated emission". *Journal of the Optical Society of America B.* **4**: 78. Bibcode:1987JOSAB...4...78F. doi:10.1364/JOSAB.4.000078.

- Saleh, Bahaa E. A. & Teich, Malvin Carl (1991). *Fundamentals of Photonics.* New York: John Wiley & Sons. ISBN 0-471-83965-5.

- Alan Corney (1977). *Atomic and Laser Spectroscopy.* Oxford: Oxford Uni. Press. ISBN 0-19-921145-0. ISBN 978-0-19-921145-6.

.3 Laser Fundamentals, William T. Silfvast

14.6 See also

- Absorption
- Active laser medium
- Laser (includes a history section)
- Laser science
- Rabi cycle
- Spontaneous emission
- STED microscopy

Chapter 15

Laser

"Laser light" redirects here. For the song, see LaserLight. For laser light show, see laser lighting display. For other uses, see Laser (disambiguation).

United States Air Force laser experiment

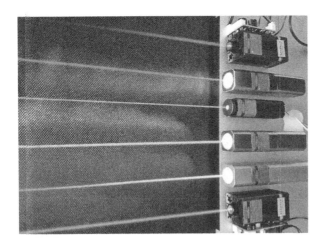

Red (660 & 635 nm), green (532 & 520 nm) and blue-violet (445 & 405 nm) lasers

A **laser** is a device that emits light through a process of optical amplification based on the stimulated emission of electromagnetic radiation. The term "laser" originated as an acronym for "**light amplification by stimulated emission of radiation**".[1][2] The first laser was built in 1960 by Theodore H. Maiman at Hughes Research Laboratories, based on theoretical work by Charles Hard Townes and Arthur Leonard Schawlow. A laser differs from other sources of light in that it emits light *coherently*. Spatial coherence allows a laser to be focused to a tight spot, enabling applications such as laser cutting and lithography. Spatial coherence also allows a laser beam to stay narrow over great distances (collimation), enabling applications such as laser pointers. Lasers can also have high temporal coherence, which allows them to emit light with a very narrow spectrum, i.e., they can emit a single color of light. Temporal coherence can be used to produce pulses of light as short as a femtosecond.

Among their many applications, lasers are used in optical disk drives, laser printers, and barcode scanners; DNA sequencing instruments, fiber-optic and free-space optical communication; laser surgery and skin treatments; cutting and welding materials; military and law enforcement devices for marking targets and measuring range and speed; and laser lighting displays in entertainment.

15.1 Fundamentals

Lasers are distinguished from other light sources by their coherence. Spatial coherence is typically expressed through the output being a narrow beam, which is diffraction-limited. Laser beams can be focused to very tiny spots, achieving a very high irradiance, or they can have very low divergence in order to concentrate their power at a great distance.

Temporal (or longitudinal) coherence implies a polarized wave at a single frequency whose phase is correlated over a relatively great distance (the coherence length) along the beam.[4] A beam produced by a thermal or other incoherent light source has an instantaneous amplitude and phase that vary randomly with respect to time and position, thus having a short coherence length.

Modern telescopes use laser technologies to compensate for the blurring effect of the Earth's atmosphere.[3]

Lasers are characterized according to their wavelength in a vacuum. Most "single wavelength" lasers actually produce radiation in several *modes* having slightly differing frequencies (wavelengths), often not in a single polarization. Although temporal coherence implies monochromaticity, there are lasers that emit a broad spectrum of light or emit different wavelengths of light simultaneously. There are some lasers that are not single spatial mode and consequently have light beams that diverge more than is required by the diffraction limit. However, all such devices are classified as "lasers" based on their method of producing light, i.e., stimulated emission. Lasers are employed in applications where light of the required spatial or temporal coherence could not be produced using simpler technologies.

15.1.1 Terminology

The word *laser* started as an acronym for "light amplification by stimulated emission of radiation". In modern usage, the term "light" includes electromagnetic radiation of any frequency, not only visible light, hence the terms *infrared*

Laser beams in fog, reflected on a car windshield

laser, *ultraviolet laser*, *X-ray laser*, *gamma-ray laser*, and so on. Because the microwave predecessor of the laser, the maser, was developed first, devices of this sort operating at microwave and radio frequencies are referred to as "masers" rather than "microwave lasers" or "radio lasers". In the early technical literature, especially at Bell Telephone Laboratories, the laser was called an **optical maser**; this term is now obsolete.[5]

A laser that produces light by itself is technically an optical oscillator rather than an optical amplifier as suggested by the acronym. It has been humorously noted that the acronym LOSER, for "light oscillation by stimulated emission of radiation", would have been more correct.[6] With the widespread use of the original acronym as a common noun, optical amplifiers have come to be referred to as "laser amplifiers", notwithstanding the apparent redundancy in that designation.

The back-formed verb *to lase* is frequently used in the field, meaning "to produce laser light,"[7] especially in reference to the gain medium of a laser; when a laser is operating it is said to be "lasing." Further use of the words *laser* and *maser* in an extended sense, not referring to laser technology or devices, can be seen in usages such as *astrophysical maser* and *atom laser*.

15.2 Design

Main article: Laser construction

A laser consists of a gain medium, a mechanism to energize it, and something to provide optical feedback.[8] The gain medium is a material with properties that allow it to amplify light by way of stimulated emission. Light of a specific wavelength that passes through the gain medium

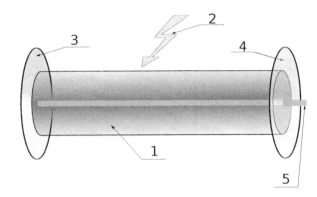

Components of a typical laser:
1. *Gain medium*
2. *Laser pumping energy*
3. *High reflector*
4. *Output coupler*
5. *Laser beam*

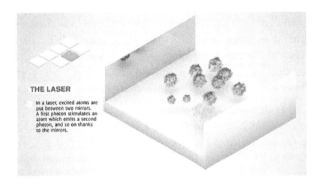

Animation explaining stimulated emission and the laser principle

is amplified (increases in power).

For the gain medium to amplify light, it needs to be supplied with energy in a process called pumping. The energy is typically supplied as an electric current or as light at a different wavelength. Pump light may be provided by a flash lamp or by another laser.

The most common type of laser uses feedback from an optical cavity—a pair of mirrors on either end of the gain medium. Light bounces back and forth between the mirrors, passing through the gain medium and being amplified each time. Typically one of the two mirrors, the output coupler, is partially transparent. Some of the light escapes through this mirror. Depending on the design of the cavity (whether the mirrors are flat or curved), the light coming out of the laser may spread out or form a narrow beam. In analogy to electronic oscillators, this device is sometimes called a *laser oscillator.*

Most practical lasers contain additional elements that affect properties of the emitted light, such as the polarization, wavelength, and shape of the beam.

15.3 Laser physics

See also: Laser science

Electrons and how they interact with electromagnetic fields are important in our understanding of chemistry and physics.

15.3.1 Stimulated emission

Main article: Stimulated emission

In the classical view, the energy of an electron orbiting an atomic nucleus is larger for orbits further from the nucleus of an atom. However, quantum mechanical effects force electrons to take on discrete positions in orbitals. Thus, electrons are found in specific energy levels of an atom, two of which are shown below:

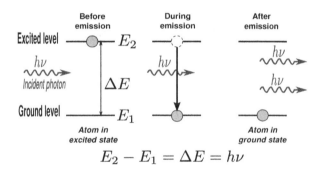

$$E_2 - E_1 = \Delta E = h\nu$$

When an electron absorbs energy either from light (photons) or heat (phonons), it receives that incident quantum of energy. But transitions are only allowed in between discrete energy levels such as the two shown above. This leads to emission lines and absorption lines.

When an electron is excited from a lower to a higher energy level, it will not stay that way forever. An electron in an excited state may decay to a lower energy state which is not occupied, according to a particular time constant characterizing that transition. When such an electron decays without external influence, emitting a photon, that is called "spontaneous emission". The phase associated with the photon that is emitted is random. A material with many atoms in such an excited state may thus result in radiation which is very spectrally limited (centered around one wavelength of light), but the individual photons would have no common phase relationship and would emanate in random directions. This is the mechanism of fluorescence and thermal emission.

An external electromagnetic field at a frequency associated with a transition can affect the quantum mechanical state of the atom. As the electron in the atom makes a transition between two stationary states (neither of which shows a dipole field), it enters a transition state which does have a dipole field, and which acts like a small electric dipole, and this dipole oscillates at a characteristic frequency. In response to the external electric field at this frequency, the probability of the atom entering this transition state is greatly increased. Thus, the rate of transitions between two stationary states is enhanced beyond that due to spontaneous emission. Such a transition to the higher state is called absorption, and it destroys an incident photon (the photon's energy goes into powering the increased energy of the higher state). A transition from the higher to a lower energy state, however, produces an additional photon; this is the process of **stimulated emission**.

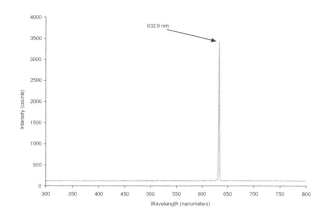

Spectrum of a helium neon laser illustrating its very high spectral purity (limited by the measuring apparatus). The 0.002 nm bandwidth of the lasing medium is well over 10,000 times narrower than the spectral width of a light-emitting diode (whose spectrum is shown **here** for comparison), with the bandwidth of a single longitudinal mode being much narrower still.

15.3.2 Gain medium and cavity

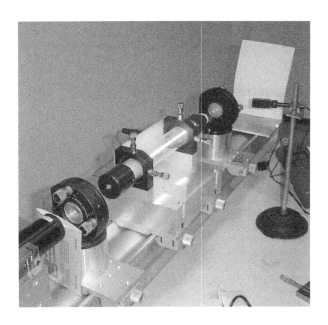

A helium–neon laser demonstration at the Kastler-Brossel Laboratory at Univ. Paris 6. The pink-orange glow running through the center of the tube is from the electric discharge which produces incoherent light, just as in a neon tube. This glowing plasma is excited and then acts as the gain medium through which the internal beam passes, as it is reflected between the two mirrors. Laser radiation output through the front mirror can be seen to produce a tiny (about 1 mm in diameter) intense spot on the screen, to the right. Although it is a deep and pure red color, spots of laser light are so intense that cameras are typically overexposed and distort their color.

The gain medium is excited by an external source of energy into an excited state. In most lasers this medium consists of a population of atoms which have been excited into such a state by means of an outside light source, or an electrical field which supplies energy for atoms to absorb and be transformed into their excited states.

The gain medium of a laser is normally a material of controlled purity, size, concentration, and shape, which amplifies the beam by the process of stimulated emission described above. This material can be of any state: gas, liquid, solid, or plasma. The gain medium absorbs pump energy, which raises some electrons into higher-energy ("excited") quantum states. Particles can interact with light by either absorbing or emitting photons. Emission can be spontaneous or stimulated. In the latter case, the photon is emitted in the same direction as the light that is passing by. When the number of particles in one excited state exceeds the number of particles in some lower-energy state, population inversion is achieved and the amount of stimulated emission due to light that passes through is larger than the amount of absorption. Hence, the light is amplified. By itself, this makes an optical amplifier. When an optical amplifier is placed inside a resonant optical cavity, one obtains a laser oscillator.[9]

In a few situations it is possible to obtain lasing with only a single pass of EM radiation through the gain medium, and this produces a laser beam without any need for a resonant or reflective cavity (see for example nitrogen laser).[10] Thus, reflection in a resonant cavity is usually required for a laser, but is not absolutely necessary.

The optical resonator is sometimes referred to as an "optical cavity", but this is a misnomer: lasers use open resonators as opposed to the literal cavity that would be employed at microwave frequencies in a maser. The resonator typically consists of two mirrors between which a coherent beam of light travels in both directions, reflecting back on itself so that an average photon will pass through the gain medium repeatedly before it is emitted from the output aperture or lost to diffraction or absorption. If the gain (amplification) in the medium is larger than the resonator losses, then the power of the recirculating light can rise exponentially. But each stimulated emission event returns an atom from its excited state to the ground state, reducing the gain of the medium. With increasing beam power the net gain (gain minus loss) reduces to unity and the gain medium is said to be saturated. In a continuous wave (CW) laser, the balance of pump power against gain saturation and cavity losses produces an equilibrium value of the laser power inside the cavity; this equilibrium determines the operating point of the laser. If the applied pump power is too small, the gain will never be sufficient to overcome the resonator losses, and laser light will not be produced. The minimum pump power needed to begin laser action is called the *lasing threshold*. The gain medium will amplify any photons passing through it, regardless of direction; but only the photons in a spatial mode supported by the resonator will pass more than once through the medium and receive substantial amplification.

15.3.3 The light emitted

The light generated by stimulated emission is very similar to the input signal in terms of wavelength, phase, and polarization. This gives laser light its characteristic coherence, and allows it to maintain the uniform polarization and often monochromaticity established by the optical cavity design.

The beam in the cavity and the output beam of the laser, when travelling in free space (or a homogeneous medium) rather than waveguides (as in an optical fiber laser), can be approximated as a Gaussian beam in most lasers; such beams exhibit the minimum divergence for a given diameter. However some high power lasers may be multimode, with the transverse modes often approximated using Hermite–Gaussian or Laguerre-Gaussian functions. It has been shown that unstable laser resonators (not used in most lasers) produce fractal shaped beams.[11] Near the beam "waist" (or focal region) it is highly *collimated*: the wavefronts are planar, normal to the direction of propagation, with no beam divergence at that point. However, due to diffraction, that can only remain true well within the Rayleigh range. The beam of a single transverse mode (gaussian beam) laser eventually diverges at an angle which varies inversely with the beam diameter, as required by diffraction theory. Thus, the "pencil beam" directly generated by a common helium–neon laser would spread out to a size of perhaps 500 kilometers when shone on the Moon (from the distance of the earth). On the other hand, the light from a semiconductor laser typically exits the tiny crystal with a large divergence: up to 50°. However even such a divergent beam can be transformed into a similarly collimated beam by means of a lens system, as is always included, for instance, in a laser pointer whose light originates from a laser diode. That is possible due to the light being of a single spatial mode. This unique property of laser light, spatial coherence, cannot be replicated using standard light sources (except by discarding most of the light) as can be appreciated by comparing the beam from a flashlight (torch) or spotlight to that of almost any laser.

15.3.4 Quantum vs. classical emission processes

The mechanism of producing radiation in a laser relies on stimulated emission, where energy is extracted from a transition in an atom or molecule. This is a quantum phenomenon discovered by Einstein who derived the relationship between the A coefficient describing spontaneous emission and the B coefficient which applies to absorption and stimulated emission. However, in the case of the free electron laser, atomic energy levels are not involved; it appears that the operation of this rather exotic device can be explained without reference to quantum mechanics.

15.4 Continuous and pulsed modes of operation

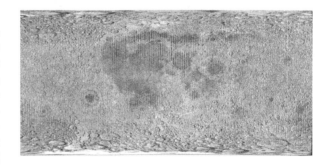

Lidar measurements of lunar topography made by Clementine mission.

A laser can be classified as operating in either continuous or pulsed mode, depending on whether the power output is essentially continuous over time or whether its output takes the form of pulses of light on one or another time scale. Of course even a laser whose output is normally continuous can be intentionally turned on and off at some rate in order to create pulses of light. When the modulation rate is on time scales much slower than the cavity lifetime and the time period over which energy can be stored in the lasing medium

15.4. CONTINUOUS AND PULSED MODES OF OPERATION

Laserlink point to point optical wireless network

Mercury Laser Altimeter (MLA) of the MESSENGER spacecraft

or pumping mechanism, then it is still classified as a "modulated" or "pulsed" continuous wave laser. Most laser diodes used in communication systems fall in that category.

15.4.1 Continuous wave operation

Some applications of lasers depend on a beam whose output power is constant over time. Such a laser is known as *continuous wave* (CW). Many types of lasers can be made to operate in continuous wave mode to satisfy such an application. Many of these lasers actually lase in several longitudinal modes at the same time, and beats between the slightly different optical frequencies of those oscillations will in fact produce amplitude variations on time scales shorter than the round-trip time (the reciprocal of the frequency spacing between modes), typically a few nanoseconds or less. In most cases these lasers are still termed "continuous wave" as their output power is steady when averaged over any longer time periods, with the very high frequency power variations having little or no impact in the intended application. (However the term is not applied to mode-locked lasers, where the *intention* is to create very short pulses at the rate of the round-trip time).

For continuous wave operation it is required for the population inversion of the gain medium to be continually replenished by a steady pump source. In some lasing media this is impossible. In some other lasers it would require pumping the laser at a very high continuous power level which would be impractical or destroy the laser by producing excessive heat. Such lasers cannot be run in CW mode.

15.4.2 Pulsed operation

Pulsed operation of lasers refers to any laser not classified as continuous wave, so that the optical power appears in pulses of some duration at some repetition rate. This encompasses a wide range of technologies addressing a number of different motivations. Some lasers are pulsed simply because they cannot be run in continuous mode.

In other cases the application requires the production of pulses having as large an energy as possible. Since the pulse energy is equal to the average power divided by the repetition rate, this goal can sometimes be satisfied by lowering the rate of pulses so that more energy can be built up in between pulses. In laser ablation for example, a small volume of material at the surface of a work piece can be evaporated if it is heated in a very short time, whereas supplying the energy gradually would allow for the heat to be absorbed into the bulk of the piece, never attaining a sufficiently high temperature at a particular point.

Other applications rely on the peak pulse power (rather than the energy in the pulse), especially in order to obtain nonlinear optical effects. For a given pulse energy, this requires creating pulses of the shortest possible duration utilizing techniques such as Q-switching.

The optical bandwidth of a pulse cannot be narrower than the reciprocal of the pulse width. In the case of extremely short pulses, that implies lasing over a considerable bandwidth, quite contrary to the very narrow bandwidths typical of CW lasers. The lasing medium in some *dye lasers* and *vibronic solid-state lasers* produces optical gain over a wide bandwidth, making a laser possible which can thus generate pulses of light as short as a few femtoseconds (10^{-15} s).

Q-switching

Main article: Q-switching

In a Q-switched laser, the population inversion is allowed to build up by introducing loss inside the resonator which exceeds the gain of the medium; this can also be described as a reduction of the quality factor or 'Q' of the cavity. Then, after the pump energy stored in the laser medium has approached the maximum possible level, the introduced loss mechanism (often an electro- or acousto-optical element) is rapidly removed (or that occurs by itself in a passive device), allowing lasing to begin which rapidly obtains the stored energy in the gain medium. This results in a short pulse incorporating that energy, and thus a high peak power.

Mode-locking

Main article: Mode-locking

A mode-locked laser is capable of emitting extremely short pulses on the order of tens of picoseconds down to less than 10 femtoseconds. These pulses will repeat at the round trip time, that is, the time that it takes light to complete one round trip between the mirrors comprising the resonator. Due to the Fourier limit (also known as energy-time uncertainty), a pulse of such short temporal length has a spectrum spread over a considerable bandwidth. Thus such a gain medium must have a gain bandwidth sufficiently broad to amplify those frequencies. An example of a suitable material is titanium-doped, artificially grown sapphire (Ti:sapphire) which has a very wide gain bandwidth and can thus produce pulses of only a few femtoseconds duration.

Such mode-locked lasers are a most versatile tool for researching processes occurring on extremely short time scales (known as femtosecond physics, femtosecond chemistry and ultrafast science), for maximizing the effect of nonlinearity in optical materials (e.g. in second-harmonic generation, parametric down-conversion, optical parametric oscillators and the like) due to the large peak power, and in ablation applications. Again, because of the extremely short pulse duration, such a laser will produce pulses which achieve an extremely high peak power.

Pulsed pumping

Another method of achieving pulsed laser operation is to pump the laser material with a source that is itself pulsed, either through electronic charging in the case of flash lamps, or another laser which is already pulsed. Pulsed pumping was historically used with dye lasers where the inverted population lifetime of a dye molecule was so short that a high energy, fast pump was needed. The way to overcome this problem was to charge up large capacitors which are then switched to discharge through flashlamps, producing an intense flash. Pulsed pumping is also required for three-level lasers in which the lower energy level rapidly becomes highly populated preventing further lasing until those atoms relax to the ground state. These lasers, such as the excimer laser and the copper vapor laser, can never be operated in CW mode.

15.5 History

15.5.1 Foundations

In 1917, Albert Einstein established the theoretical foundations for the laser and the maser in the paper *Zur Quantentheorie der Strahlung* (On the Quantum Theory of Radiation) via a re-derivation of Max Planck's law of radiation, conceptually based upon probability coefficients (Einstein coefficients) for the absorption, spontaneous emission, and stimulated emission of electromagnetic radiation.[12] In 1928, Rudolf W. Ladenburg confirmed the existence of the phenomena of stimulated emission and negative absorption.[13] In 1939, Valentin A. Fabrikant predicted the use of stimulated emission to amplify "short" waves.[14] In 1947, Willis E. Lamb and R. C. Retherford found apparent stimulated emission in hydrogen spectra and effected the first demonstration of stimulated emission.[13] In 1950, Alfred Kastler (Nobel Prize for Physics 1966) proposed the method of optical pumping, experimentally confirmed, two years later, by Brossel, Kastler, and Winter.[15]

15.5.2 Maser

Main article: Maser

In 1951, Joseph Weber submitted a paper on using stimulated emissions to make a microwave amplifier to the June 1952 Institute of Radio Engineers Vacuum Tube Research Conference at Ottawa.[16] After this presentation, RCA asked Weber to give a seminar on this idea, and Charles Hard Townes asked him for a copy of the paper.[17]

In 1953, Charles Hard Townes and graduate students James P. Gordon and Herbert J. Zeiger produced the first mi-

15.5. HISTORY

Aleksandr Prokhorov

crowave amplifier, a device operating on similar principles to the laser, but amplifying microwave radiation rather than infrared or visible radiation. Townes's maser was incapable of continuous output. Meanwhile, in the Soviet Union, Nikolay Basov and Aleksandr Prokhorov were independently working on the quantum oscillator and solved the problem of continuous-output systems by using more than two energy levels. These gain media could release stimulated emissions between an excited state and a lower excited state, not the ground state, facilitating the maintenance of a population inversion. In 1955, Prokhorov and Basov suggested optical pumping of a multi-level system as a method for obtaining the population inversion, later a main method of laser pumping.

Townes reports that several eminent physicists—among them Niels Bohr, John von Neumann, and Llewellyn Thomas—argued the maser violated Heisenberg's uncertainty principle and hence could not work. Others such as Isidor Rabi and Polykarp Kusch expected that it would be impractical and not worth the effort.[18] In 1964 Charles H. Townes, Nikolay Basov, and Aleksandr Prokhorov shared the Nobel Prize in Physics, "for fundamental work in the field of quantum electronics, which has led to the construction of oscillators and amplifiers based on the maser–laser principle".

15.5.3 Laser

In 1957, Charles Hard Townes and Arthur Leonard Schawlow, then at Bell Labs, began a serious study of the infrared laser. As ideas developed, they abandoned infrared radiation to instead concentrate upon visible light. The concept originally was called an "optical maser". In 1958, Bell Labs filed a patent application for their proposed optical maser; and Schawlow and Townes submitted a manuscript of their theoretical calculations to the *Physical Review*, published that year in Volume 112, Issue No. 6.

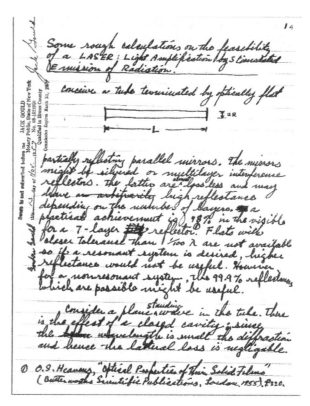

LASER notebook: *First page of the notebook wherein Gordon Gould coined the LASER acronym, and described the elements for constructing the device.*

Simultaneously, at Columbia University, graduate student Gordon Gould was working on a doctoral thesis about the energy levels of excited thallium. When Gould and Townes met, they spoke of radiation emission, as a general subject; afterwards, in November 1957, Gould noted his ideas for a "laser", including using an open resonator (later an essential laser-device component). Moreover, in 1958, Prokhorov independently proposed using an open resonator, the first published appearance (the USSR) of this idea. Elsewhere, in the U.S., Schawlow and Townes had agreed to an open-resonator laser design – apparently unaware of Prokhorov's publications and Gould's unpublished laser work.

At a conference in 1959, Gordon Gould published the term LASER in the paper *The LASER, Light Amplification by Stimulated Emission of Radiation*.[1][6] Gould's linguistic intention was using the "-aser" word particle as a suffix – to accurately denote the spectrum of the light emitted by the LASER device; thus x-rays: *xaser*, ultraviolet: *uvaser*, et cetera; none established itself as a discrete term, although "raser" was briefly popular for denoting radio-frequency-emitting devices.

Gould's notes included possible applications for a laser, such as spectrometry, interferometry, radar, and nuclear fusion. He continued developing the idea, and filed a patent application in April 1959. The U.S. Patent Office denied his application, and awarded a patent to Bell Labs, in 1960. That provoked a twenty-eight-year lawsuit, featuring scientific prestige and money as the stakes. Gould won his first minor patent in 1977, yet it was not until 1987 that he won the first significant patent lawsuit victory, when a Federal judge ordered the U.S. Patent Office to issue patents to Gould for the optically pumped and the gas discharge laser devices. The question of just how to assign credit for inventing the laser remains unresolved by historians.[19]

On May 16, 1960, Theodore H. Maiman operated the first functioning laser [20][21] at Hughes Research Laboratories, Malibu, California, ahead of several research teams, including those of Townes, at Columbia University, Arthur Schawlow, at Bell Labs,[22] and Gould, at the TRG (Technical Research Group) company. Maiman's functional laser used a solid-state flashlamp-pumped synthetic ruby crystal to produce red laser light, at 694 nanometers wavelength; however, the device only was capable of pulsed operation, because of its three-level pumping design scheme. Later that year, the Iranian physicist Ali Javan, and William R. Bennett, and Donald Herriott, constructed the first gas laser, using helium and neon that was capable of continuous operation in the infrared (U.S. Patent 3,149,290); later, Javan received the Albert Einstein Award in 1993. Basov and Javan proposed the semiconductor laser diode concept. In 1962, Robert N. Hall demonstrated the first *laser diode* device, made of gallium arsenide and emitted at 850 nm the near-infrared band of the spectrum. Later that year, Nick Holonyak, Jr. demonstrated the first semiconductor laser with a visible emission. This first semiconductor laser could only be used in pulsed-beam operation, and when cooled to liquid nitrogen temperatures (77 K). In 1970, Zhores Alferov, in the USSR, and Izuo Hayashi and Morton Panish of Bell Telephone Laboratories also independently developed room-temperature, continual-operation diode lasers, using the heterojunction structure.

15.5.4 Recent innovations

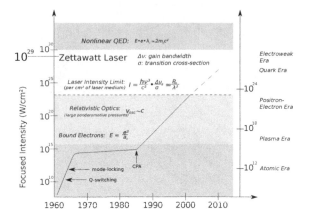

Graph showing the history of maximum laser pulse intensity throughout the past 40 years.

Since the early period of laser history, laser research has produced a variety of improved and specialized laser types, optimized for different performance goals, including:

- new wavelength bands
- maximum average output power
- maximum peak pulse energy
- maximum peak pulse power
- minimum output pulse duration
- maximum power efficiency
- minimum cost

and this research continues to this day.

Lasing without maintaining the medium excited into a population inversion was discovered in 1992 in sodium gas and again in 1995 in rubidium gas by various international teams. This was accomplished by using an external maser to induce "optical transparency" in the medium by introducing and destructively interfering the ground electron transitions between two paths, so that the likelihood for the ground electrons to absorb any energy has been cancelled.

15.6 Types and operating principles

For a more complete list of laser types see this list of laser types.

15.6. TYPES AND OPERATING PRINCIPLES

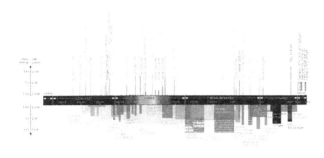

Wavelengths of commercially available lasers. Laser types with distinct laser lines are shown above the wavelength bar, while below are shown lasers that can emit in a wavelength range. The color codifies the type of laser material (see the figure description for more details).

15.6.1 Gas lasers

Main article: Gas laser

Following the invention of the HeNe gas laser, many other gas discharges have been found to amplify light coherently. Gas lasers using many different gases have been built and used for many purposes. The helium–neon laser (HeNe) is able to operate at a number of different wavelengths, however the vast majority are engineered to lase at 633 nm; these relatively low cost but highly coherent lasers are extremely common in optical research and educational laboratories. Commercial carbon dioxide (CO_2) lasers can emit many hundreds of watts in a single spatial mode which can be concentrated into a tiny spot. This emission is in the thermal infrared at 10.6 μm; such lasers are regularly used in industry for cutting and welding. The efficiency of a CO_2 laser is unusually high: over 30%.[23] Argon-ion lasers can operate at a number of lasing transitions between 351 and 528.7 nm. Depending on the optical design one or more of these transitions can be lasing simultaneously; the most commonly used lines are 458 nm, 488 nm and 514.5 nm. A nitrogen transverse electrical discharge in gas at atmospheric pressure (TEA) laser is an inexpensive gas laser, often home-built by hobbyists, which produces rather incoherent UV light at 337.1 nm.[24] Metal ion lasers are gas lasers that generate deep ultraviolet wavelengths. Helium-silver (HeAg) 224 nm and neon-copper (NeCu) 248 nm are two examples. Like all low-pressure gas lasers, the gain media of these lasers have quite narrow oscillation linewidths, less than 3 GHz (0.5 picometers),[25] making them candidates for use in fluorescence suppressed Raman spectroscopy.

Chemical lasers

Chemical lasers are powered by a chemical reaction permitting a large amount of energy to be released quickly. Such very high power lasers are especially of interest to the military, however continuous wave chemical lasers at very high power levels, fed by streams of gasses, have been developed and have some industrial applications. As examples, in the hydrogen fluoride laser (2700–2900 nm) and the deuterium fluoride laser (3800 nm) the reaction is the combination of hydrogen or deuterium gas with combustion products of ethylene in nitrogen trifluoride.

Excimer lasers

Excimer lasers are a special sort of gas laser powered by an electric discharge in which the lasing medium is an excimer, or more precisely an exciplex in existing designs. These are molecules which can only exist with one atom in an excited electronic state. Once the molecule transfers its excitation energy to a photon, therefore, its atoms are no longer bound to each other and the molecule disintegrates. This drastically reduces the population of the lower energy state thus greatly facilitating a population inversion. Excimers currently used are all noble gas compounds; noble gasses are chemically inert and can only form compounds while in an excited state. Excimer lasers typically operate at ultraviolet wavelengths with major applications including semiconductor photolithography and LASIK eye surgery. Commonly used excimer molecules include ArF (emission at 193 nm), KrCl (222 nm), KrF (248 nm), XeCl (308 nm), and XeF (351 nm).[26] The molecular fluorine laser, emitting at 157 nm in the vacuum ultraviolet is sometimes referred to as an excimer laser, however this appears to be a misnomer inasmuch as F_2 is a stable compound.

15.6.2 Solid-state lasers

Solid-state lasers use a crystalline or glass rod which is "doped" with ions that provide the required energy states. For example, the first working laser was a ruby laser, made from ruby (chromium-doped corundum). The population inversion is actually maintained in the dopant. These materials are pumped optically using a shorter wavelength than the lasing wavelength, often from a flashtube or from another laser. The usage of the term "solid-state" in laser physics is narrower than in typical use. Semiconductor lasers (laser diodes) are typically *not* referred to as solid-state lasers.

Neodymium is a common dopant in various solid-state laser crystals, including yttrium orthovanadate (Nd:YVO$_4$), yttrium lithium fluoride (Nd:YLF) and yttrium aluminium garnet (Nd:YAG). All these lasers can produce high powers in the infrared spectrum at 1064 nm. They are used for cutting, welding and marking of metals and other materi-

A 50 W FASOR, based on a Nd:YAG laser, used at the Starfire Optical Range.

als, and also in spectroscopy and for pumping dye lasers. These lasers are also commonly frequency doubled, tripled or quadrupled to produce 532 nm (green, visible), 355 nm and 266 nm (UV) beams, respectively. Frequency-doubled diode-pumped solid-state (DPSS) lasers are used to make bright green laser pointers.

Ytterbium, holmium, thulium, and erbium are other common "dopants" in solid-state lasers. Ytterbium is used in crystals such as Yb:YAG, Yb:KGW, Yb:KYW, Yb:SYS, Yb:BOYS, Yb:CaF_2, typically operating around 1020–1050 nm. They are potentially very efficient and high powered due to a small quantum defect. Extremely high powers in ultrashort pulses can be achieved with Yb:YAG. Holmium-doped YAG crystals emit at 2097 nm and form an efficient laser operating at infrared wavelengths strongly absorbed by water-bearing tissues. The Ho-YAG is usually operated in a pulsed mode, and passed through optical fiber surgical devices to resurface joints, remove rot from teeth, vaporize cancers, and pulverize kidney and gall stones.

Titanium-doped sapphire (Ti:sapphire) produces a highly tunable infrared laser, commonly used for spectroscopy. It is also notable for use as a mode-locked laser producing ultrashort pulses of extremely high peak power.

Thermal limitations in solid-state lasers arise from unconverted pump power that heats the medium. This heat, when coupled with a high thermo-optic coefficient (dn/dT) can cause thermal lensing and reduce the quantum efficiency. Diode-pumped thin disk lasers overcome these issues by having a gain medium that is much thinner than the diameter of the pump beam. This allows for a more uniform temperature in the material. Thin disk lasers have been shown to produce beams of up to one kilowatt.[27]

15.6.3 Fiber lasers

Main article: Fiber laser

Solid-state lasers or laser amplifiers where the light is guided due to the total internal reflection in a single mode optical fiber are instead called fiber lasers. Guiding of light allows extremely long gain regions providing good cooling conditions; fibers have high surface area to volume ratio which allows efficient cooling. In addition, the fiber's waveguiding properties tend to reduce thermal distortion of the beam. Erbium and ytterbium ions are common active species in such lasers.

Quite often, the fiber laser is designed as a double-clad fiber. This type of fiber consists of a fiber core, an inner cladding and an outer cladding. The index of the three concentric layers is chosen so that the fiber core acts as a single-mode fiber for the laser emission while the outer cladding acts as a highly multimode core for the pump laser. This lets the pump propagate a large amount of power into and through the active inner core region, while still having a high numerical aperture (NA) to have easy launching conditions.

Pump light can be used more efficiently by creating a fiber disk laser, or a stack of such lasers.

Fiber lasers have a fundamental limit in that the intensity of the light in the fiber cannot be so high that optical nonlinearities induced by the local electric field strength can become dominant and prevent laser operation and/or lead to the material destruction of the fiber. This effect is called photodarkening. In bulk laser materials, the cooling is not so efficient, and it is difficult to separate the effects of photodarkening from the thermal effects, but the experiments in fibers show that the photodarkening can be attributed to the formation of long-living color centers.

15.6.4 Photonic crystal lasers

Photonic crystal lasers are lasers based on nano-structures that provide the mode confinement and the density of optical states (DOS) structure required for the feedback to take place. They are typical micrometer-sized and tunable on the bands of the photonic crystals.[28]

15.6.5 Semiconductor lasers

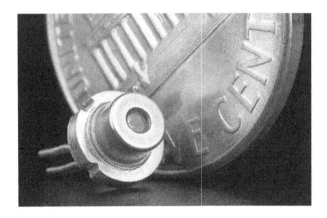

A 5.6 mm 'closed can' commercial laser diode, probably from a CD or DVD player

Semiconductor lasers are diodes which are electrically pumped. Recombination of electrons and holes created by the applied current introduces optical gain. Reflection from the ends of the crystal form an optical resonator, although the resonator can be external to the semiconductor in some designs.

Commercial laser diodes emit at wavelengths from 375 nm to 3500 nm.[29] Low to medium power laser diodes are used in laser pointers, laser printers and CD/DVD players. Laser diodes are also frequently used to optically pump other lasers with high efficiency. The highest power industrial laser diodes, with power up to 10 kW (70 dBm), are used in industry for cutting and welding. External-cavity semiconductor lasers have a semiconductor active medium in a larger cavity. These devices can generate high power outputs with good beam quality, wavelength-tunable narrow-linewidth radiation, or ultrashort laser pulses.

In 2012, Nichia and OSRAM developed and manufactured commercial high-power green laser diodes (515/520 nm), which compete with traditional diode-pumped solid-state lasers.[30][31]

Vertical cavity surface-emitting lasers (VCSELs) are semiconductor lasers whose emission direction is perpendicular to the surface of the wafer. VCSEL devices typically have a more circular output beam than conventional laser diodes. As of 2005, only 850 nm VCSELs are widely available, with 1300 nm VCSELs beginning to be commercialized,[32] and 1550 nm devices an area of research. VECSELs are external-cavity VCSELs. Quantum cascade lasers are semiconductor lasers that have an active transition between energy *sub-bands* of an electron in a structure containing several quantum wells.

The development of a silicon laser is important in the field of optical computing. Silicon is the material of choice for integrated circuits, and so electronic and silicon photonic components (such as optical interconnects) could be fabricated on the same chip. Unfortunately, silicon is a difficult lasing material to deal with, since it has certain properties which block lasing. However, recently teams have produced silicon lasers through methods such as fabricating the lasing material from silicon and other semiconductor materials, such as indium(III) phosphide or gallium(III) arsenide, materials which allow coherent light to be produced from silicon. These are called hybrid silicon laser. Another type is a Raman laser, which takes advantage of Raman scattering to produce a laser from materials such as silicon.

15.6.6 Dye lasers

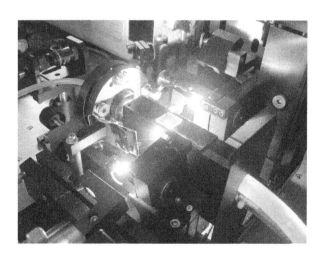

Close-up of a table-top dye laser based on Rhodamine 6G

Dye lasers use an organic dye as the gain medium. The wide gain spectrum of available dyes, or mixtures of dyes, allows these lasers to be highly tunable, or to produce very short-duration pulses (on the order of a few femtoseconds). Although these tunable lasers are mainly known in their liquid form, researchers have also demonstrated narrow-linewidth tunable emission in dispersive oscillator configurations incorporating solid-state dye gain media.[33] In their most prevalent form these solid state dye lasers use dye-doped polymers as laser media.

15.6.7 Free-electron lasers

The free-electron laser FELIX at the FOM Institute for Plasma Physics Rijnhuizen, Nieuwegein

Free-electron lasers, or FELs, generate coherent, high power radiation that is widely tunable, currently ranging in wavelength from microwaves through terahertz radiation and infrared to the visible spectrum, to soft X-rays. They have the widest frequency range of any laser type. While FEL beams share the same optical traits as other lasers, such as coherent radiation, FEL operation is quite different. Unlike gas, liquid, or solid-state lasers, which rely on bound atomic or molecular states, FELs use a relativistic electron beam as the lasing medium, hence the term *free-electron*.

15.6.8 Exotic media

The pursuit of a high-quantum-energy laser using transitions between isomeric states of an atomic nucleus has been the subject of wide-ranging academic research since the early 1970s. Much of this is summarized in three review articles.[34][35][36] This research has been international in scope, but mainly based in the former Soviet Union and the United States. While many scientists remain optimistic that a breakthrough is near, an operational gamma-ray laser is yet to be realized.[37]

Some of the early studies were directed toward short pulses of neutrons exciting the upper isomer state in a solid so the gamma-ray transition could benefit from the line-narrowing of Mössbauer effect.[38][39] In conjunction, several advantages were expected from two-stage pumping of a three-level system.[40] It was conjectured that the nucleus of an atom, embedded in the near field of a laser-driven coherently-oscillating electron cloud would experience a larger dipole field than that of the driving laser.[41][42] Furthermore, nonlinearity of the oscillating cloud would produce both spatial and temporal harmonics, so nuclear transitions of higher multipolarity could also be driven at multiples of the laser frequency.[43][44][45][46][47][48][49]

In September 2007, the BBC News reported that there was speculation about the possibility of using positronium annihilation to drive a very powerful gamma ray laser.[50] Dr. David Cassidy of the University of California, Riverside proposed that a single such laser could be used to ignite a nuclear fusion reaction, replacing the banks of hundreds of lasers currently employed in inertial confinement fusion experiments.[50]

Space-based X-ray lasers pumped by a nuclear explosion have also been proposed as antimissile weapons.[51][52] Such devices would be one-shot weapons.

Living cells have been used to produce laser light.[53][54] The cells were genetically engineered to produce green fluorescent protein (GFP). The GFP is used as the laser's "gain medium", where light amplification takes place. The cells were then placed between two tiny mirrors, just 20 millionths of a meter across, which acted as the "laser cavity" in which light could bounce many times through the cell. Upon bathing the cell with blue light, it could be seen to emit directed and intense green laser light.

15.7 Uses

Main article: List of applications for lasers

When lasers were invented in 1960, they were called "a solution looking for a problem".[55] Since then, they have become ubiquitous, finding utility in thousands of highly varied applications in every section of modern society, including consumer electronics, information technology, science, medicine, industry, law enforcement, entertainment, and the military. Fiber-optic communication using lasers is a key technology in modern communications, allowing services such as the Internet.

The first use of lasers in the daily lives of the general population was the supermarket barcode scanner, introduced in 1974. The laserdisc player, introduced in 1978, was the first successful consumer product to include a laser but the compact disc player was the first laser-equipped device to become common, beginning in 1982 followed shortly by laser printers.

Some other uses are:

- Medicine: Bloodless surgery, laser healing, surgical treatment, kidney stone treatment, eye treatment, dentistry.
- Industry: Cutting, welding, material heat treatment, marking parts, non-contact measurement of parts.

15.7. USES

Lasers range in size from microscopic diode lasers (top) with numerous applications, to football field sized neodymium glass lasers (bottom) used for inertial confinement fusion, nuclear weapons research and other high energy density physics experiments.

- Military: Marking targets, guiding munitions, missile defence, electro-optical countermeasures (EOCM), alternative to radar, blinding troops.

- Law enforcement: used for latent fingerprint detection in the forensic identification field[56][57]

- Research: Spectroscopy, laser ablation, laser annealing, laser scattering, laser interferometry, lidar, laser capture microdissection, fluorescence microscopy, metrology.

- Product development/commercial: laser printers, optical discs (e.g. CDs and the like), barcode scanners, thermometers, laser pointers, holograms, bubblegrams.

- Laser lighting displays: Laser light shows.

- Cosmetic skin treatments: acne treatment, cellulite and striae reduction, and hair removal.

In 2004, excluding diode lasers, approximately 131,000 lasers were sold with a value of US$2.19 billion.[58] In the same year, approximately 733 million diode lasers, valued at $3.20 billion, were sold.[59]

15.7.1 Examples by power

Laser application in astronomical adaptive optics imaging

Different applications need lasers with different output powers. Lasers that produce a continuous beam or a series of short pulses can be compared on the basis of their average power. Lasers that produce pulses can also be characterized based on the *peak* power of each pulse. The peak power of a pulsed laser is many orders of magnitude greater than its average power. The average output power is always less than the power consumed.

Examples of pulsed systems with high peak power:

- 700 TW (700×10^{12} W) – National Ignition Facility, a 192-beam, 1.8-megajoule laser system adjoining a 10-meter-diameter target chamber.[63]

- 1.3 PW (1.3×10^{15} W) – world's most powerful laser as of 1998, located at the Lawrence Livermore Laboratory[64]

15.7.2 Hobby uses

In recent years, some hobbyists have taken interests in lasers. Lasers used by hobbyists are generally of class IIIa or IIIb (see Safety), although some have made their own class IV types.[65] However, compared to other hobbyists, laser hobbyists are far less common, due to the cost and potential dangers involved. Due to the cost of lasers, some hobbyists use inexpensive means to obtain lasers, such as salvaging laser diodes from broken DVD players (red), Blu-ray players (violet), or even higher power laser diodes from CD or DVD burners.[66]

Hobbyists also have been taking surplus pulsed lasers from retired military applications and modifying them for pulsed holography. Pulsed Ruby and pulsed YAG lasers have been used.

15.7.3 As weapons

The US-Israeli Tactical High Energy weapon has been used to shoot down rockets and artillery shells.

Lasers of all but the lowest powers can potentially be used as incapacitating weapons, through their ability to produce temporary or permanent vision loss in varying degrees when aimed at the eyes. The degree, character, and duration of vision impairment caused by eye exposure to laser light varies with the power of the laser, the wavelength(s), the collimation of the beam, the exact orientation of the beam, and the duration of exposure. Lasers of even a fraction of a watt in power can produce immediate, permanent vision loss under certain conditions, making such lasers potential non-lethal but incapacitating weapons. The extreme handicap that laser-induced blindness represents makes the use of lasers even as non-lethal weapons morally controversial, and weapons designed to cause blindness have been banned by the Protocol on Blinding Laser Weapons. Incidents of pilots being exposed to lasers while flying have prompted aviation authorities to implement special procedures to deal with such hazards.[67]

Laser weapons capable of directly damaging or destroying a target in combat are still in the experimental stage. The general idea of laser-beam weaponry is to hit a target with a train of brief pulses of light. The rapid evaporation and expansion of the surface causes shockwaves that damage the target. The power needed to project a high-powered laser beam of this kind is beyond the limit of current mobile power technology, thus favoring chemically powered gas dynamic lasers. Example experimental systems include MIRACL and the Tactical High Energy Laser.

Boeing YAL-1. The laser system is mounted in a turret attached to the aircraft nose

Throughout the 2000s, the United States Air Force worked on the Boeing YAL-1, an airborne laser mounted in a Boeing 747. It was intended to be used to shoot down incoming ballistic missiles over enemy territory. In March 2009, Northrop Grumman claimed that its engineers in Redondo Beach had successfully built and tested an electrically powered solid state laser capable of producing a 100-kilowatt beam, powerful enough to destroy an airplane. According to Brian Strickland, manager for the United States Army's Joint High Power Solid State Laser program, an electrically powered laser is capable of being mounted in an aircraft, ship, or other vehicle because it requires much less space for its supporting equipment than a chemical laser.[68] However, the source of such a large electrical power in a mobile application remained unclear. Ultimately, the project was deemed to be infeasible,[69][70][71] and was cancelled in December 2011,[72] with the Boeing YAL-1 prototype being stored and eventually dismantled.

The United States Navy is developing a laser weapon referred to as the Laser Weapon System or LaWS.[73]

15.7.4 Telecommunications in space

Main article: Laser communication in space

Recent technology has allowed prototypes for laser communications and visible light communication in outer space. The communication range of free-space optical communication is currently of the order of several thousand kilometers,[74] but has the potential to bridge interplanetary distances of millions of kilometers, using optical telescopes as beam expanders.[75][76]

15.8 Safety

European laser warning symbol required for Class 2 lasers and higher. Right: US laser warning label, in this case for a Class 3B laser
Main article: Laser safety

Even the first laser was recognized as being potentially dangerous. Theodore Maiman characterized the first laser as having a power of one "Gillette" as it could burn through one Gillette razor blade. Today, it is accepted that even low-power lasers with only a few milliwatts of output power can be hazardous to human eyesight when the beam hits the eye directly or after reflection from a shiny surface. At wavelengths which the cornea and the lens can focus well, the coherence and low divergence of laser light means that it can be focused by the eye into an extremely small spot on the retina, resulting in localized burning and permanent damage in seconds or even less time.

Lasers are usually labeled with a safety class number, which identifies how dangerous the laser is:

- Class 1 is inherently safe, usually because the light is contained in an enclosure, for example in CD players.

- Class 2 is safe during normal use; the blink reflex of the eye will prevent damage. Usually up to 1 mW power, for example laser pointers.

- Class 3R (formerly IIIa) lasers are usually up to 5 mW and involve a small risk of eye damage within the time of the blink reflex. Staring into such a beam for several seconds is likely to cause damage to a spot on the retina.

- Class 3B can cause immediate eye damage upon exposure.

- Class 4 lasers can burn skin, and in some cases, even scattered light can cause eye and/or skin damage. Many industrial and scientific lasers are in this class.

The indicated powers are for visible-light, continuous-wave lasers. For pulsed lasers and invisible wavelengths, other power limits apply. People working with class 3B and class 4 lasers can protect their eyes with safety goggles which are designed to absorb light of a particular wavelength.

Infrared lasers with wavelengths longer than about 1.4 micrometers are often referred to as "eye-safe", because the cornea tends to absorb light at these wavelengths, protecting the retina from damage. The label "eye-safe" can be misleading, however, as it applies only to relatively low power continuous wave beams; a high power or Q-switched laser at these wavelengths can burn the cornea, causing severe eye damage, and even moderate power lasers can injure the eye.

15.9 See also

- Bessel beam
- Coherent perfect absorber
- Dazzler (weapon)
- Homogeneous broadening
- Induced gamma emission
- Injection seeder
- International Laser Display Association
- Laser accelerometer
- Lasers and aviation safety
- Laser beam profiler
- Laser bonding
- Laser converting
- Laser cooling
- Laser engraving
- Laser medicine
- Laser scalpel
- 3D scanner
- Laser turntable
- Laser beam welding

- List of laser articles
- List of light sources
- Mercury laser
- Nanolaser
- Nonlinear optics
- Reference beam
- Rytov number
- Sound amplification by stimulated emission of radiation
- Selective laser sintering
- Spaser
- Speckle pattern
- Tophat beam
- Vortex laser

15.10 References

[1] Gould, R. Gordon (1959). "The LASER, Light Amplification by Stimulated Emission of Radiation". In Franken, P.A.; Sands R.H. (Eds.). *The Ann Arbor Conference on Optical Pumping, the University of Michigan, 15 June through 18 June 1959*. p. 128. OCLC 02460155.

[2] "laser". Reference.com. Retrieved May 15, 2008.

[3] "Four Lasers Over Paranal". *www.eso.org*. European Southern Observatory. Retrieved 9 May 2016.

[4] *Conceptual physics*, Paul Hewitt, 2002

[5] "Schawlow and Townes invent the laser". Lucent Technologies. 1998. Archived from the original on October 17, 2006. Retrieved October 24, 2006.

[6] Chu, Steven; Townes, Charles (2003). "Arthur Schawlow". In Edward P. Lazear (ed.),. *Biographical Memoirs*. vol. 83. National Academy of Sciences. p. 202. ISBN 0-309-08699-X.

[7] "lase". Dictionary.reference.com. Retrieved December 10, 2011.

[8] Siegman, Anthony E. (1986). *Lasers*. University Science Books. p. 2. ISBN 0-935702-11-3.

[9] Siegman, Anthony E. (1986). *Lasers*. University Science Books. p. 4. ISBN 0-935702-11-3.

[10] "Nitrogen Laser". *Light and Its Uses*. Scientific American. June 1974. pp. 40–43. ISBN 0-7167-1185-0.

[11] G. P. Karman, G. S. McDonald, G. H. C. New, J. P. Woerdman, "Laser Optics: Fractal modes in unstable resonators", *Nature*, Vol. 402, 138, November 11, 1999.

[12] Einstein, A (1917). "Zur Quantentheorie der Strahlung". *Physikalische Zeitschrift*. **18**: 121–128. Bibcode:1917PhyZ...18..121E.

[13] Steen, W. M. "Laser Materials Processing", 2nd Ed. 1998.

[14] Batani, Dimitri (2004). "Il rischio da laser: cosa è e come affrontarlo; analisi di un problema non così lontano da noi" [The risk from laser: what it is and what it is like facing it; analysis of a problem which is thus not far away from us]. *wwwold.unimib.it*. Programma Corso di Formazione Obbligatorio (in Italian). University of Milano-Bicocca. p. 12. Archived from the original (Powerpoint presentation) on June 14, 2007. Retrieved January 1, 2007.

[15] The Nobel Prize in Physics 1966 Presentation Speech by Professor Ivar Waller. Retrieved January 1, 2007.

[16] "American Institute of Physics Oral History Interview with Joseph Weber".

[17] Bertolotti, Mario (2015), *Masers and Lasers, Second Edition: An Historical Approach*, CRC Press, pp. 89–91, ISBN 9781482217803, retrieved March 15, 2016

[18] Townes, Charles H. (1999). *How the Laser Happened: Adventures of a Scientist*, Oxford University Press, ISBN 9780195122688, pp. 69-70.

[19] Joan Lisa Bromberg, *The Laser in America, 1950–1970* (1991), pp. 74–77 online

[20] Maiman, T. H. (1960). "Stimulated optical radiation in ruby". *Nature*. **187** (4736): 493–494. Bibcode:1960Natur.187..493M. doi:10.1038/187493a0.

[21] Townes, Charles Hard. "The first laser". University of Chicago. Retrieved May 15, 2008.

[22] Hecht, Jeff (2005). *Beam: The Race to Make the Laser*. Oxford University Press. ISBN 0-19-514210-1.

[23] Nolen, Jim; Derek Verno. "The Carbon Dioxide Laser". Davidson Physics. Retrieved 17 August 2014.

[24] Csele, Mark (2004). "The TEA Nitrogen Gas Laser". *Homebuilt Lasers Page*. Archived from the original on September 11, 2007. Retrieved September 15, 2007.

[25] "Deep UV Lasers" (PDF). Photon Systems, Covina, Calif. Archived from the original (PDF) on 2007-07-01. Retrieved May 27, 2007.

[26] Schuocker, D. (1998). *Handbook of the Eurolaser Academy*. Springer. ISBN 0-412-81910-4.

[27] C. Stewen, M. Larionov, and A. Giesen, "Yb:YAG thin disk laser with 1 kW output power", in OSA Trends in Optics and Photonics, Advanced Solid-State Lasers, H. Injeyan, U. Keller, and C. Marshall, ed. (Optical Society of America, Washington, D.C.., 2000) pp. 35–41.

[28] Wu, X.; et al. (October 25, 2004). "Ultraviolet photonic crystal laser". *Applied Physics Letters.* **85** (17): 3657. arXiv:physics/0406005. Bibcode:2004ApPhL..85.3657W. doi:10.1063/1.1808888.

[29] "Laser Diode Market". Hanel Photonics. Retrieved Sep 26, 2014.

[30] "LASER Diode". *nichia.co.jp*.

[31] "Green Laser". *osram-os.com*. August 19, 2015.

[32] "Picolight ships first 4-Gbit/s 1310-nm VCSEL transceivers". *Laser Focus World Online.* December 9, 2005. Archived from the original on March 13, 2006. Retrieved May 27, 2006.

[33] F. J. Duarte, *Tunable Laser Optics*, 2nd Edition (CRC, New York, 2015).

[34] Baldwin, G. C.; Solem, J. C.; Gol'danskii, V. I. (1981). "Approaches to the development of gamma-ray lasers". *Reviews of Modern Physics.* **53**: 687–744. Bibcode:1981RvMP...53..687B. doi:10.1103/RevModPhys.53.687.

[35] Baldwin, G. C.; Solem, J. C. (1995). "Recent proposals for gamma-ray lasers". *Laser Physics.* **5** (2): 231–239.

[36] Baldwin, G. C.; Solem, J. C. (1997). "Recoilless gamma-ray lasers". *Reviews of Modern Physics.* **69** (4): 1085–1117. Bibcode:1997RvMP...69.1085B. doi:10.1103/RevModPhys.69.1085.

[37] Baldwin, G. C.; Solem, J. C. (1982). "Is the time ripe? Or must we wait so long for breakthroughs?". *Laser Focus.* **18** (6): 6&8.

[38] Solem, J. C. (1979). "On the feasibility of an impulsively driven gamma-ray laser". *Los Alamos Scientific Laboratory Report LA-7898*.

[39] Baldwin, G. C.; Solem, J. C. (1979). "Maximum density and capture rates of neutrons moderated from a pulsed source". *Nuclear Science & Engineering.* **72** (3): 281–289.

[40] Baldwin, G. C.; Solem, J. C. (1980). "Two-stage pumping of three-level Mössbauer gamma-ray lasers". *Journal of Applied Physics.* **51**: 2372–2380. Bibcode:1980JAP....51.2372B. doi:10.1063/1.328007.

[41] Solem, J. C. (1986). "Interlevel transfer mechanisms and their application to grasers". *Proceedings of Advances in Laser Science-I, First International Laser Science Conference, Dallas, TX 1985 (American Institute of Physics, Optical Science and Engineering, Series 6).* **146**: 22–25.

[42] Biedenharn, L. C.; Boyer, K.; Solem, J. C. (1986). "Possibility of grasing by laser-driven nuclear excitation". *Proceedings of AIP Advances in Laser Science-I, Dallas, TX, November 18–22, 1985.* **146**: 50–51.

[43] Rinker, G. A.; Solem, J. C.; Biedenharn, L. C. (1988). "Calculation of harmonic radiation and nuclear coupling arising from atoms in strong laser fields". *Proceedings of SPIE 0875, Short and Ultrashort Wavelength Lasers, Los Angeles, CA, January 11, 1988 (International Society for Optics and Photonics).* **146**: 92–101.

[44] Rinker, G. A.; Solem, J. C.; Biedenharn, L. C. (1987). "Nuclear interlevel transfer driven by collective outer shell electron excitations". *Proceedings of the Second International Laser Science Conference, Seattle, WA (Advances in Laser Science-II) Lapp, M.; Stwalley, W. C.; Kenney-Wallace G.A., eds. (American Institute of Physics, New York).* **160**: 75–86.

[45] Solem, J. C. (1988). "Theorem relating spatial and temporal harmonics for nuclear interlevel transfer driven by collective electronic oscillation". *Journal of Quantitative Spectroscopy and Radiative Transfer.* **40** (6): 713–715. Bibcode:1988JQSRT..40..713S. doi:10.1016/0022-4073(88)90067-2.

[46] Solem, J. C.; Biedenharn, L. C. (1987). "Primer on coupling collective electronic oscillations to nuclei" (PDF). *Los Alamos National Laboratory Report LA-10878*.

[47] Solem, J. C.; Biedenharn, L. C. (1988). "Laser coupling to nuclei via collective electronic oscillations: A simple heuristic model study". *Journal of Quantitative Spectroscopy and Radiative Transfer.* **40** (6): 707–712. Bibcode:1988JQSRT..40..707S. doi:10.1016/0022-4073(88)90066-0.

[48] Boyer, K.; Java, H.; Luk, T. S.; McIntyre, I. A.; McPherson, A.; Rosman, R.; Solem, J. C.; Rhodes, C. K.; Szöke, A. (1987). "Discussion of the role of many-electron motions in multiphoton ionization and excitation". *Proceedings of International Conference On Multiphoton Processes (ICOMP) IV, July 13–17, 1987, Boulder, CA, Smith, S.; Knight, P.; eds. (Cambridge University Press, Cambridge, England)*: 58.

[49] Biedenharn, L. C.; Rinker, G. A.; Solem, J. C. (1989). "A solvable approximate model for the response of atoms subjected to strong oscillatory electric fields". *Journal of the Optical Society of America B.* **6** (2): 221–227. Bibcode:1989JOSAB...6..221B. doi:10.1364/JOSAB.6.000221.

[50] Fildes, Jonathan (September 12, 2007). "Mirror particles form new matter". *BBC News.* Retrieved May 22, 2008.

[51] Hecht, Jeff (May 2008). "The history of the x-ray laser". *Optics and Photonics News.* Optical Society of America. **19** (5): 26–33. Bibcode:2008OptPN..19R..26H. doi:10.1364/opn.19.5.000026.

[52] Robinson, Clarence A. (1981). "Advance made on high-energy laser". *Aviation Week & Space Technology* (February 23, 1981): 25–27.

[53] Palmer, Jason (June 13, 2011). "Laser is produced by a living cell". *BBC News.* Retrieved June 13, 2011.

[54] Malte C. Gather & Seok Hyun Yun (June 12, 2011). "Single-cell biological lasers". *Nature Photonics*. Retrieved June 13, 2011.

[55] Charles H. Townes (2003). "The first laser". In Laura Garwin; Tim Lincoln. *A Century of Nature: Twenty-One Discoveries that Changed Science and the World*. University of Chicago Press. pp. 107–12. ISBN 0-226-28413-1. Retrieved February 2, 2008.

[56] Dalrymple B. E., Duff J. M., Menzel E. R. "Inherent fingerprint luminescence – detection by laser". *Journal of Forensic Sciences*, 22(1), 1977, 106–115

[57] Dalrymple B. E. "Visible and infrared luminescence in documents : excitation by laser". *Journal of Forensic Sciences*, 28(3), 1983, 692–696

[58] Kincade, Kathy; Anderson, Stephen (January 1, 2005). "Laser Marketplace 2005: Consumer applications boost laser sales 10%". *Laser Focus World*. Vol. 41 no. 1.

[59] Steele, Robert V. (February 1, 2005). "Diode-laser market grows at a slower rate". *Laser Focus World*. Vol. 41 no. 2.

[60] "Green Laser 400 mW burn a box CD in 4 second". *YouTube*. Retrieved December 10, 2011.

[61] "Laser Diode Power Output Based on DVD-R/RW specs". elabz.com. Retrieved December 10, 2011.

[62] Peavy, George M. "How to select a surgical veterinary laser". *Aesculight*. Retrieved March 30, 2016.

[63] Heller, Arnie, "Orchestrating the world's most powerful laser." *Science and Technology Review*. Lawrence Livermore National Laboratory, July/August 2005. URL accessed May 27, 2006.

[64] Schewe, Phillip F.; Stein, Ben (November 9, 1998). "Physics News Update 401". American Institute of Physics. Archived from the original on June 14, 2008. Retrieved March 15, 2008.

[65] PowerLabs CO_2 LASER! Sam Barros June 21, 2006. Retrieved January 1, 2007.

[66] Maks, Stephanie. "Howto: Make a DVD burner into a high-powered laser". *Transmissions from Planet Stephanie*. Retrieved April 6, 2015.

[67] "Police fight back on laser threat". *BBC News*. April 8, 2009. Retrieved April 4, 2010.

[68] Peter, Pae (March 19, 2009). "Northrop Advance Brings Era Of The Laser Gun Closer". *Los Angeles Times*. p. B2.

[69] "Missile Defense Umbrella?". Center for Strategic and International Studies.

[70] "Schwartz: Get those AF boots off the ground". airforcetimes.com.

[71] Hodge, Nathan (February 11, 2011). "Pentagon Loses War To Zap Airborne Laser From Budget". *Wall Street Journal*.

[72] Butler, Amy (December 21, 2011). "Lights Out For The Airborne Laser". *Aviation Week*.

[73] Luis Martinez (9 Apr 2013). "Navy's New Laser Weapon Blasts Bad Guys From Air, Sea". *ABC*. Retrieved 9 April 2013.

[74] "Another world first for Artemis: a laser link with an aircraft". European Space Agency. December 18, 2006. Retrieved June 28, 2011.

[75] Boroson, Don M. (2005), *Optical Communications: A Compendium of Signal Formats, Receiver Architectures, Analysis Mathematics, and Performance Characteristics*, retrieved 8 Jan 2013

[76] Steen Eiler Jørgensen (October 27, 2003). "Optisk kommunikation i deep space– Et feasibilitystudie i forbindelse med Bering-missionen" (PDF). Dansk Rumforskningsinstitut. Retrieved June 28, 2011. (Danish) Optical Communications in Deep Space, University of Copenhagen

15.11 Further reading

Books

- Bertolotti, Mario (1999, trans. 2004). *The History of the Laser*. Institute of Physics. ISBN 0-7503-0911-3.

- Bromberg, Joan Lisa (1991). *The Laser in America, 1950–1970*. MIT Press. ISBN 978-0-262-02318-4.

- Csele, Mark (2004). *Fundamentals of Light Sources and Lasers*. Wiley. ISBN 0-471-47660-9.

- Koechner, Walter (1992). *Solid-State Laser Engineering*. 3rd ed. Springer-Verlag. ISBN 0-387-53756-2.

- Siegman, Anthony E. (1986). *Lasers*. University Science Books. ISBN 0-935702-11-3.

- Silfvast, William T. (1996). *Laser Fundamentals*. Cambridge University Press. ISBN 0-521-55617-1.

- Svelto, Orazio (1998). *Principles of Lasers*. 4th ed. Trans. David Hanna. Springer. ISBN 0-306-45748-2.

- Taylor, Nick (2000). *LASER: The inventor, the Nobel laureate, and the thirty-year patent war*. New York: Simon & Schuster. ISBN 0-684-83515-0.

- Wilson, J. & Hawkes, J.F.B. (1987). *Lasers: Principles and Applications*. Prentice Hall International Series in Optoelectronics, Prentice Hall. ISBN 0-13-523697-5.

- Yariv, Amnon (1989). *Quantum Electronics*. 3rd ed. Wiley. ISBN 0-471-60997-8.

Periodicals

- *Applied Physics B: Lasers and Optics* (ISSN 0946-2171)
- *IEEE Journal of Lightwave Technology* (ISSN 0733-8724)
- *IEEE Journal of Quantum Electronics* (ISSN 0018-9197)
- *IEEE Journal of Selected Topics in Quantum Electronics* (ISSN 1077-260X)
- *IEEE Photonics Technology Letters* (ISSN 1041-1135)
- *Journal of the Optical Society of America B: Optical Physics* (ISSN 0740-3224)
- *Laser Focus World* (ISSN 0740-2511)
- *Optics Letters* (ISSN 0146-9592)
- *Photonics Spectra* (ISSN 0731-1230)

15.12 External links

- Encyclopedia of laser physics and technology by Dr. Rüdiger Paschotta
- A Practical Guide to Lasers for Experimenters and Hobbyists by Samuel M. Goldwasser
- Homebuilt Lasers Page by Professor Mark Csele
- Powerful laser is 'brightest light in the universe' – The world's most powerful laser as of 2008 might create supernova-like shock waves and possibly even antimatter (*New Scientist*, April 9, 2008)
- "Laser Fundamentals" an online course by Prof. F. Balembois and Dr. S. Forget. *Instrumentation for Optics*, 2008, (accessed January 17, 2014)
- Northrop Grumman's Press Release on the Firestrike 15kw tactical laser product.
- Website on Lasers 50th anniversary by APS, OSA, SPIE
- Advancing the Laser anniversary site by SPIE: Video interviews, open-access articles, posters, DVDs
- Bright Idea: The First Lasers history of the invention, with audio interview clips.
- Free software for Simulation of random laser dynamics
- Video Demonstrations in Lasers and Optics Produced by the Massachusetts Institute of Technology (MIT). Real-time effects are demonstrated in a way that would be difficult to see in a classroom setting.
- Virtual Museum of Laser History, from the touring exhibit by SPIE
- website with animations, applications and research about laser and other quantum based phenomena Universite Paris Sud

Chapter 16

Photon structure function

The quark content of the photon is described in quantum field theory by the *photon structure function* defined by the process $e + \gamma \to e +$ hadrons. It is uniquely characterized by the linear increase in the logarithm of the electronic momentum transfer $\log Q^2$ and by the approximately linear rise in x, the fraction of the quark momenta within the photon. These characteristics are borne out by the experimental analyses of the photon structure function.

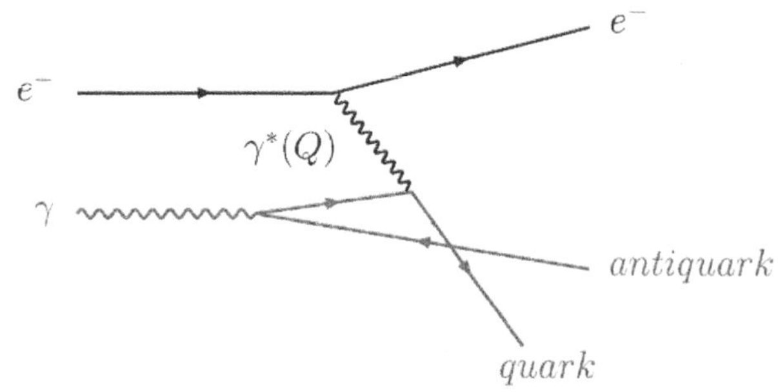

Figure 1. Electron-photon scattering generic Feynman diagram.

16.1 Theoretical basis

High energy photons can transform in quantum mechanics to lepton and quark pairs, the latter fragmented subsequently to jets of hadrons, i.e. protons, pions etc. At high energies E the lifetime t of such quantum fluctuations of mass M becomes nearly macroscopic: $t \approx E/M^2$; this amounts to flight lengths as large as one micrometer for electron pairs in a 100 GeV photon beam, and still 10 fermi, i.e. the tenfold radius of a proton, for light hadrons. High energy photon beams have been generated by photon radiation off electron beams in e^-e^+ circular beam facilities such as PETRA at DESY in Hamburg and LEP at CERN in Geneva. Exceedingly high photon energies may be generated in the future by shining laser light on TeV electron beams in a linear collider facility.

The classical technique for analyzing the virtual particle content of photons is provided by scattering electrons off the photons. In high-energy, large-angle scattering the experimental facility can be viewed as an electron microscope of very high resolution Q, corresponding to the momentum transfer in the scattering process according to Heisenberg's uncertainty principle. The intrinsic quark structure of the target photon beam is revealed by observing characteristic patterns of the scattered electrons in the final state.

The incoming target photon splits into a nearly collinear quark-antiquark pair. The impinging electron is scattered off the quark to large angles, the scatter pattern revealing the internal quark structure of the photon. Quark and antiquark finally transform to hadrons. Most exciting is the theoretical analysis of the quark content of the photon, termed "photon structure function". The analysis can be described quantitatively in quantum chromodynamics (QCD), the theory of quarks as constituents of the strongly interacting elementary particles, which are bound together by gluonic forces. The primary splitting of photons to quark pairs, cf. Fig.1, regulates the essential characteristica of the photon structure function, the number and the energy spectrum of the quark constituents within the photon.[1] QCD refines the picture [2][3] by modifying the shape of the spectrum, to order unity unlike the small modifications naively expected as a result of asymptotic freedom.

Quantum mechanics predicts the number of quark pairs in the photon splitting process to increase logarithmically with the resolution Q, and (approximately) linearly with the momenta x. The characteristic behavior

$$F_{2,B}^{\gamma}(x, Q^2) = f_B(x) \log Q^2/\Lambda^2 + ...$$

with

$$f_B(x) = \frac{3\alpha}{2\pi} \sum_{q,\bar{q}} e_q^4 x [x^2 + (1-x)^2]$$

is predicted for the photon structure function in the quark model to leading logarithmic behavior, the Sommerfeld fine-structure constant denoted by α=1/137 and the fractional quark charges by eq; the factor 3 counting the quark color degrees. Turning on the radiation of gluon quanta off quarks in QCD, the quark momenta are reshuffled partly from large to small x values with increasing resolution. At the same time the radiation is damped moderately due to asymptotic freedom. The delicate interplay between photon splitting and damped gluon radiation re-normalizes the photon structure function

$$F_{2,B}^{\gamma}(x,Q^2) \to F_2^{\gamma}(x,Q^2) = f(x) \log Q^2/\Lambda^2$$

to order unity, leaving the logarithmic behavior in the resolution Q untouched apart from superficially introducing the fundamental QCD scale Λ, but tilting the shape of the structure function $fB(x) \to f(x)$ by damping the momentum spectrum at large x. These characteristica, dramatically different from the proton parton density, are unique features of the photon structure function within QCD. They are the origin of the excitement associated with the photon structure function.[4]

While electron scattering off photons maps out the quark spectra, the electrically neutral gluon content of the photons can best be analyzed by jet pair production in photon-proton scattering. Gluons as components of the photon may scatter off gluons residing in the proton, and generate two hadron jets in the final state. The complexity of these scattering processes, due to the superposition of many subprocesses, renders the analysis of the gluon content of the photon quite complicated.

The quantitative representation of the photon structure function introduced above is strictly valid only for asymptotically high resolution Q, i.e. the logarithm of Q being much larger than the logarithm of the quark masses. However, the asymptotic behavior is approached steadily with increasing Q for x away from zero as demonstrated next. In this asymptotic regime the photon structure function is predicted uniquely in QCD to logarithmic accuracy.

16.2 Experimental analyses

Up to now the photon structure function has only been investigated experimentally by electron scattering off a beam of quasi-real photons. The experiments utilize the so-called two-photon reactions at electron-positron colliders $e^-e^+ \to e^-e^+ + h$, where h includes all hadrons of the final state. The kinematics chosen is characterized by the electron scattered at large angles and the positron at very small angles thus providing a calculable flux of quasi-real photons (Weizsäcker-Williams approximation). The cross section for electron-photon scattering is then analyzed in terms of the photon structure function quite analogously to studies of the nucleon structure in electron-nucleon scattering.

In order to ensure a small virtual mass of the target photon one uses the so-called anti-tagging. Special forward detectors are arranged down to small angles close to the beam pipe. Events with a positron signal in these detectors are eliminated from the analysis. By contrast, events with the positrons travelling undetected down the beam pipe, are accepted. The energy of the emitted quasi-real target photon is unknown. Whereas the four-momentum transfer squared Q^2 can be determined alone from the energy and angle of the scattered electron, x has to be calculated from Q^2 and the invariant mass W of the hadronic system using $x = Q^2/(Q^2+W^2)$. The experimental situation is thus comparable to neutrino-nucleon scattering where the unknown energy of the incoming neutrino also requires the determination of W for calculating the kinematical parameters of the neutrino quark scattering process.

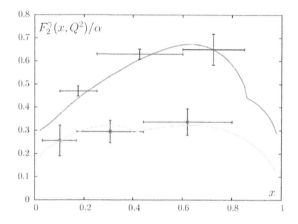

Fig2: Photon structure function versus x for $Q^2 = 4.3$ GeV^2 (blue crosses) and 39.7 GeV^2 (black crosses) compared to the QCD prediction explained in the text.

The hadronic system produced in two-photon reactions has in general a rather high momentum along the beam direction resulting in small hadronic scattering angles. This kinematic feature again requires special forward detectors. A high efficiency in reconstructing hadronic events is now also essential. Nevertheless losses of hadronic energy are practically unavoidable and the real hadronic energy is therefore determined using sophisticated unfolding techniques.[5] [6]

The first measurement of the photon structure function has been performed using the detector PLUTO at the DESY storage ring PETRA [7] followed subsequently by many investigations at all large electron-positron colliders. A comprehensive discussion of data and theory can be found in reviews of 2000 [6] and 2014.[8] It is customary to display the structure function in units of the fine-structure constant α. The basic theoretical features discussed above are impressively verified by the data. The increase of $F_2^\gamma(x,Q^2)$ with x, shown in Fig. 2 at $Q^2 = 4.3$ GeV^2 and 39.7 GeV^2, is obviously quite different from the behaviour of the proton structure function, which falls with rising x, and it demonstrates nicely the influence of the photon splitting to quark pairs. The predicted log Q^2 dependence of $F_2(x,Q^2)$ is clearly demonstrated in Fig. 3, here plotted for data with $0.3 < x < 0.5$.

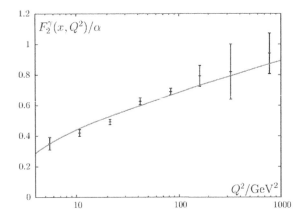

Fig 3: *Photon structure function versus log Q^2 for $0.3 < x < 0.5$ compared to the QCD prediction explained in the text.*

In both figures the data are compared to theoretical calculations, the curves representing the analysis of photon structure function data based on the standard higher-order QCD prediction for three light quarks [9] supplemented by the charm quark contribution and a residual hadronic component accounted for by vector meson dominance. The numerical values were calculated using $\Lambda = 0.338$ GeV and a charm quark mass of 1.275 GeV. See[8] for details of the data selection and the theoretical model.

One might be tempted to use the data for a precision measurement of Λ. However, while the asymptotic solution, defined properly at higher order, appears superficially very sensitive to Λ, spurious singularities at small x require either technical ad-hoc regularizations or the switching to the evolution from pre-fixed initial conditions at small Q^2. Both techniques reduce the sensitivity to Λ. Nevertheless, values of

$$\alpha_s(M_Z) = 0.1198 \pm 0.0028(ex) \pm 0.0040(th)$$

in analyses of the QCD coupling along these lines [10] agree well with other experimental methods.

It is remarkable to realize that even a single parameter (Λ) fit performed to all data with[10] $x > 0.45$, $Q^2 > 59$ GeV^2 or to all data with[8] $x > 0.1$ leads to very similar results for $\alpha S (M_Z)$.

16.3 Conclusion

In summary, the prediction for the number of quarks and their momentum spectrum in high-energy photons, with characteristics very much different from the proton, together with the value of the QCD coupling constant, are borne out nicely by the experimental analyses—a fascinating success of QCD.

16.4 References

[1] T. F. Walsh and P. M. Zerwas, "Two-photon processes in the parton model", *Physics Letters* B44 (1973) 195.

[2] E. Witten, "Anomalous cross-section for photon-photon scattering in gauge theories", *Nuclear Physics* B120 (1977) 189.

[3] W. A. Bardeen and A. J. Buras, "Higher order asymptotic freedom corrections to photon-photon scattering", *Physical Review* D20 (1979) 166 [E-ibid. D21 (1980) 2041].

[4] A. J. Buras, "Photon structure functions: 1978 and 2005", *Acta Physica Polonica* B37 (2006) 609, arXiv:hep-ph/0512238v2.

[5] Ch. Berger and W.Wagner, Photon-photon reactions, Physics Reports 146 (1987) 1

[6] R. Nisius, The photon structure from deep inelastic electron photon scattering, Physics Report 332 (2000) 165

[7] PLUTO Collaboration Ch. Berger et al., First measurement of the photon structure function, Physics Letters B107 (1981) 168

[8] Ch. Berger, Photon structure function revisited , arXiv: 1404.3551, http://de.arxiv.org/abs/1404.3551

[9] M. Glück, E. Reya and A. Vogt, Parton structure of the photon beyond the leading order, Physical Review D45 (1992) 3986

[10] S.Albino, M.Klasen and S.Söldner-Rembold, Strong coupling constant from the photon structure function, Physical Review Letters 89 (2002) 122004, [hep-ph/0205069].

Chapter 17

Ballistic photon

Ballistic photons are the light photons that travel through a scattering (turbid) medium in a straight line. Also known as **ballistic light**. If laser pulses are sent through a turbid medium such as fog or body tissue, most of the photons are either randomly scattered or absorbed. However, across short distances, a few photons pass through the scattering medium in straight lines. These coherent photons are referred to as ballistic photons. Photons that are slightly scattered, retaining some degree of coherence, are referred to as *snake* photons.

If efficiently detected, there are many applications for ballistic photons especially in coherent high resolution medical imaging systems. Ballistic scanners (using ultrafast time gates) and optical coherence tomography (OCT) (using the interferometry principle) are just two of the popular imaging systems that rely on ballistic photon detection to create diffraction-limited images. Due to the exponential reduction (with respect to distance) of ballistic photons in a scattering medium, often image processing techniques are applied to the raw captured ballistic images, to reconstruct high quality ones.

17.1 References

- K. Yoo and R. R. Alfano, "Time-resolved coherent and incoherent components of forward light scattering in random media," Optics Letters 15, 320–322 (1990).

- S. Farsiu, J. Christofferson, B. Eriksson, P. Milanfar, B. Friedlander, A. Shakouri, R. Nowak, "Statistical detection and imaging of objects hidden in turbid media using ballistic photons", *Applied Optics*, vol. 46, no. 23, pp. 5805–5822, Aug. 2007.

Chapter 18

Photonic molecule

Photonic molecules are a natural form of matter which can also be made artificially in which photons bind together to form "molecules". These type of particles are found in sunlight. According to Mikhail Lukin, individual (massless) photons "interact with each other so strongly that they act as though they have mass". The effect is analogous to refraction. The light enters another medium, transferring part of its energy to the medium. Inside the medium, it exists as coupled light and matter, but it exits as light.[1]

Researchers drew analogies between the phenomenon and the fictional "lightsaber" from *Star Wars*.[1][2]

18.1 Construction

Gaseous rubidium atoms were pumped into a vacuum chamber. The cloud was cooled using lasers to just a few degrees above absolute zero. Using weak laser pulses, small numbers of photons were fired into the cloud.[1]

As the photons entered the cloud, their energy excited atoms along their path, causing them to lose speed. Inside the cloud medium the photons dispersively coupled to strongly interacting atoms in highly excited Rydberg states. This caused the photons to behave as massive particles with strong mutual attraction (photon molecules). Eventually the photons exited the cloud together as normal photons (often entangled in pairs).[1]

The effect is caused by a so-called Rydberg blockade, which, in the presence of one excited atom, prevents nearby atoms from being excited to the same degree. In this case, as two photons enter the atomic cloud, the first excites an atom, but must move forward before the second can excite nearby atoms. In effect the two photons push and pull each other through the cloud as their energy is passed from one atom to the next, forcing them to interact. This photonic interaction is mediated by the electromagnetic interaction between photons and atoms.[1]

18.2 Possible applications

The interaction of the photons suggests that the effect could be employed to build a system that can preserve quantum information, and process it using quantum logic operations.[1]

The system could also be useful in classical computing, given the much-lower power required to manipulate photons than electrons.[1]

It may be possible to arrange the photonic molecules in such a way within the medium that they form larger three-dimensional structures (similar to crystals).[1]

18.3 Interacting microcavities

The term photonic molecule has been also used since 1998 for an unrelated phenomenon involving electromagnetically-interacting optical microcavities. The properties of quantized confined photon states in optical micro- and nanocavities are very similar to those of confined electron states in atoms.[3] Owing to this similarity, optical microcavities can be termed 'photonic atoms'. Taking this analogy even further, a cluster of several mutually-coupled photonic atoms forms a photonic molecule.[4] When individual photonic atoms are brought into close proximity, their optical modes interact and give rise to a spectrum of hybridized super-modes of photonic molecules.[5]

"A micrometer-sized piece of semiconductor can trap photons inside it in such a way that they act like electrons in an atom. Now the 21 September PRL describes a way to link two of these "photonic atoms" together. The result of such a close relationship is a "photonic molecule," whose optical modes bear a strong resemblance to the electronic states of a diatomic molecule like hydrogen."[6]

"Photonic molecules, named by analogy with chemical molecules, are clusters of closely located electromagnetically interacting microcavities or "photonic atoms"."[7]

"Optically coupled microcavities have emerged as photonic structures with promising properties for investigation of fundamental science as well as for applications."[8]

The first demonstration of a lithographically-fabricated photonic molecule was inspired by an analogy with a simple diatomic molecule.[9] However, other nature-inspired PM structures (such as 'photonic benzene') have been proposed and shown to support confined optical modes closely analogous to the ground-state molecular orbitals of their chemical counterparts.[10]

Photonic molecules offer advantages over isolated photonic atoms in a variety of applications, including bio(chemical) sensing,[11][12] cavity optomechanics,[13][14] and microlasers,[15][16][17][18] Photonic molecules can also be used as quantum simulators of many-body physics and as building blocks of future optical quantum information processing networks.[19]

In complete analogy, clusters of metal nanoparticles - which support confined surface plasmon states - have been termed 'plasmonic molecules."[20][21][22][23][24]

Finally, hybrid photonic-plasmonic (or opto-plasmonic) molecules have also been proposed and demonstrated.,[25][26][27][28]

18.4 References

[1] "Seeing light in a new light: Scientists create never-before-seen form of matter". Sciencedaily.com. Retrieved 2013-09-27.

[2] Firstenberg, O.; Peyronel, T.; Liang, Q. Y.; Gorshkov, A. V.; Lukin, M. D.; Vuletić, V. (2013). "Attractive photons in a quantum nonlinear medium". *Nature*. Bibcode:2013Natur.502...71F. doi:10.1038/nature12512.

[3] Benson, T. M.; Boriskina, S. V.; Sewell, P.; Vukovic, A.; Greedy, S. C.; Nosich, A. I. (2006). "Micro-Optical Resonators for Microlasers and Integrated Optoelectronics". *Frontiers in Planar Lightwave Circuit Technology*. NATO Science Series II: Mathematics, Physics and Chemistry. **216**. p. 39. doi:10.1007/1-4020-4167-5_02. ISBN 1-4020-4164-0.

[4] Boriskina, S. V. (2010). "Photonic Molecules and Spectral Engineering". *Photonic Microresonator Research and Applications*. Springer Series in Optical Sciences. **156**. p. 393. doi:10.1007/978-1-4419-1744-7_16. ISBN 978-1-4419-1743-0.

[5] Rakovich, Y.; Donegan, J.; Gerlach, M.; Bradley, A.; Connolly, T.; Boland, J.; Gaponik, N.; Rogach, A. (2004). "Fine structure of coupled optical modes in photonic molecules". *Physical Review A*. **70** (5). Bibcode:2004PhRvA..70e1801R. doi:10.1103/PhysRevA.70.051801.

[6] doi:10.1103/PhysRevFocus.2.14

[7] arXiv:0704.2154

[8] doi:10.1038/lsa.2013.38

[9] Bayer, M.; Gutbrod, T.; Reithmaier, J.; Forchel, A.; Reinecke, T.; Knipp, P.; Dremin, A.; Kulakovskii, V. (1998). "Optical Modes in Photonic Molecules". *Physical Review Letters*. **81** (12): 2582. Bibcode:1998PhRvL..81.2582B. doi:10.1103/PhysRevLett.81.2582.

[10] Lin, B. (2003). "Variational analysis for photonic molecules: Application to photonic benzene waveguides". *Physical Review E*. **68** (3). Bibcode:2003PhRvE..68c6611L. doi:10.1103/PhysRevE.68.036611.

[11] Boriskina, S. V. (2006). "Spectrally engineered photonic molecules as optical sensors with enhanced sensitivity: A proposal and numerical analysis". *Journal of the Optical Society of America B*. **23** (8): 1565. arXiv:physics/0603228. Bibcode:2006JOSAB..23.1565B. doi:10.1364/JOSAB.23.001565.

[12] Boriskina, S. V.; Dal Negro, L. (2010). "Self-referenced photonic molecule bio(chemical)sensor". *Optics Letters*. **35** (14): 2496–8. Bibcode:2010OptL....35.2496B. doi:10.1364/OL.35.002496. PMID 20634875.

[13] Jiang, X.; Lin, Q.; Rosenberg, J.; Vahala, K.; Painter, O. (2009). "High-Q double-disk microcavities for cavity optomechanics". *Optics Express*. **17** (23): 20911–9. doi:10.1364/OE.17.020911. PMID 19997328.

[14] Hu, Y. W.; Xiao, Y. F.; Liu, Y. C.; Gong, Q. (2013). "Optomechanical sensing with on-chip microcavities". *Frontiers of Physics*. **8** (5): 475. Bibcode:2013FrPhy...8..475H. doi:10.1007/s11467-013-0384-y.

[15] Hara, Y.; Mukaiyama, T.; Takeda, K.; Kuwata-Gonokami, M. (2003). "Photonic molecule lasing". *Optics Letters*. **28** (24): 2437–9. Bibcode:2003OptL....28.2437H. doi:10.1364/OL.28.002437. PMID 14690107.

[16] Nakagawa, A.; Ishii, S.; Baba, T. (2005). "Photonic molecule laser composed of GaInAsP microdisks". *Applied Physics Letters*. **86** (4): 041112. Bibcode:2005ApPhL..86d1112N. doi:10.1063/1.1855388.

[17] Boriskina, S. V. (2006). "Theoretical prediction of a dramatic Q-factor enhancement and degeneracy removal of whispering gallery modes in symmetrical photonic molecules". *Optics Letters*. **31** (3): 338–40. Bibcode:2006OptL....31..338B. doi:10.1364/OL.31.000338. PMID 16480201.

[18] Smotrova, E. I.; Nosich, A. I.; Benson, T. M.; Sewell, P. (2006). "Threshold reduction in a cyclic photonic molecule laser composed of identical microdisks with whispering-gallery modes". *Optics Letters*. **31** (7): 921–3. Bibcode:2006OptL....31..921S. doi:10.1364/OL.31.000921. PMID 16599212.

[19] Hartmann, M.; Brandão, F.; Plenio, M. (2007). "Effective Spin Systems in Coupled Microcavities". *Physical Review Letters.* **99** (16). arXiv:0704.3056. Bibcode:2007PhRvL..99p0501H. doi:10.1103/PhysRevLett.99.160501.

[20] Nordlander, P.; Oubre, C.; Prodan, E.; Li, K.; Stockman, M. I. (2004). "Plasmon Hybridization in Nanoparticle Dimers". *Nano Letters.* **4** (5): 899. Bibcode:2004NanoL...4..899N. doi:10.1021/nl049681c.

[21] Fan, J. A.; Bao, K.; Wu, C.; Bao, J.; Bardhan, R.; Halas, N. J.; Manoharan, V. N.; Shvets, G.; Nordlander, P.; Capasso, F. (2010). "Fano-like Interference in Self-Assembled Plasmonic Quadrumer Clusters". *Nano Letters.* **10** (11): 4680–5. Bibcode:2010NanoL..10.4680F. doi:10.1021/nl1029732. PMID 20923179.

[22] Liu, N.; Mukherjee, S.; Bao, K.; Brown, L. V.; Dorfmüller, J.; Nordlander, P.; Halas, N. J. (2012). "Magnetic Plasmon Formation and Propagation in Artificial Aromatic Molecules". *Nano Letters.* **12** (1): 364–9. Bibcode:2012NanoL..12..364L. doi:10.1021/nl203641z. PMID 22122612.

[23] Yan, B.; Boriskina, S. V.; Reinhard, B. R. M. (2011). "Optimizing Gold Nanoparticle Cluster Configurations (n≤ 7) for Array Applications". *The Journal of Physical Chemistry C.* **115** (11): 4578. doi:10.1021/jp112146d.

[24] Yan, B.; Boriskina, S. V.; Reinhard, B. R. M. (2011). "Design and Implementation of Noble Metal Nanoparticle Cluster Arrays for Plasmon Enhanced Biosensing". *The Journal of Physical Chemistry C.* **115** (50): 24437. doi:10.1021/jp207821t.

[25] Boriskina, S. V.; Reinhard, B. M. (2011). "Spectrally and spatially configurable superlenses for optoplasmonic nanocircuits". *Proceedings of the National Academy of Sciences.* **108** (8): 3147. arXiv:1110.6822. Bibcode:2011PNAS..108.3147B. doi:10.1073/pnas.1016181108.

[26] Boriskina, S. V.; Reinhard, B. R. M. (2011). "Adaptive on-chip control of nano-optical fields with optoplasmonic vortex nanogates". *Optics Express.* **19** (22): 22305–15. arXiv:1111.0022. Bibcode:2011OExpr..1922305B. doi:10.1364/OE.19.022305. PMC 3298770. PMID 22109072.

[27] Hong, Y.; Pourmand, M.; Boriskina, S. V.; Reinhard, B. R. M. (2013). "Enhanced Light Focusing in Self-Assembled Optoplasmonic Clusters with Subwavelength Dimensions". *Advanced Materials.* **25**: 115. doi:10.1002/adma.201202830.

[28] Ahn, W.; Boriskina, S. V.; Hong, Y.; Reinhard, B. R. M. (2012). "Photonic–Plasmonic Mode Coupling in On-Chip Integrated Optoplasmonic Molecules". *ACS Nano.* **6** (1): 951–60. doi:10.1021/nn204577v. PMID 22148502.

18.5 External links

- http://prl.aps.org/abstract/PRL/v81/i12/p2582_1

Chapter 19

Two-photon physics

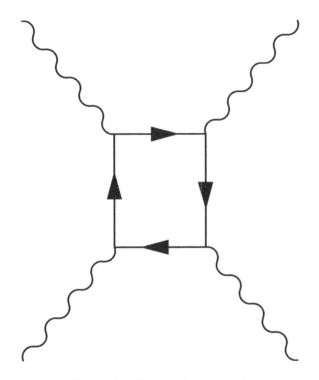

A *Feynman diagram* (box diagram) *for photon–photon scattering, one photon scatters from the transient vacuum charge fluctuations of the other*

Two-photon physics, also called **gamma–gamma physics**, is a branch of particle physics that describes the interactions between two photons. Normally, beams of light pass through each other unperturbed. Inside an optical material, and if the intensity of the beams is high enough, the beams may affect each other through a variety of non-linear effects. In pure vacuum, some weak scattering of light by light will exist if the center-of-mass energy of the system of the two photons is large enough. Also, above some threshold of this center-of-mass energy of the system of the two photons, matter can be created.

19.1 Astronomy

Photon–photon scattering limits the spectrum of observed gammas to below 80 TeV. The other photon is one of the many photons of the cosmic microwave background. In the frame of reference where the invariant mass of the two photons is at rest, both photons are gammas with just enough energy to pair-produce an electron–positron pair.

19.2 Experiments

Two-photon physics can be studied with high-energy particle accelerators, where the accelerated particles are not the photons themselves but charged particles that will radiate photons. The most significant studies so far were performed at the Large Electron–Positron Collider (LEP) at CERN. If the transverse momentum transfer and thus the deflection is large, one or both electrons can be detected; this is called tagging. The other particles that are created in the interaction are tracked by large detectors to reconstruct the physics of the interaction.

Frequently, photon-photon interactions will be studied via ultraperipheral collisions (UPCs) of heavy ions, such as gold or lead. These are collisions in which the colliding nuclei do not touch each other; i.e., the impact parameter b is larger than the sum of the radii of the nuclei. The strong interaction between the quarks composing the nuclei is thus greatly suppressed, making the weaker electromagnetic $\gamma\gamma$ interaction much more visible. In UPCs, because the ions are heavily charged, it is possible to have two independent interactions between a single ion pair, such as production of two electron-positron pairs. UPCs are studied with the STARlight simulation code.

Light-by-light scattering has not been directly observed so far. As of 2012, the best constraint on the elastic photon–photon scattering cross section belongs to PVLAS, which reports an upper limit far above the level predicted by the Standard Model.[1] Proposals have been made to measure elastic light-by-light scattering using the strong electromagnetic fields of the hadrons collided at the LHC.[2] Observation of a cross section larger than that predicted by the Standard Model could signify new physics such as axions, the search of which is the primary goal of PVLAS and several similar experiments.

19.3 Processes

From quantum electrodynamics it can be found that photons cannot couple directly to each other, since they carry no charge, but they can interact through higher-order processes: A photon can, within the bounds of the uncertainty principle, fluctuate into a virtual charged fermion–antifermion pair, to either of which the other photon can couple. This fermion pair can be leptons or quarks. Thus, two-photon physics experiments can be used as ways to study the photon structure, or, somewhat metaphorically, what is "inside" the photon.

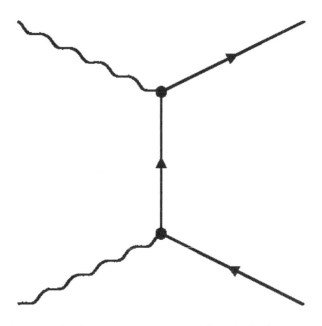

Creation of a fermion–antifermion pair through the direct two-photon interaction. These drawings are Feynman diagrams.

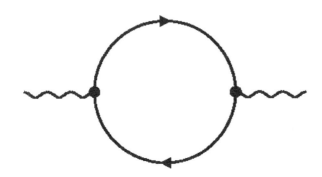

The photon fluctuates into a fermion–antifermion pair.

We distinguish three interaction processes:

- *Direct* or *pointlike*: The photon couples directly to a quark inside the target photon.[3] If a lepton–antilepton pair is created, this process involves only quantum electrodynamics (QED), but if a quark–antiquark pair is created, it involves both QED and perturbative quantum chromodynamics (QCD).[4][5] The intrinsic quark content of the photon is described by the photon structure function, experimentally analyzed in deep-inelastic electron–photon scattering.[6][7]

- *Single resolved*: The quark pair of the target photon form a vector meson. The probing photon couples to a constituent of this meson.

- *Double resolved*: Both target and probe photon have formed a vector meson. This results in an interaction between two hadrons.

For the latter two cases, the scale of the interaction is such as the strong coupling constant is large. This is called *Vector Meson Dominance* (VMD) and has to be modelled in non-perturbative QCD.

19.4 See also

- Channelling radiation has been considered as a method to generate polarized high energy photon beams for gamma–gamma colliders.

- Matter creation

- Pair production

19.5 References

[1] G. Zavattini et al., "Measuring the magnetic birefringence of vacuum: the PVLAS experiment", Accepted for publication in the *Proceedings of the QFEXT11 Benasque Conference*,

[2] D. d'Enterria, G. G. da Silveira, "Observing Light-by-Light Scattering at the Large Hadron Collider", *Phys. Rev. Lett.*, **111** (2013) 080405

[3] T.F.Walsh and P.M.Zerwas, "Two photon processes in the parton model", Phys. Lett. B44 (1973) 195.

[4] E.Witten, "Anomalous Cross-Section for Photon – Photon Scattering in Gauge Theories", Nucl. Phys. B120} (1977) 189.

[5] W.A.Bardeen and A.J.Buras, "Higher Order Asymptotic Freedom Corrections to Photon–Photon Scattering", Phys. Rev. D20 (1979) 166, [Erratum-ibid. D21 (1980) 2041].

[6] L3 Collaboration, Measurement of the photon structure function F_2^γ with the L3 detector at LEP, *Phys. Lett. B 622, 249 (2005)*

[7] R. Nisius, The photon structure from deep inelastic electron photon scattering, Physics Report 332 (2000) 165

19.6 External links

- Lauber,J A, 1997, A small tutorial in gamma–gamma Physics
- Two-photon physics at LEP
- Two-photon physics at CESR

19.7 Text and image sources, contributors, and licenses

19.7.1 Text

- **Photon** *Source:* https://en.wikipedia.org/wiki/Photon?oldid=735591138 *Contributors:* AxelBoldt, WojPob, Mav, Bryan Derksen, The Anome, Tarquin, Koyaanis Qatsi, Ap, Josh Grosse, Ben-Zin~enwiki, Heron, Youandme, Spiff~enwiki, Bdesham, Michael Hardy, Ixfd64, TakuyaMurata, NuclearWinner, Looxix~enwiki, Snarfies, Ahoerstemeier, Stevenj, Julesd, Glenn, AugPi, Mxn, Smack, Pizza Puzzle, Wikiborg, Reddi, Lfh, Jitse Niesen, Kbk, Laussy, Bevo, Shizhao, Raul654, Jusjih, Donarreiskoffer, Robbot, Hankwang, Fredrik, Eman, Sanders muc, Altenmann, Bkalafut, Merovingian, Gnomon Kelemen, Hadal, Wereon, Anthony, Wjbeaty, Giftlite, Art Carlson, Herbee, Xerxes314, Everyking, Dratman, Michael Devore, Bensaccount, Foobar, Jaan513, DÅ‚ugosz, Zeimusu, LucasVB, Beland, Setokaiba, Kaldari, Vina, RetiredUser2, Icairns, Lumidek, Zondor, Randwicked, Eep², Chris Howard, Zowie, Naryathegreat, Discospinster, Rich Farmbrough, Yuval madar, Pjacobi, Vsmith, Ivan Bajlo, Dbachmann, Mani1, SpookyMulder, Bender235, Kbh3rd, RJHall, Ben Webber, El C, Edwinstearns, Laurascudder, RoyBoy, Spoon!, Dalf, Drhex, Bobo192, Foobaz, I9Q79oL78KiL0QTFHgyc, Chbarts, La goutte de pluie, Zr40, Apostrophe, Minghong, Rport, Alansohn, Gary, Sade, Corwin8, PAR, UnHoly, Hu, Caesura, Wtmitchell, Bucephalus, Max rspct, BanyanTree, Cal 1234, Count Iblis, Egg, Dominic, Gene Nygaard, Ghirlandajo, Kazvorpal, UTSRelativity, Falcorian, Drag09, Boothy443, Richard Arthur Norton (1958-), Woohookitty, Linas, Gerd Breitenbach, StradivariusTV, Oliphaunt, Cleonis, Pol098, Ruud Koot, Mpatel, Nakos2208~enwiki, Dbl2010, Ch'marr, SDC, CharlesC, Alan Canon, Reddwarf2956, Mandarax, BD2412, Kbdank71, Zalasur, Sjakkalle, Rjwilmsi, Саша Стефановић, Strait, MarSch, Dennis Estenson II, Trlovejoy, Mike Peel, HappyCamper, Bubba73, Brighterorange, Cantorman, Egopaint, Noon, Godzatswing, FlaBot, RobertG, Arnero, Mathbot, Nihiltres, Fresheneesz, TeaDrinker, Srleffler, BradBeattie, Chobot, Jaraalbe, DVdm, Elfguy, EamonnPKeane, YurikBot, Bambaiah, Splintercellguy, Jimp, RussBot, Supasheep, JabberWok, Wavesmikey, KevinCuddeback, Stephenb, Gaius Cornelius, Salsb, Trovatore, Długosz, Tailpig, Joelr31, SCZenz, Randolf Richardson, Ravedave, Tony1, Roy Brumback, Gadget850, Dna-webmaster, Enormousdude, Lt-wiki-bot, Oysteinp, JoanneB, Ligart, John Broughton, GrinBot~enwiki, Sbyrnes321, Itub, SmackBot, Moeron, Incnis Mrsi, KnowledgeOfSelf, CelticJobber, Melchoir, Rokfaith, WilyD, Jagged 85, Jab843, Cessator, AnOddName, Skizzik, Dauto, JSpudeman, Robin Whittle, Ati3414, Persian Poet Gal, MK8, Jprg1966, Complexica, Sbharris, Colonies Chris, Ebertek, WordLife565, V1adis1av, RWincek, Aces lead, Stangbat, Cybercobra, Valenciano, EVula, A.R., Mini-Geek, AEM, DMacks, N Shar, Sadi Carnot, FlyHigh, The Fwanksta, Drunken Pirate, Yevgeny Kats, Lambiam, Harryboyles, IronGargoyle, Ben Moore, A. Parrot, Mr Stephen, Fbartolom, Dicklyon, SandyGeorgia, Mets501, Ceeded, Ambuj.Saxena, Ryulong, Vincecate, Astrobayes, Newone, J Di, Lifeverywhere, Tawkerbot2, JRSpriggs, Chetvorno, Luis A. Veguilla-Berdecia, CalebNoble, Xod, Gregory9, CmdrObot, Wafulz, Van helsing, John Riemann Soong, Rwflammang, Banedon, Wquester, Outriggr (2006-2009), Logical2u, Myasuda, Howardsr, Cydebot, Krauss, Kanags, A876, WillowW, Bvcrist, Hyperdeath, Hkyriazi, Rracecarr, Difluoroethene, Edgerck, Michael C Price, Tawkerbot4, Christian75, Ldussan, RelHistBuff, Waxigloo, Kozuch, Thijs!bot, Epbr123, Opabinia regalis, Markus Pössel, Mglg, 24fan24, Headbomb, Newton2, John254, J.christianson, Dawnseeker2000, Escarbot, Stannered, AntiVandalBot, Luna Santin, Jtrain4469, Normanmargolus, Tyco.skinner, TimVickers, NSH001, Dodecahedron~enwiki, Tim Shuba, Gdo01, Lfstevens, Sluzzelin, Abyssoft, CosineKitty, Plantsurfer, AndyBloch, Bryanv, ScottStearns, WolfmanSF, Hroðulf, Bongwarrior, VoABot II, B&W Anime Fan, SHCarter, Lgoger, I JethroBT, Dirac66, Hveziris, Maliz, Lord GaleVII, TRWBW, Shijualex, Glen, DerHexer, Patstuart, Gwern, Taborgate, MartinBot, MNAdam, Jay Litman, HEL, Ralf 58, J.delanoy, DrKay, Trusilver, C. Trifle, AstroHurricane001, Numbo3, Pursey, CMDadabo, Kevin aylward, UchihaFury, Pirate452, H4xx0r, Iamthewalrus35, Iamthewalrus36, Gee Eff, Chimpy07, Dirkdiggler69, Lk69, Hallamfm, Annoying editter, Yehoodig, Acalamari, Foreigner1, McSly, Samtheboy, Tarotcards, Rominandreu, Ontarioboy, ARTE, Tanaats, Potatoswatter, Y2H, Divad89, Scott Illini, Stack27, THEblindwarrior, VolkovBot, AlnoktaBOT, Hyperlinker, DoorsAjar, TXiKiBoT, Oshwah, Cosmic Latte, The Original Wildbear, Davehi1, Chiefwaterfall, Vipinhari, Hqb, Anonymous Dissident, HansMair, Predator24, BotKung, Luuva, Calvin4986, Improve~enwiki, Kmhkmh, Richwil, Antixt, Gorank4, Falcon8765, GlassFET, Cryptophile, MattiasAndersson, AlleborgoBot, Carlodn6, FlyingLeopard2014, Relilles~enwiki, Tpb, SieBot, Timb66, Graham Beards, WereSpielChequers, ToePeu.bot, JerrySteal, Android Mouse, Likebox, RadicalOne, Paolo.dL, Lightmouse, PbBot, Spartan-James, Duae Quartunciae, Hamiltondaniel, StewartMH, Dstebbins, ClueBot, Bobathon71, The Thing That Should Not Be, Mwengler, EoGuy, Jagun, RODERICKMOLASAR, Wwheaton, Dmlcyal8er, Razimantv, Mild Bill Hiccup, Feebas factor, J8079s, Rotational, MaxwellsLight, Awickert, Excirial, PixelBot, Sun Creator, NuclearWarfare, PhySusie, El bot de la dieta, DerBorg, Shamanchill, PoofyPeter99, J1.grammar natz, Laserheinz, TimothyRias, XLinkBot, Jovianeye, Interferometrist, Petedskier, Hess88, Addbot, Mathieu Perrin, DOI bot, DougsTech, Download, James thirteen, AndersBot, LinkFA-Bot, Barak Sh, AgadaUrbanit, Тивериополник, Dayewalker, Quantumobserver, Kein Einstein, Legobot, Luckas-bot, Yobot, Kilom691, Allowgolf~enwiki, AnomieBOT, Cantanchorus, Ratul2000, Kingpin13, Materialscientist, Citation bot, Xqbot, Ambujarind69, Mananay, Sharhalakis, Shirik, RibotBOT, Rickproser, SongRenKai, Max derner, Merrrr, A. di M., ??, CES1596, Paine Ellsworth, Gsthae with tempo!, Nageh, TimonyCrickets, WurzelT, Steve Quinn, Spacekid99, Radeksonic, Citation bot 1, Pinethicket, I dream of horses, HRoestBot, Tanweer Morshed, Eno crux, Tom.Reding, Jschnur, RedBot, IVAN3MAN, Gamewizard71, FoxBot, TobeBot, Earthandmoon, PleaseStand, Marie Poise, RjwilmsiBot, Антон Глинисты, Ripchip Bot, Ofercomay, Chemyanda, EmausBot, Bookalign, WikitanvirBot, Roxbreak, Word2need, Gcastellanos, Tommy2010, Dcirovic, K6ka, Hhhippo, Cogiati, 1howardsr1, StringTheory11, Waperkins, Jojojlj, Midas02, Access Denied, Quondum, AManWithNoPlan, Raynor42, L Kensington, Chrisman62, Maschen, HCPotter, Haiti333, RockMagnetist, Rocketrod1960, ClueBot NG, JASMEET SINGH HAFIST, Schicagos, Snotbot, Vinícius Machado Vogt, Helpful Pixie Bot, SzMithrandir, Bibcode Bot, BG19bot, Roberticus, Paolo Lipparini, Wzrd1, Rifath119, Davidiad, Mark Arsten, Peter.sujak, Wikarchitect, Zedshort, Hamish59, Caypartisbot, Penguinstorm300, Amphibio, Millennium bug, Sfarney, KSI ROX, Bhargavuk1997, Chromastone1998, TheJJJunk, Nimmo1859, EagerToddler39, RockBoxt, EZas3pt14, Webclient101, Chrisanion, Vanquisher.UA, Tony Mach, PREMDASKANNAN, Meghas, Reatlas, Profb39, Zerberos, Thesuperseo, The User 111, Eyesnore, Ybidzian, Tentinator, Illusterati, JustBerry, Celso ad, Quenhitran, Manul, DrMattV, Anrnusna, Wyn.junior, K0RTD, Monkbot, Yikkayaya, Vieque, Tigercompanion25, BethNaught, Markmizzi, Garfield Garfield, Dhm44444, Smokey2022, Zargol Rejerfree, RAL2014, Shahriar Kabir Pavel, Sdjncskdjnfskje, Anshul1908, Professor Flornoy, Thatguytestw, Tetra quark, Harshit100, KasparBot, Chinta 01, Geek3, Leisor2, SSTflyer, TheKingOfPhysics, Chemistry1111, Bassam.Mak, Daniel kenneth, FishmrWiki, Fmadd, Artm7777 and Anonymous: 520

- **Light** *Source:* https://en.wikipedia.org/wiki/Light?oldid=734728200 *Contributors:* AxelBoldt, TwoOneTwo, Kpjas, Trelvis, Lee Daniel Crocker, Brion VIBBER, Bryan Derksen, Zundark, The Anome, Koyaanis Qatsi, Rjstott, Andre Engels, Xaonon, XJaM, Fredbauder, SJK, William Avery, DrBob, Panairjdde~enwiki, Heron, Ewen, Olivier, Someone else, MimirZero, Stevertigo, Lir, Patrick, Michael Hardy, Fred Bauder, Dante Alighieri, Ixfd64, Sannse, Qaz, Shoaler, Minesweeper, Dgrant, Looxix~enwiki, Ahoerstemeier, Mac, 5ko, Kingturtle, Athypique~enwiki, Julesd, Glenn, AugPi, Andres, Evercat, David Stewart, Mxn, Pizza Puzzle, Emperorbma, Greenrd, WhisperToMe, Saltine, Carol

Fenijn, SEWilco, Omegatron, Wernher, Dpbsmith, Jusjih, Robbot, Mulberry~enwiki, Hankwang, Tomchiukc, Chris 73, Altenmann, Kowey, Lowellian, Merovingian, Yosri, Sverdrup, Henrygb, Academic Challenger, Blainster, Sunray, Hadal, UtherSRG, Wereon, Wile E. Heresiarch, FrankSier, Alan Liefting, Giftlite, Crculver, DocWatson42, Harp, Seabhcan, Ævar Arnfjörð Bjarmason, Netoholic, Lupin, Ayman, Everyking, Curps, Frencheigh, AJim, Al-khowarizmi, Bobblewik, Bact, Zeimusu, LucasVB, Antandrus, Beland, ClockworkLunch, Karol Langner, DragonflySixtyseven, RetiredUser2, Icairns, Sayeth, Arcturus, WpZurp, Neutrality, Joyous!, Kevin Rector, Deglr6328, Mike Rosoft, Smokris, JTN, Discospinster, Rich Farmbrough, Pak21, Thematicunity, Vsmith, Chub~enwiki, Roybb95~enwiki, Dbachmann, Paul August, Horsten, Bender235, ESkog, Andrejj, Danny B-), RJHall, Dkroll2, El C, Robert P. O'Shea, Lankiveil, Laurascudder, Shanes, Art LaPella, West London Dweller, Causa sui, Bobo192, Army1987, NetBot, Fir0002, AnyFile, Smalljim, Reinyday, AllyUnion, Elipongo, I9Q79oL78KiL0QTFHgyc, Man vyi, Jojit fb, Nk, PeterisP, Sam Korn, Krellis, Pearle, Alansohn, Gary, Arthena, Atlant, Ricky81682, Phiddipus, CiceronI, Batmanand, Malo, Stillnotelf, Snowolf, Blobglob, L33th4x0rguy, Almafeta, Yuckfoo, Stephan Leeds, Jwinius, Gpvos, Sciurinæ, LFaraone, Kusma, DV8 2XL, Gene Nygaard, Netkinetic, Forderud, Feezo, Snowmanmelting, Fred Condo, Simetrical, OwenX, Whitemanburntmyland, TigerShark, LOL, Jersyko, Uncle G, Borb, Jacobolus, Kzollman, Davidkazuhiro, JeremyA, Sengkang, Mike McGovern, Pinkgothic, Prashanthns, Dysepsion, Mandarax, MassGalactusUniversum, RichardWeiss, Ashmoo, BD2412, Kane5187, Rjwilmsi, Phileas, Vary, Strait, Captain Disdain, Nneonneo, Quietust, Thechamelon, Oblivious, Ligulem, The wub, Yamamoto Ichiro, Kevmitch, Ravidreams, Titoxd, FlaBot, Splarka, AED, Dan Guan, Doc glasgow, Pumeleon, Nihiltres, Crazycomputers, Andy85719, RexNL, Ewlyahoocom, Gurch, Hja, KFP, Fresheneesz, Alphachimp, Srleffler, King of Hearts, Chobot, DVdm, Bomb319, EamonnPKeane, YurikBot, Wavelength, Pip2andahalf, RussBot, SpuriousQ, Akamad, Stephenb, Bill52270, Shell Kinney, Gaius Cornelius, Alex Bakharev, Pseudomonas, Salsb, Wimt, Bullzeye, David R. Ingham, NawlinWiki, SEWilcoBot, Wiki alf, Grafen, Matticus78, RL0919, Voidxor, Xompanthy, BOT-Superzerocool, Paulwesterman, DeadEyeArrow, Bota47, Oliverdl, Wknight94, Enormousdude, 2over0, Where next Columbus?, Closedmouth, KGasso, Petri Krohn, GraemeL, Back ache, Kungfuadam, Junglecat, RG2, Serendipodous, Chrismith, DVD R W, SmackBot, RDBury, MattieTK, Elonka, Sergeymk, Lestrade, Urania3, KnowledgeOfSelf, CelticJobber, Ma8thew, Melchoir, Unyoyega, C.Fred, The Photon, Bomac, Jagged 85, ScaldingHotSoup, Delldot, Eskimbot, Rojomoke, Hardyplants, Frymaster, Gilliam, Ohnoitsjamie, Hmains, Betacommand, Saros136, KaragouniS, Persian Poet Gal, Oli Filth, MalafayaBot, Fluri, SchfiftyThree, The Rogue Penguin, Afasmit, Deli nk, JavaJake, DHN-bot~enwiki, Sbharris, Colonies Chris, Mkamensek, Hallenrm, Darth Panda, Sct72, Audriusa, Can't sleep, clown will eat me, Shalom Yechiel, PeteShanosky, Onorem, Geoffrey Gibson, Rrburke, VMS Mosaic, Neopergoss, RedHillian, Stiangk, Stevenmitchell, Popsup, COMPFUNK2, Jmlk17, Cybercobra, Khukri, Nakon, PatelRahul, Invincible Ninja, Mini-Geek, Trieste, Tinclon, DMacks, Salamurai, Sayden, Vina-iwbot~enwiki, Kukini, The undertow, SashatoBot, EMan32x, Quendus, Vanished user 9i39j3, Kuru, Philosophus, J 1982, JohnCub, Antonielly, Peterlewis, Scetoaux, The Man in Question, Dicklyon, Waggers, Intranetusa, Ryulong, Novangelis, Thrindel, Christian Roess, KJS77, Politepunk, Levineps, Iridescent, Ronius, JacekW, Izapuff, Robbie Cook, GDallimore, Tony Fox, Sam Li, Courcelles, Jjedmond, Tawkerbot2, Hammer Raccoon, Conrad.Irwin, Slmader, JForget, Rnn2walls, Liam Skoda, Tanthalas39, Mattbr, Jackzhp, Unionhawk, BoneyKing, Linus M., Runningonbrains, GHe, THF, Nmacu, Mattbuck, Utsav.schurmans, Mblumber, Jonathan Tweet, WillowW, Hyperdeath, Meno25, Gogo Dodo, Anonymi, JFreeman, Flowerpotman, Llort, Stone.samuel, Tkynerd, Benjiboi, Tawkerbot4, Doug Weller, DumbBOT, SteveMcCluskey, Kansas Sam, NMChico24, Dischdeldritsch~enwiki, Thijs!bot, Epbr123, Martin Hogbin, Mojo Hand, Oliver202, Headbomb, Trevyn, Dtgriscom, Marek69, Second Quantization, 1234sam, Tellyaddict, Pcbene, D.H, TarkusAB, Dawnseeker2000, Escarbot, Darknesscrazyman, AntiVandalBot, Majorly, Gioto, Luna Santin, Widefox, EdgarCarpenter, Seaphoto, Tweesdad, Opelio, QuiteUnusual, Paste, Tyco.skinner, Tlabshier, Farosdaughter, Mikeryz, Flod logic, Myanw, Canadian-Bacon, JAnDbot, Vivelequebeclibre, Leuko, Kaobear, MER-C, Sonicsuns, Dmbstudio, Hamsterlopithecus, Dcooper, Hut 8.5, 100110100, LittleOldMe, Angelofdeath275, Penubag, Jaysweet, Bongwarrior, VoABot II, Fusionmix, Wikidudeman, Kuyabribri, Shahin-e-iqbal, Think outside the box, Lucyin, Rich257, Bobcat64, Tonyfaull, The Anomebot2, Catgut, Theroadislong, Khuz, Animum, First Harmonic, 28421u2232nfenfcenc, LorenzoB, DerHexer, Esanchez7587, Waninge, WLU, The Sanctuary Sparrow, Lunakeet, 0612, MartinBot, CRACKER66, Hellspawn123, STBot, Nishantsah, Vigyani, Butterfly reflections, NAHID, Poeloq, Kostisl, R'n'B, CommonsDelinker, AlexiusHoratius, J.delanoy, Captain panda, Pharaoh of the Wizards, AAA!, Numbo3, Anas Salloum, Uncle Dick, Maurice Carbonaro, All Is One, Stephanwehner, Pumpknhd, Ijustam, M C Y 1008, Rod57, It Is Me Here, H8jd5, McSly, Mikael Häggström, Engunneer, Manofsteel32, Maxberners, GhostPirate, Raining girl, Belovedfreak, Darrendeng, Cadwaladr, KCinDC, Kraftlos, FJPB, DigitallyBorn, Cmichael, Juliancolton, Vanished user 39948282, Micro01, Craklyn, Bonadea, Idioma-bot, Wikieditor06, Lights, X!, McNoddy~enwiki, VolkovBot, Jeff G., TheMindsEye, Nthndude23, DavidBrahm, Ryan032, ChaNy123, Philip Trueman, TXiKiBoT, Zidonuke, PercyWaldram, Outpt, Medlat, Comtraya, Z.E.R.O., Anonymous Dissident, Karan bob, Asama~enwiki, Karanbob 123456789, BURDA, Rebornsoldier, JayC, Qxz, Lradrama, Clarince63, NattraN, Elphion, Leafyplant, Sirkad, Setreset, Abdullais4u, Jackfork, LeaveSleaves, Ripepette, ^demonBot2, Wingray2, Heavenskid, Waycool27, LordLincoln8494, Arjunatgv, Flaw600, Kabuzol, Happy5214, Synthebot, Falcon8765, VanishedUserABC, Enviroboy, RaseaC, Agüeybaná, Itachi8009, AlleborgoBot, JoeSeale, Masterofpsi, Legoktm, CT Cooper, Justbabychub, Jayman65, EmxBot, Starkrm, Lilfiregal, Demmy, Technion, SieBot, Gago16, Timb66, Ttonyb1, Justinritter, Hertz1888, Viskonsas, Caltas, Hyper oxtane, Edmund Patrick, Lennartgoosens, The high and mighty, Triwbe, Keilana, Tiptoety, Radon210, Exert, Jhsbulldog, Heislegend9111, Minnir45, Arbor to SJ, Stanleymilgram, Jonbean13, Paolo.dL, Aruton, Oxymoron83, Faradayplank, Avnjay, Pac72, AnonGuy, Tombomp, Rajeevtco, Macy, NBS, Maelgwnbot, Cyfal, Anchor Link Bot, Mygerardromance, Reubentg, Wisemannn, Sphilbrick, Nn123645, Into The Fray, Aadgray, Dig deeper, ElectronicsEnthusiast, Faithlessthewonderboy, Atif.t2, Kevin Langendyk, Church, Loren.wilton, Tanvir Ahmmed, Beeblebrox, ClueBot, Andrew Nutter, Foxj, The Thing That Should Not Be, Guelao, Vacio, Gawaxay, Loathsome Dog, Gaia Octavia Agrippa, Razimantv, Dvratnam~enwiki, Mild Bill Hiccup, Uncle Milty, J8079s, Boing! said Zebedee, Gartree5, Rotational, Mattw91, House13, Mspraveen, Srguy, DragonBot, Djr32, Stuart.clayton.22, Alexbot, Wikibobspider, Black dragon lucifer, Ludwigs2, SpikeToronto, Vivio Testarossa, Lartoven, Simon D M, Aks818guy, Superduperboard, ParisianBlade, Brews ohare, NuclearWarfare, Arjayay, Promethean, M.O.X, Bludshotinc, Razorflame, Huntthetroll, SchreiberBike, Saebjorn, Dogcutter, Sallicio, Thehelpfulone, Existentialistcowboy, Thingg, Marsbarz, Aitias, Versus22, LieAfterLie, Amaltheus, Jb dodo, Ibliz, Dhanisha, XLinkBot, Lights2, Pichpich, Rror, Avoided, Ethanthepimp, SilvonenBot, Joearmacost, JinJian, HeartOnShelf, NonNobisSolum, Stormoffire, Wikiwkiz, Addbot, PhilT2, Willking1979, Manic19, El cangri386, Glassneko, Travis002, FeRD NYC, Ronhjones, Njaelkies Lea, Fieldday-sunday, Laurinavicius, Startstop123, W1k1p3diaf0rt3hw1n, NjardarBot, Cst17, Lukester94, Download, Chamal N, Glane23, Glass Sword, AndersBot, Debresser, Favonian, Bettafish2hamsters, LinkFABot, Quercus solaris, AgadaUrbanit, Numbo3-bot, Tide rolls, Bloodcited, Bfigura's puppy, Lightbot, Ajtc14, Gail, Slgcat, The only one who knew, Coolio98, Angrysockhop, Frehley, Legobot, Drdonzi, Yobot, Aliasmk, Les boys, Marez512, Bykgardner, Amble, Mr.tennisman37, KamikazeBot, Mj fan1995, 8ung3st, Tempodivalse, Synchronism, Dragos 85, AnomieBOT, A More Perfect Onion, Floquenbeam, Kristen Eriksen, Yourself1, Galoubet, Piano non troppo, Ufim, Sfvace, NS96091, Ulric1313, Flewis, Materialscientist, The High Fin Sperm Whale, Citation bot, Andrewalsterda, GB fan, The Firewall, Stephentucker, MauritsBot, Xqbot, Capricorn42, Emezei, Wperdue, Renaissancee, Ilexministrator, Gatorgirl7563, GreenWood86, Crzer07, Hoobladoobla, RadiX, HolgerFiedler, Omnipaedista, Blindwolf88, Igloowiki, Bashar, Shubinator, Der

Falke, Doulos Christos, Saralmao, Merpre, SobaNoodleForYou, Sjsclass, A. di M., Endothermic, Sesu Prime, Dougofborg, Legobot III, Bekus, Bout2gohuntin, Prari, FrescoBot, LucienBOT, Paine Ellsworth, Pepper, Iamaditya, Steve Quinn, Wifione, Tlork Thunderhead, Craig Pemberton, HamburgerRadio, Citation bot 1, DigbyDalton, Þjóðólfr, Alipson, Pinethicket, Elockid, WikiAntPedia, Hard Sin, A8UDI, Lars Washington, Raisusi, ActivExpression, Irbisgreif, TobeBot, Lotje, Comet Tuttle, Oracleofottawa, Vrenator, Zvn, Jssa1995, Sosonat, Jaybag91, Cardinality, Smarty02, Tbhotch, Reach Out to the Truth, Marie Poise, Karatekidd10, DARTH SIDIOUS 2, Shooosh, Jw079232, Ripchip Bot, Rdema13, Dukeofnewmexico, Jonlegere, Narayanan20092009, EmausBot, John of Reading, Ksaranme, Gfoley4, JawsBrody, Antony11031989, Ajraddatz, Niuhaibiao, Calland10, Ericyang1337, Pens98, CarlowGraphics, Rajkiandris, Inspector Soumik, Tommy2010, Dcirovic, K6ka, Evanh2008, Shadow one eight seven, Boblikesrob, Vinne2, Access Denied, Hgetnet, GianniG46, Wayne Slam, Looscan, Mattblythe, Resprinter123, L Kensington, HCPotter, Shashank artemis fowl, Crazyboom2, Teapeat, Rememberway, ClueBot NG, 66mat66, Lhimec, Rajayuv, Candy853, C4100, Matthiaspaul, JohnsonL623, Hazhk, Vinícius Machado Vogt, MerlIwBot, Helpful Pixie Bot, Thisthat2011, Calidum, Bibcode Bot, Tiscando, AvocatoBot, Zyxwv99, Jontyla, Cadiomals, Jmccormick927, TehJayden, Nitrobutane, Comfr, RudolfRed, BattyBot, Evolvo365247, Yavor18, Tandrum, SD5bot, Khazar2, BrightStarSky, Dexbot, Dissident93, Mogism, Hwangrox99, LMANSH, Reatlas, Rkaup, Jellyfrank, Evergreen-Fir, DavRosen, Wyn.junior, Rhlius, Maxwell Verbeek, Bentsutomu1234, SarahTehCat, Yukeshrbs, Tetra quark, Awjfi, MacPoli1, Lshaw93, KasparBot, Jmc76, Kafishabbir, C.Gesualdo, Plantlady223, Robot psychiatrist, Fmadd, Motivação and Anonymous: 1199

- **Photon energy** *Source:* https://en.wikipedia.org/wiki/Photon_energy?oldid=725410026 *Contributors:* Berek, Bearcat, Ilyak, AdiJapan, A.R., Andrarias, Answer~hewiki and Anonymous: 1

- **Photon polarization** *Source:* https://en.wikipedia.org/wiki/Photon_polarization?oldid=733423468 *Contributors:* Michael Hardy, Sebastian-Helm, William M. Connolley, Nv8200pa, AJim, Egg, Rjwilmsi, Bhadani, Srleffler, Kri, Phantomsteve, Długosz, Cojoco, SmackBot, Kmarinas86, RDBrown, Droll, Complexica, Rwh555, Dicklyon, Spiel496, JdH, Jackzhp, Cydebot, SyntaxError55, RJ4, Four Dog Night, Magioladitis, Rettetast, R'n'B, CommonsDelinker, BigrTex, Gillwill2000, Hennessey, Patrick, IanBushfield, DnetSvg, HHHEB3, Interferometrist, Addbot, Fgnievinski, Skippy le Grand Gourou, AnomieBOT, Citation bot, Dave3457, Vladimir.manea, Tom.Reding, John of Reading, ClamDip, Wi4hic, NULL, Bibcode Bot, Srodrig, Mark Arsten, Historyphysics and Anonymous: 41

- **Photon counting** *Source:* https://en.wikipedia.org/wiki/Photon_counting?oldid=654970513 *Contributors:* DavidCary, Srleffler, Xiaphias, Zwiller, Grantmidnight, John of Reading, MrNiceGuy1113 and Anonymous: 2

- **Photonics** *Source:* https://en.wikipedia.org/wiki/Photonics?oldid=729756827 *Contributors:* WojPob, Bryan Derksen, Oliver Pereira, Ahoerstemeier, Mac, Sergiusz Patela, Raul654, Drxenocide, Robbot, Sanders muc, Lowellian, DHN, Graeme Bartlett, Jason Quinn, Karol Langner, Kevin Rector, Rich Farmbrough, Pjacobi, Vsmith, Mani1, Dmr2, Bender235, Laurascudder, Bobo192, Stesmo, Giraffedata, Cavrdg, Cigno, Passw0rd, Anthony Appleyard, RPaschotta, Arthena, Keenan Pepper, Sp00n17, Ceyockey, Distantbody, MONGO, JonBirge, Toussaint, Dwward, Vegaswikian, FlaBot, Vclaw, Gordonfu, Ewlyahoocom, Intgr, Srleffler, Rlee1185, Smithbrenon, Chobot, Abarenbo, Nehalem, YurikBot, Stephenb, Gaius Cornelius, Kkmurray, Dan Austin, SmackBot, Twistedcritique, J-beda, The Photon, EncMstr, Mithaca, DHN-bot~enwiki, A. B., Dce194, Pyo, JzG, Dicklyon, Nikvist, Hu12, DabMachine, Iridescent, BeenAroundAWhile, Hyperdeath, Meno25, Waxigloo, Thijs!bot, AMCDawes, Martin Hedegaard, Funny-phani, MichaelMaggs, Escarbot, Austin Maxwell, Zylorian, Spencer, Jabeles, JAnDbot, MER-C, Crmrmurphy, Swpb, Laserboy1969, *smb, Dima373, R'n'B, CommonsDelinker, ArcAngel, Ctroy36, Tgeairn, Sr903, JCarlos, Rod57, Afluegel, CUDOS, Celos, WinterSpw, Squids and Chips, VolkovBot, Chanetsa, Phfromspie.org, TXiKiBoT, Pearsallt, Rdsherwood, Mawkish1983, Chuck Sirloin, Barkeep, Stgean53, Photonicscenter, Ptr123, Joel2013, Maxq 2006, ClueBot, 7Piguine, DanielleJ, Edknol, Rubin joseph 10, A3camero, Renamed user 3, Hdorren, Adrory, Addbot, Mseyfang, Cpia, Fieldday-sunday, Chamal N, AndersBot, SpBot, Lightbot, OlEnglish, Luckas-bot, Yobot, AnomieBOT, Redarmy101, Piano non troppo, 2001hal9000, Materialscientist, Corrigendas, Xqj37, Jhbdel, Sheeson, Much noise, Photonicsuka, Pinethicket, Sydney20, Dgiltner, Opticalgirl, Lotje, JackFrozen, Mannaro85, WikitanvirBot, Beatnik8983, ZéroBot, Quantumavik, Donner60, Damirgraffiti, Petrb, ClueBot NG, IOPhysics, Widr, Metricopolus, Jayashri fegade, S Larctia, Omegaoptical, Durdham, Pratyya Ghosh, JYBot, LalahGrace, Aymankamelwiki, Lugia2453, ScienceofLight, Djoka.panama, Lagoset, Bigdaddybrabantio, RadermacherSKT, KasparBot and Anonymous: 160

- **Electromagnetic radiation** *Source:* https://en.wikipedia.org/wiki/Electromagnetic_radiation?oldid=735687668 *Contributors:* AxelBoldt, Tobias Hoevekamp, Bryan Derksen, Timo Honkasalo, The Anome, AdamW, Youssefsan, Fredbauder, PierreAbbat, Ray Van De Walker, DrBob, TomCerul, Heron, Olivier, Stevertigo, Lir, Patrick, Tim Starling, LenBudney, Gabbe, Looxix~enwiki, Ahoerstemeier, William M. Connolley, Den fjättrade ankan~enwiki, Julesd, Glenn, Jeandré du Toit, Mxn, Smack, Pizza Puzzle, Reddi, Nv8200pa, SEWilco, Omegatron, Phoebe, EikwaR, Denelson83, Phil Boswell, Donarreiskoffer, Robbot, Hankwang, Agilulfe~enwiki, Blainster, Wikibot, Wereon, Aetheling, Ramir, Enochlau, Srtxg, Wjbeaty, Giftlite, Snags, Peruvianllama, Anville, Bensaccount, Ssd, AJim, Saaga, Bobblewik, Edcolins, Louis Labrèche, Utcursch, Clinton reece, Antandrus, Aulis Eskola, Karol Langner, Rdsmith4, Icairns, Mozzerati, Craig Currier, Jcw69, Buickid, Deglr6328, Adashiel, Canterbury Tail, The stuart, Maestrosync, Mike Rosoft, Discospinster, Guanabot, Pak21, Vsmith, Jpk, CODOR, Mykhal, Kbh3rd, Kaisershatner, Dkroll2, El C, Anphanax, Rgdboer, Hayabusa future, Laurascudder, Edward Z. Yang, Bobo192, Smalljim, I9Q79oL78KiL0QTFHgyc, Sparkgap, Nk, Franl, Deryck Chan, Ranveig, Storm Rider, Msh210, Alansohn, Jamyskis, The RedBurn, Atlant, Snowolf, KJK::Hyperion, Wtmitchell, Kdau, RainbowOfLight, DV8 2XL, Akidd dublin, Mpatel, Prashanthns, Graham87, Magister Mathematicae, Abach, Chun-hian, Sjö, Rjwilmsi, JVz, Pleiotrop3, The wub, Bhadani, Oliverkeenan, FlaBot, Michaelbluejay, Ysangkok, Gurch, Fresheneesz, Lmatt, Srleffler, Chobot, DVdm, The Rambling Man, YurikBot, Wavelength, Crotalus horridus, Sceptre, Arado, Bhny, JabberWok, Kerowren, Rintrah, Salsb, SEWilcoBot, Grafen, Jpowell, Killdevil, Scottfisher, Bota47, Kkmurray, Ms2ger, Tigershrike, Light current, Enormousdude, 2over0, C h fleming, Sagsaw, Closedmouth, Zerodamage, Modify, JoanneB, Sizarieldor, GrinBot~enwiki, Serendipodous, Mejor Los Indios, Sbyrnes321, Veinor, SmackBot, Melchoir, Pgk, C.Fred, The Photon, KocjoBot~enwiki, Delldot, Binarypower, HalfShadow, Yamaguchi先生, Gilliam, Dauto, Simsea, Rmosler2100, Chris the speller, Keegan, MK8, Complexica, The Rogue Penguin, VincenzoAmpolo~enwiki, Octahedron80, DHN-bot~enwiki, Sbharris, Can't sleep, clown will eat me, JustUser, Chlewbot, Pax85, Shadow1, Drphilharmonic, Hammer1980, Mwtoews, Daniel.Cardenas, Tfl, DJIndica, Nmnogueira, SashatoBot, Lambiam, Chazchaz101, Vasu123, Tefnut~enwiki, Calum MacÙisdean, JorisvS, Bjankuloski06en~enwiki, Melody Concerto, Jmorkel, Noah Salzman, Alethiophile, Topazg, Rogerbrent, Dicklyon, Macellarius, Doczilla, Jose77, Caiaffa, Shezhenting, Smin0, R~enwiki, GDallimore, IanOfNorwich, Jp0186, Rgjm, Tawkerbot2, JRSpriggs, G-W, Chetvorno, JForget, Sakurambo, Armin T, GHe, WeggeBot, Eecon, Bvcrist, FIL (usurped), Raomap, Yeanold Viskersenn, Odie5533, Aajaja, Doug Weller, Christian75, Dchristle, NMChico24, Omicronpersei8, RDates, Thijs!bot, Epbr123, Barticus88, Mbell, Yy-bo, Martin Hogbin, N5iln, Headbomb, KSSA, Marek69, Nick Number, Rriegs, SvenAERTS, Handface, David D., KrakatoaKatie, AntiVandalBot, Luna Santin, Seaphoto, QuiteUnusual, RapidR, Chill doubt, Rico402, Lfstevens, Andreazy, Arx Fortis, Golgofrinchian, IrishFBall32, JAnDbot, MER-C, Andonic, 100110100, Acroterion, Bongwarrior,

19.7. TEXT AND IMAGE SOURCES, CONTRIBUTORS, AND LICENSES

VoABot II, Jetstreamer, JNW, JamesBWatson, Mclay1, Nyttend, Catgut, Allstarecho, V 1993, InvertRect, RisingStick, Ashishbhatnagar72, NatureA16, Oren0, Hdt83, MartinBot, Mermaid from the Baltic Sea, Rettetast, Federico Benitez Conte, Kostisl, Kateshortforbob, Lcabanel, Nono64, J.delanoy, Troyboy53, Uncle Dick, Javawizard, Kar.ma, Eliz81, Extransit, StalinsLoveChild, Cpiral, St.daniel, Darkspots, McSly, Tarotcards, Gurchzilla, NewEnglandYankee, ARTE, Mufka, Fylwind, Atropos235, KylieTastic, Cometstyles, Ibrasg, Sheliak, Sokratesla, Black Kite, VolkovBot, Thedjatclubrock, AlnoktaBOT, Philip Trueman, Yakitoriman, TXiKiBoT, GLPeterson, Paine, Nxavar, Qxz, Seraphim, Bass fishing physicist, Jackfork, Mishlai, Wiae, WikiCantona, Venny85, Stamulevich, MADe, Andy Dingley, Lamro, Editorpark, John David Wright, Rhopkins8, Dianneknight, Goodwill289, Bsayusd, Logan, EmxBot, Deconstructhis, Ratsbew, The Random Editor, EJF, SieBot, Spammerman, Demologian, Scarian, WereSpielChequers, Jim E. Black, Jauerback, Gerakibot, Tigerdragon, The way, the truth, and the light, LeadSongDog, Likebox, Flyer22 Reborn, Paolo.dL, JSpung, Antonio Lopez, Hatster301, Dtvjho, Iknowyourider, Thinghy, Mike2vil, Mygerardromance, Dust Filter, BentzyCo, Denisarona, Randy Kryn, Sbacle, Lascorz, Jlc0023, Rickcandell, ClueBot, LAX, Trojancowboy, Binksternet, Vladkornea, PipepBot, Snigbrook, The Thing That Should Not Be, Amen316, Thubing, Kanhef, CrazieXninja, Catintehbox, Razimantv, Boing! said Zebedee, Ravirathore1984, Niceguyedc, VandalCruncher, Yongy, Vql, DragonBot, Djr32, Excirial, Jusdafax, Eeekster, Abrech, Rubin joseph 10, SpikeToronto, Brews ohare, PhySusie, M.O.X, Kaiba, Aitias, Jonverve, Amaltheus, CorpITGuy, SoxBot III, Anon126, Editorofthewiki, Jytdog, BodhisattvaBot, Dthomsen8, Nicoguaro, Shieber, SilvonenBot, WikiDao, ZooFari, Shikasannin, HexaChord, Addbot, Eric Drexler, Some jerk on the Internet, Dharmendra srivastva, Tcncv, Betterusername, Ukberry, Ronhjones, TutterMouse, Njaelkies Lea, Fieldday-sunday, Ironholds, Laurinavicius, GyroMagician, Churibo, Redheylin, Epzcaw, AndersBot, Favonian, Doniago, LinkFA-Bot, 5 albert square, Tide rolls, Lightbot, Teles, Meisam, Genius101, Legobot, Luckas-bot, Yobot, THEN WHO WAS PHONE?, KamikazeBot, Linktex, Tempodivalse, AnomieBOT, Jim1138, Galoubet, Bluerasberry, Materialscientist, Legofreak2008, Citation bot, ArthurBot, Xqbot, Phazvmk, Cureden, Romanfall, Capricorn42, Emezei, Nickkid5, The Original Economist, Br77rino, Loiskristellemum, GrouchoBot, Redpanda900, RibotBOT, Victamonn, Www.ca, Mathonius, Maplestory101, Doulos Christos, Sophus Bie, January2009, Shadowjams, Erik9, Kierkkadon, Niaoulibloodelf, Earwax09, Prari, ImaFirinMaLazor, Lookang, Amilnerwhite, Vuldoraq, Steve Quinn, JameKelly, Austria156, HamburgerRadio, Citation bot 1, Alipson, Redrose64, Pinethicket, MJ94, A8UDI, Jschnur, MondalorBot, Serols, Merlion444, December21st2012Freak, Jauhienij, Ronak abna, Trappist the monk, Yunshui, Sumone10154, J-p-fm, Reaper Eternal, Suffusion of Yellow, Reach Out to the Truth, Marie Poise, 360flip360, Onel5969, RjwilmsiBot, TjBot, Rollins83, EmausBot, John of Reading, Orphan Wiki, Acather96, Bio watcher, WikitanvirBot, 8lak3st3r, Immunize, Poi830, Racerx11, Yt95, Sp33dyphil, CoincidentalBystander, Tommy2010, Netheril96, Wikipelli, Dcirovic, Bdjwww, Thedoctor123, Susfele, 1howardsr1, Danilomath, Quinnd16, AManWithNoPlan, OrdinaryFattyAcid, Sky380, Otuguldur, Maxrokatanski, L Kensington, ꜟꜟꜟꜟ~enwiki, Donner60, Zueignung, ChuispastonBot, RockMagnetist, ResearchRave, ClueBot NG, Twihard123, Lcdrovers, Encycloshave, Waistlesselk, Frietjes, Braincricket, Mesoderm, HazelAB, Marechal Ney, Sina-chemo, Go Phightins!, Widr, MerlIwBot, Diyar se, Helpful Pixie Bot, Calabe1992, Bibcode Bot, DBigXray, Lowercase sigmabot, BG19bot, Ryanross43, Vivek Verma 38, AvocatoBot, Robert the Devil, TROPtastic, DARIO SEVERI, Sparkie82, YVSREDDY, Zedshort, Shaun, Tomohama, BattyBot, LeviathanPMS, ChrisGualtieri, BlazeNinja418, Khazar2, MSUGRA, Dániel I fiz, Gdrg22, JYBot, TopGarbageCollector, Agumonkey, BrightStarSky, Dexbot, Tharvey100, AyaLovesAmjad, Webclient101, CuriousMind01, Mark viking, Epicgenius, Secondhand Work, Sevınti faıv, Mrsquirrel dh, Teeth69, Kharkiv07, Ugog Nizdast, Jordanvandijk, DavRosen, Gillemc, Hanthoec, Sam Sailor, Jianhui67, SpecialPiggy, Param Mudgal, Linuxjava, CharlesIJ1948, Lakun.patra, Johansen.fred, Csutric, Monkbot, Yikkayaya, Starteller, AKS.9955, Gauracho, Vieque, Joeleoj123, Quarter2002, Nelsonaugust3, Amortias, Tris1313, Indranil1993, WyattAlex, Anmikmore, Easy Secrets, SageGreenRider, Electric Toast, Telkomsel013, GeneralizationsAreBad, Username123123123, KasparBot, Jmc76, Eat me, I'm an azuki, Chrisemblhh, Loudandrews123, Lantolar, Redzemp, Lm800001064, 1416domination, Asma.sharief.aanhtt, UrveshPatel741, GABRIEL600$13, Jsb71, Fmadd and Anonymous: 926

- **Photoelectric effect** Source: https://en.wikipedia.org/wiki/Photoelectric_effect?oldid=735323570 Contributors: AxelBoldt, Brion VIBBER, Bryan Derksen, Rjstott, Css, XJaM, Fredbauder, William Avery, DrBob, Heron, Ewen, Stevertigo, Michael Hardy, Looxix~enwiki, Ahoerstemeier, Mac, Stevenj, William M. Connolley, Theresa knott, Julesd, Glenn, AugPi, Tristanb, Pizza Puzzle, Hashar, Emperorbma, Reddi, Itai, Omegatron, Pstudier, Robbot, Fredrik, Npettiaux, Chris 73, Chancemill, Hadal, SpellBott, Enochlau, Giftlite, Ferkelparade, Bensaccount, Glenn Koenig, ConradPino, Csmiller, Karol Langner, Unquantum, Mikko Paananen, Zfr, Urhixidur, Joyous!, Deglr6328, M1ss1ontomars2k4, Brianjd, Twinxor, Bosteen, Pjacobi, Roo72, Pavel Vozenilek, Nabla, Mwanner, Gilgamesh he, Robarnler, Sietse Snel, David kitson, Honeycake, Atlant, Axl, Pion, Vedantm, Wtshymanski, RainbowOfLight, Lerdsuwa, Forteblast, Dennis Bratland, Linas, StradivariusTV, Kzollman, Urod, Lensovet, Eilthireach, SeventyThree, Palica, Jan.bannister, Li-sung, Volland, Saperaud~enwiki, Rjwilmsi, Zbxgscqf, Vegaswikian, Nneonneo, InFairness, Sferrier, Ems57fcva, Sjlegg, Yamamoto Ichiro, Anurup, Arnero, Doc glasgow, RexNL, Srleffler, Chobot, DaGizza, DVdm, YurikBot, Wavelength, Sceptre, Jimp, Wolfmankurd, Lucinos~enwiki, CambridgeBayWeather, NawlinWiki, Wiki alf, Grafen, Kdkeller, E2mb0t~enwiki, STufaro, Ospalh, Kkmurray, Ott2, Georgewilliamherbert, Light current, Lt-wiki-bot, Imaninjapirate, Ageekgal, Petri Krohn, GraemeL, CWenger, Cjfsyntropy, Junglecat, SmackBot, InverseHypercube, KnowledgeOfSelf, Joonhon, Melchoir, Nickst, Delldot, Gary2863, Yamaguchi⁈⁈, Gilliam, JSpudeman, Saros136, SMP, Jprg1966, SchfiftyThree, Droll, Complexica, KJie.Neo, CMacMillan, VirtualSteve, ThePromenader, Chlewbot, TheKMan, GeorgeMoney, PsychoCola, Waprap, Cybercobra, Nakon, James084, A5b, Sadi Carnot, Chymicus, The undertow, SashatoBot, Zahid Abdassabur, Kuru, John, Loodog, OutSales, JorisvS, Mgiganteus1, Cielomobile, Ben Moore, Dicklyon, ChadyWady, Tawkerbot2, CmdrChon, Tarchon, Mattbr, Van helsing, Laplacian, Harriemkali, Tex, Kanags, A876, Gravitroid, Jedonnelley, Michael C Price, Tawkerbot4, Roberta F., Branclem, Thijs!bot, Poorleno, Mojo Hand, Headbomb, West Brom 4ever, Tapir Terrific, Brichcja, Hcobb, D.H, Escarbot, AntiVandalBot, KP Botany, Mrshaba, LibLord, Rico402, JAnDbot, Montparnasse, IJMacD, Gumby600, Magioladitis, Pedro, Bongwarrior, VoABot II, Delta107~enwiki, WODUP, Giggy, Dirac66, Cpl Syx, Khalid Mahmood, Gwern, Gjd001, MartinBot, Nono64, J.delanoy, Trusilver, Yabbadabbadoo, AstroHurricane001, Numbo3, Mostlymostly, Tdadamemd, McSly, AntiSpamBot, Luke FM, Uberdude85, Treisijs, Deor, Tourbillon, Meaningful Username, Coldplasma, Philip Trueman, TXiKiBoT, The Original Wildbear, Red Act, Calwiki, FDominec, Nxavar, Anna Lincoln, Martin451, Jackfork, Tarun06071987, UnitedStatesian, Zain Ebrahim111, Entropy1963, Andy Dingley, Meters, Enviroboy, Junkinbomb, Planet-man828, Neparis, SieBot, AquaDTRS, YonaBot, Krawi, Matthew Yeager, The way, the truth, and the light, JerrySteal, Dattebayo321, Happysailor, Proofhand, Janfri, CultureDrone, Tesi1700, Dolphin51, Sfan00 IMG, ClueBot, Rumping, Binksternet, Yaleks, The Thing That Should Not Be, Mriya, Frdayeen, Niceguyedc, DragonBot, Djr32, Jamespitt, ChrisHodgesUK, DumZiBoT, XLinkBot, BodhisattvaBot, Azaz129, Avoided, Mifter, Padfoot79, Jht4060, SkyLined, Truthnlove, Tayste, Addbot, Yousou, ProjectTux, CarsracBot, Debresser, Norman21, Numbo3-bot, Tide rolls, Lightbot, QuadrivialMind, Gail, Legobot, Luckas-bot, Yobot, Tamtamar, THEN WHO WAS PHONE?, AmeliorationBot, KamikazeBot, Mhmolitor, Tempodivalse, AnomieBOT, Galoubet, EryZ, Materialscientist, Citation bot, Soldarat, ArthurBot, Xqbot, Capricorn42, Pandaninja91, Idegmcsa, Turk oğlan, GrouchoBot, Sujitmahj, SobaNoodleForYou, ⁈⁈, FrescoBot, Cwtiyar, Paine Ellsworth, VI, Amadeus666, Dr John Kalien, Endofskull, Steve Quinn, Left Coast Bernard, Dscraggs, Metastabil01, Louperibot, Citation bot 1, Pinethicket, I dream of horses, Serols, Clivebeale, Boobarkee, Jeans.a, Earthandmoon, Marie Poise, Mean as custard,

RjwilmsiBot, Rajettan, EmausBot, Dewritech, Racerx11, Mitartep, Kiran Gopi, Old nic, Hhhippo, M1arvin, Harddk, Druzhnik, AManWithNoPlan, Wayne Slam, Looscan, RaptureBot, Donner60, Surajt88, Damirgraffiti, Orange Suede Sofa, ChuispastonBot, RockMagnetist, ClueBot NG, Elcubano91, Lmoding, A520, SusikMkr, Crlance, Prateek.sondhi, Fkhwang, SolidStateDD, NuclearEnergy, Helpful Pixie Bot, Mulhollant, Wbm1058, Bibcode Bot, BG19bot, Shesinastro, KlausWilhelm, Jarad619, Zedshort, Shawn Worthington Laser Plasma, TheGoodBadWorst, Kisokj, Abdulllahumar, Cyberbot II, ChrisGualtieri, JYBot, Sflintg, Dexbot, Erinepwright, Lugia2453, Mohpof, Me, Myself, and I are Here, Kptech, NeapleBerlina, SJ Defender, Torqu3e, Diegonolovich, Sreynoldsbros2, SkateTier, HiYahhFriend, Kanawishi, Sofia Koutsouveli, Trackteur, Pentazoid, Jarodtp3, LemarqueSadler, TerryAlex, Laxman z, LauraIsabelleDB, Kush bansal, Saltimoore, Katgirl2000, KasparBot, Jmc76, Aaseeshdatla, Frvfrferv, Jeffwu64, Shaik sunain, K;jassfhshfkfsh;, GreenC bot, Sadeesha7, Bender the Bot, Kolbyiskolbo34 and Anonymous: 560

- **Wave–particle duality** Source: https://en.wikipedia.org/wiki/Wave%E2%80%93particle_duality?oldid=735759857 Contributors: AxelBoldt, Tobias Hoevekamp, Derek Ross, MarXidad, The Anome, Manning Bartlett, Wayne Hardman, Andre Engels, Josh Grosse, Miguel~enwiki, ChangChienFu, DrBob, Heron, KF, Stevertigo, Michael Hardy, Gabbe, Dgrant, William M. Connolley, Pizza Puzzle, Hike395, Charles Matthews, Timwi, Geoff, Reddi, El~enwiki, ErikStewart, Timc, Maximus Rex, Populus, BenRG, Francs2000, Phil Boswell, SJRubenstein, Robbot, Hankwang, Cdang, Owain, Fredrik, Chris 73, Bkalafut, Sverdrup, Roscoe x, Anthony, Giftlite, BenFrantzDale, Lethe, AJim, Jason Quinn, Jorge Stolfi, Quamaretto, Eequor, Pcarbonn, Antandrus, Unquantum, Lumidek, Chadernook, Adashiel, Trevor MacInnis, Eep^2, Chris Howard, Rich Farmbrough, Filthybutter, Hidaspal, Vsmith, Mjpieters, Bender235, Cyclopia, JustinWick, Pt, El C, Laurascudder, RoyBoy, Spoon!, Army1987, Enric Naval, Atlant, Ricky81682, AzaToth, Fritzpoll, PAR, BRW, Jheald, Gene Nygaard, Afshar, Oleg Alexandrov, Velho, Woohookitty, Linas, Davidkazuhiro, Zealander, Pol098, Ruud Koot, Cbdorsett, Ch'marr, Mandarax, BD2412, Qwertyus, Thierry Dugnolle~enwiki, Drbogdan, Rjwilmsi, KYPark, Strait, Jmcc150, Cjpuffin, Arnero, Musical Linguist, Nihiltres, Alfred Centauri, Fresheneesz, Srleffler, Kri, Snailwalker, CiaPan, Chobot, DVdm, Siddhant, Vyroglyph, YurikBot, Wavelength, X42bn6, Wolfmankurd, Hede2000, Gaius Cornelius, Salsb, Schlafly, Expensivehat, SCZenz, Jb849, Danlaycock, Chichui, Zwobot, Ospalh, Wknight94, Ott2, Rwxrwxrwx, Smkolins, Light current, Enormousdude, StuRat, Oysteinp, Keithd, Mavaddat, Sbyrnes321, That Guy, From That Show!, Luk, KnightRider~enwiki, SmackBot, Ashenai, Dmccaig, Joonhon, Delldot, Eskimbot, Gilliam, Hmains, Rrscott, Melburnian, Complexica, DHN-bot~enwiki, Colonies Chris, Fiziker, Duncombe, Lesnail, Kittybrewster, ThreeAnswers, King Vegita, Rich.lewis, Turms, DMacks, Bentreuherz~enwiki, SciBrad, Lambiam, ArglebargleIV, Petr Kopač, Feraudyh, Jpawloski, SMasters, Phancy Physicist, Dicklyon, Spiel496, Hetar, Wizard191, Newone, IRevLinas, Pathosbot, Garrettcobb, JRSpriggs, Chetvorno, Hairyfairycarpetfluff, CRGreathouse, BeenAroundAWhile, Linus M., John courtneidge, LPerson, Myasuda, Act333, Yzphub, Krauss, Nick Y., Bvcrist, Peterdjones, Michael C Price, Viridae, Abtract, Raoul NK, Letranova, Thijs!bot, Michael D. Wolok, Mbell, Headbomb, JustAGal, GordonRoss, Thadius856, Nipisiquit, JAnDbot, MER-C, .anacondabot, VoABot II, SHCarter, Docduke, AllenDowney, Tserton, Keith D, R'n'B, Andrej.westermann, Rieffel, J.delanoy, AstroHurricane001, Maurice Carbonaro, Kevin aylward, TimLong2001, Azus~enwiki, Hakufu Sonsaku, Tarotcards, Aqm2241, Fountains of Bryn Mawr, Laurenpass, SirHolo, Lseixas, TraceyR, Sheliak, Club house, Maniaphobic, DParlevliet, AlnoktaBOT, Macspaunday, Philip Trueman, Revilo314, TXiKiBoT, The Original Wildbear, Red Act, Tom239, Voorlandt, RedAndr, Robert1947, Kpedersen1, Rshob, YohanN7, SieBot, ShiftFn, Hertz1888, Ergateesuk, Caltas, Yuefairchild, Julianva, Anfieldman, Android Mouse, Flyer22 Reborn, Oxymoron83, Janfri, Szalaglora, Hamiltondaniel, Randy Kryn, Velvetron, ImageRemovalBot, MenoBot, Martarius, ClueBot, Binksternet, Razimantv, Mild Bill Hiccup, Isirr, DanielDeibler, Djr32, TheUNOFFICIALvandalpolice, Tyler, NuclearWarfare, Iohannes Animosus, Zilliput, MelonBot, YouRang?, Mchaddock, James Kanjo, Ost316, Wyatt915, Addbot, DOI bot, Captain-tucker, Moshamoot, WMdeMuynck, CarsracBot, AnnaFrance, Favonian, LinkFA-Bot, ATOE, Luckas-bot, Yobot, TaBOT-zerem, THEN WHO WAS PHONE?, Silca678, AnomieBOT, Piano non troppo, EryZ, Andaza, Danno uk, Citation bot, KT-2500, Peterdx, J JMesserly, Vonharris, RibotBOT, Flaviusvulso, LazyMapleSunday, ??, Chjoaygame, Tank.hasmukh, Machine Elf 1735, Ysyoon, Citation bot 1, Jonesey95, Vanzac11, Cnwilliams, Trappist the monk, Wdanbae, Vrenator, Fivedoughnut, RjwilmsiBot, Jonlegere, EmausBot, John of Reading, Dewritech, Tommy2010, Dcirovic, Slawekb, AvicBot, 1howardsr1, H3llBot, Fizicist, Chrisman62, Maschen, Bulwersator, Chewings72, Sudozero, Will Beback Auto, ClueBot NG, Jostikas, Andybiddulph, Widr, Helpful Pixie Bot, Verberate, Jubobroff, Bibcode Bot, Bm gub2, BG19bot, Guy vandegrift, Vokesk, PhnomPencil, MusikAnimal, F=q(E+v^B), Jivey81, Elemenat, BattyBot, Samanthaclark11, Mdann52, Thojuf, Dexbot, Hmainsbot1, Purplearc, CuriousMind01, Nishantarya98, Pdecalculus, Iztwoz, Perseus.3,14, Jsresearch, My name is not dave, Mdominguez611, Johnfranciscollins, Monkbot, PlaidPolarity, Master Pok, Rhynhardtk, Pasten2, Isambard Kingdom, Softsay, Vespro Latuna, Knife-in-the-drawer, DireNeed, Shane Gregor Ducharme, Ezra Kirkpatrick and Anonymous: 371

- **Squeezed coherent state** Source: https://en.wikipedia.org/wiki/Squeezed_coherent_state?oldid=735652719 Contributors: Jordi Burguet Castell, Sanders muc, Sho Uemura, Lumidek, Rich Farmbrough, Matt McIrvin, Keenan Pepper, Alai, Falcorian, Oleg Alexandrov, Bobrayner, Linas, Gerd Breitenbach, BD2412, Nanite, Vegaswikian, Srleffler, Gaius Cornelius, SmackBot, Chris the speller, TimBentley, Hadmack, Second Quantization, Nick Number, Steelpillow, JamesBWatson, Cuzkatzimhut, Deciwill, SchreiberBike, DumZiBoT, Phys0111, Addbot, AgadaUrbanit, Yobot, AnomieBOT, JackieBot, Xqbot, Mchcmc963369, Thinking of England, Sijothankam, Quantanew, Dewritech, Quantumavik, Mattedia, RockMagnetist, NULL, Gilderien, Bibcode Bot, BG19bot, Nje1987, Hctrmycss, Ashraf Katnah, Chengkai Zhang, Geek3, AntonyRichardLee, Jbohnet and Anonymous: 29

- **Uncertainty principle** Source: https://en.wikipedia.org/wiki/Uncertainty_principle?oldid=731110355 Contributors: AxelBoldt, LC~enwiki, CYD, Bryan Derksen, Zundark, MarXidad, Slrubenstein, Mirwin, XJaM, Miguel~enwiki, Roadrunner, SimonP, Fredb, Youandme, Stevertigo, Edward, Bdesham, JohnOwens, PhilipMW, Michael Hardy, Qaz, TakuyaMurata, Fwappler, Karada, JeremyR, SebastianHelm, Looxix~enwiki, Stevenj, J-Wiki, Theresa knott, JWSchmidt, Julesd, Glenn, Victor Gijsbers, Kimiko, Nikai, Iorsh, Schneelocke, Tom Peters, Disdero, Charles Matthews, Joshk, Patrick0Moran, Taxman, Fibonacci, Phys, Omegatron, Shizhao, Fvw, Phil Boswell, Vt-aoe, Robbot, Owain, Sverdrup, Rursus, Texture, DHN, ElBenevolente, JerryFriedman, Tea2min, Giftlite, Netoholic, Lethe, Fastfission, Everyking, No Guru, Dratman, Alison, Christopherlin, Chowbok, Geni, R. fiend, Beland, Karol Langner, CSTAR, DragonflySixtyseven, Erik Garrison, TobinFricke, Mschlindwein, JohnArmagh, BrianWilloughby, Mh, Chris Howard, DanielCD, Dablaze, Pyrop, Noisy, 4pq1injbok, Rich Farmbrough, TedPavlic, Pjacobi, Ericamick, Zazou, Bender235, El C, NTiOzymandias, Kwamikagami, Vinsci, Laurascudder, Edward Z. Yang, RoyBoy, Theshowmecanuck, Kine, Bobo192, Army1987, Irrbloss, Clawson, I9Q79oL78KiL0QTFHgyc, Slicky, Kjkolb, Rje, Helix84, SPUI, Danski14, Alansohn, BlackJava, Ashley Pomeroy, WhiteC, PAR, GeorgeStepanek, Cortonin, Velella, RJFJR, DV8 2XL, Ssammu, Galaxiaad, Kenyon, Ott, Linas, Localh77, ToddFincannon, Kzollman, GregorB, Btyner, Marudubshinki, RuM, Mandarax, Graham87, Thierry Dugnolle~enwiki, Rjwilmsi, Coemgenus, Joel D. Reid, Tawker, Mike Peel, Bubba73, FlaBot, Xaque, Mathbot, Gparker, Elmer Clark, Goeagles4321, Fresheneesz, Aeroknight, EronMain, Sodin, Srleffler, Snailwalker, Theshibboleth, Chobot, Benlisquare, Bgwhite, YurikBot, Wavelength, Maelin, Anuran, Charles Gaudette, RussBot, JabberWok, Jengelh, SpuriousQ, JohnJSal, Gaius Cornelius, Varnav, RadioKirk, CarlHewitt, Robertvan1, Captain Yesterday, Harksaw, Lepi-

19.7. TEXT AND IMAGE SOURCES, CONTRIBUTORS, AND LICENSES

doptera, SCZenz, Mike Lin, Fs, Chichui, Syrthiss, Bota47, Dna-webmaster, Light current, Enormousdude, Wtlegis, Deville, 2over0, Confuted, Zzuuzz, Oysteinp, Closedmouth, Tabish q, Mumuwenwu, Ephilei, RG2, Teply, GrinBot~enwiki, Segv11, Sbyrnes321, Tom Morris, That Guy, From That Show!, Treesmill, SmackBot, InverseHypercube, Melchoir, KocjoBot~enwiki, Motorneuron, BugoK, Gint'ire, Edgar181, Skizzik, Chris the speller, Bduke, MK8, Jprg1966, Oli Filth, MalafayaBot, Timneu22, Complexica, Dustimagic, Nbarth, Adibob, DHN-bot~enwiki, Sbharris, Can't sleep, clown will eat me, Pasachoff, Tschwenn, Yidisheryid, Voyajer, Clinkophonist, Landon, Jmnbatista, Soosed, Wen D House, Khukri, Kntrabssi, Jimbatka, TheMaster42, Jon Awbrey, Sadi Carnot, Vina-iwbot~enwiki, Kukini, Andrei Stroe, Tesseran, Jonnty, DJIndica, SashatoBot, Lambiam, Nathanael Bar-Aur L., Harryboyles, Richard L. Peterson, UberCryxic, Mike1901, Anthony Almighty, Mgiganteus1, Zarniwoot, Neathway, Treyp, Kvng, Quaeler, Dead3y3, Simon12, Iridescent, Clarityfiend, UncleDouggie, Woodshed, Tawkerbot2, DKqwerty, Phil Christs, Joechuck, Ale jrb, Sir Vicious, Memetics, CBM, BKalesti, ShelfSkewed, WeggeBot, Gregbard, Quackquackquack, Cydebot, Ash.furrow, WillowW, Gtxfrance, Gagueci, Damiancorrigan, Lugnuts, Miguel de Servet, David edwards, Michael C Price, M a s, Waxigloo, Gonzo fan2007, Maziotis, Epbr123, Barticus88, Mbell, Rcq, Mojo Hand, Headbomb, Marek69, Bruce H. McCosar, Bobblehead, FelixP~enwiki, Bulldogsully, Second Quantization, Hcobb, Duncan McB, CharlotteWebb, Gregoryloyse, Dawnseeker2000, Hryun, Mentifisto, AntiVandalBot, Horal~enwiki, Boravi, Orionus, Notaxi, Promsi, Doc Tropics, Farringo, Bikkan, Jantoy, Fermant, Darklilac, Tim Shuba, Themark, Spartaz, Mmyotis, Byrgenwulf, Qwerty Binary, Kainino, Austinsanity, JAnDbot, Riesz, Cvkline, Dp76764, Charlesrkiss, Wikidudeman, SHCarter, Tajik24, Couki, Dirac66, DerHexer, Edward321, Infovarius, Kyle824, ClansOfIntrigue, JCraw, TechnoFaye, Zouavman Le Zouave, J.delanoy, Richiekim, Maurice Carbonaro, Kevin aylward, Mike.lifeguard, Extransit, George963 au, McSly, Petersec, Janus Shadowsong, Tarotcards, DjScrawl, 123qwertyasdf, NewEnglandYankee, Teol~enwiki, Policron, 83d40m, Sarregouset, Samlyn.josfyn, Lseixas, Kgoff, Madblueplanet, Sheliak, Cuzkatzimhut, Malik Shabazz, Deor, DParlevliet, LokiClock, AlnoktaBOT, Thurth, ContributorX, The Original Wildbear, Davehi1, Mieszko the first, Michael H 34, Perrymc, Srinivasvuk, Wiae, Lejarrag, Madhero88, Kpedersen1, Spiral5800, Greswik, Entropy1963, Jakerussell777, Windigo65, Vsst, Kbrose, Quietbritishjim, SieBot, WereSpielChequers, Peeter.joot, Djayjp, Likebox, Eurion, ScottColemanKC, Jmwise, Prestonmag, Thehotelambush, Jdaloner, KathrynLybarger, Sanya3, Hatster301, Autumn Wind, OKBot, Zdilli, Frank121629, Neurophysics, Slapazoid, StewartMH, Myrvin, InceInce, Martarius, ClueBot, SummerWithMorons, Pateblen, Badger Drink, Pakaraki, Products1234567, VsBot, Edfredkin, Franamax, Der Golem, Alisandi~enwiki, Furtherpale, Schuermann~enwiki, LizardJr8, Mike0001, GregIngram, Somno, DragonBot, That john chap, Winston365, Wiki libs, Jotterbot, Vojtech.sidorin, Maheshexp, SchreiberBike, Jonverve, Straw07, Chadoh, Sparklicious, Shamanchill, Finalnight, YouRang?, SHostetler, XLinkBot, Jossysayir, Anthony Nolan O'Nymous, Onlinereal, MarcM1098, Mmaman12, Stephen Poppitt, Addbot, Deepmath, Mortense, Dunhere, DOI bot, Betterusername, Gregz08, WMdeMuynck, Goddord, Aboctok, Kyle1278, FutureKingMark, LinkFA-Bot, AgadaUrbanit, Tassedethe, MeritBadge, Tide rolls, Lightbot, Ettrig, Zuxbeinyufoo, Legobot, Artichoke-Boy, Yobot, Enemyunknown, Ptbotgourou, The Earwig, Bosonichadron, AnomieBOT, AndrooUK, DemocraticLuntz, Rubinbot, Jim1138, AdjustShift, Materialscientist, Geekyguild, Citation bot, LilHelpa, Bdmy, Tomwsulcer, Gap9551, GrouchoBot, Omnipaedista, RibotBOT, A. di M., Constructive editor, Nixón, George585, FrescoBot, Tobby72, Zacksc, Sławomir Biały, Xhaoz, Red33410, Quantum 235, OgreBot, Citation bot 1, Pinethicket, I dream of horses, Abhi2005singh, Stpasha, RedBot, Serols, FratleyLionheart, IVAN3MAN, Meier99, Trappist the monk, DixonDBot, Beta Orionis, Manuelferreria, Jazz2y, Remuson, Keith Cascio, Wikifan798, Cplusplusian, Diannaa, WikiTome, Poizned, Shuvo915, Xphileprof, Mikejones2255, Tesseract2, EmausBot, Dewritech, Kierany5, Faolin42, GoingBatty, Slightsmile, Dcirovic, Slawekb, Thecheesykid, Hhhippo, ZéroBot, Fæ, Fintil, Ὁ οἶστρος, Quondum, Zetafun, Maschen, Zueignung, Tls60, Gjshisha, Puffin, BoredextraWorkvidid, Mr ket, Davikrehalt, RockMagnetist, Patrolboat, JamesRJohnston, Iketsi, Xronon, Unomio, Shi Hou, Isocliff, Rkyriakakis, ClueBot NG, PoincareHenri, Quandle, Tox1c harlequ1n, Parcly Taxel, Jorge Gomes Sharky, ThatAMan, Davidcarfi, Timflutre, Helpful Pixie Bot, Bibcode Bot, BG19bot, Absconded Northerner, Herpyderpy1234, MusikAnimal, Reebopareebop, Stormx39, Mark Arsten, Dwightboone, Drift chambers, Pattalur, QPhysics137, Toccata quarta, Metallic org, TubesUntil, Profesores, B1naryatr0phy, BattyBot, DLetourneau, Timothy Gu, ChrisGualtieri, Electricmuffin11, Dexbot, GyaroMaguus, Hablab, Colinddelia, Degaram, TJ-Grite, Rohitarura, Fidelledo, Katherine Pendleton, I am One of Many, GOHORN, Mearns1952, CensoredScribe, Penitence, Zenibus, Zilmino, AtticTapestry, FabForrest, Elytper9865, Anrnusna, Mathphysman, Dolzikid, Dfoverdx, 22merlin, Monkbot, Pokharel1996, RegistryKey, TheHecster, Samrajdwivedi, Benjir21, KasparBot, Stiglich, Theoretical.physicstdx, Miar32345, Datboi(me) and Anonymous: 709

- **De Broglie–Bohm theory** *Source:* https://en.wikipedia.org/wiki/De_Broglie%E2%80%93Bohm_theory?oldid=731905976 *Contributors:* LC~enwiki, The Anome, Roadrunner, David spector, Edward, Michael Hardy, Tim Starling, Anders Feder, J-Wiki, Loren Rosen, Charles Matthews, Timwi, Rednblu, The Anomebot, Zoicon5, Patrick0Moran, Mir Harven, Goethean, Sverdrup, Jfire, Mattflaschen, Giftlite, DocWatson42, Barbara Shack, Pretzelpaws, Michael Devore, Andycjp, Yafufidie, CSTAR, Togo~enwiki, Lumidek, Eyv, Chris Howard, Freakofnurture, Rich Farmbrough, Guanabot, Dbachmann, Swiftly, Dmr2, Bender235, ESkog, Floorsheim, Evand, Tjic, L33tminion, Scentoni, AshtonBenson, M0rph, Kuratowski's Ghost, Jason Davies, Andrewpmk, Plumbago, SlimVirgin, DV8 2XL, Falcorian, Linas, Kkchang, GregorB, Pfalstad, Rjwilmsi, MarSch, Smithfarm, Mike Peel, Vegaswikian, Vuen, TiagoTiago, Alfred Centauri, GangofOne, DVdm, NTBot~enwiki, KSchutte, Anomalocaris, Benja, Holon, Olleicua, DomenicDenicola, Tonymec, Cojoco, Profero, Alessandro70~enwiki, SmackBot, RDBury, Chris the speller, SuezanneC Baskerville, Emurphy42, John Reaves, Jefffire, RJN, Murf42, Ebitnet, Michael Rogers, L0rents, Scottie 000, Vgy7ujm, JorisvS, Extremophile, Noah Salzman, Hypnosifl, Kripkenstein, Dl2000, Stephen B Streater, Dan Gluck, Iridescent, Jambaugh, IRevLinas, Duduong, Agger~enwiki, Banedon, Editorius, Gregbard, Dragon's Blood, Cydebot, Peterdjones, Michael C Price, Christian75, Thijs!bot, Headbomb, Basilo, RogueNinja, Tomixdf, Dhemm, WolfmanSF, JamesBWatson, Duendeverde, DinoBot, Vandermude, Rickard Vogelberg, Lexivore, Were-Bunny, Dataweaver, The Anonymous One, Kevin aylward, AoS1014, Ross Fraser, Borat fan, Ummonk, Efgé, Itangalo, Voorlandt, Leafyplant, Venny85, Nsomnia03, SQL, W1k13rh3nry, Pierre-Alain Gouanvic, SiegeLord, Likebox, Arjen Dijksman, Weierstrass, RandomHumanoid, Sfwild, Felixaldonso, Phys-demystifier, Randy Kryn, A.C. Norman, Myrvin, ClueBot, Razimantv, Fjados, Skewyou, Mastertek, 1ForTheMoney, Alphatronic, Crowsnest, Addbot, Grayfell, DOI bot, Kalonymos, Lightbot, Quantumobserver, Legobot, Yobot, Linket, Wireader, Dickdock, AnomieBOT, Archon 2488, Materialscientist, Citation bot, Deadly Nut, Jostylr, Andersæøå, Jmundo, Patlatus, Cmdulya, Waleswatcher, Dvtausk, Shadowjams, FrescoBot, Tnorsen, Zicovich, Ilja Schmelzer, MistySpock, Trappist the monk, Jordgette, OlderIgor, ZRPerry, Tumulka, Slon02, Quantanew, AvicBot, ZéroBot, Mpc755, Quantholic, Jim A. Wilson, Clearlyfakeusername, Helpful Pixie Bot, Bibcode Bot, BG19bot, Bronzmash, Slippingspy, Mcarmier, Malyszkz, BattyBot, Plutoniumjesus, ChrisGualtieri, Dobie80, AnInformedDude, Trompedo, Master Lenman, Anrnusna, Monkbot, Aerosheet, Epigogue and Anonymous: 144

- **Bose gas** *Source:* https://en.wikipedia.org/wiki/Bose_gas?oldid=690168049 *Contributors:* SimonP, Michael Hardy, Schneelocke, Giftlite, PAR, Tom davis, Nanite, Chobot, Tony1, Kmarinas86, Colonies Chris, Lambiam, OS2Warp, Headbomb, Dirac66, Venny85, Ectomaniac, Hugepeak, MystBot, Addbot, DOI bot, LaaknorBot, Barak Sh, Omnipaedista, A. di M., Citation bot 1, RedBot, Korepin, Bibcode Bot, Monkbot and Anonymous: 6

- **Stimulated emission** *Source:* https://en.wikipedia.org/wiki/Stimulated_emission?oldid=733185495 *Contributors:* AxelBoldt, CYD, Zundark, DrBob, B4hand, Michael Hardy, Adam Bishop, Topbanana, Raul654, Robbot, Sanders muc, Rorro, Lupo, Troyrock, Wmahan, Neffk, Nabla, Cedders, I9Q79oL78KiL0QTFHgyc, Wrs1864, Zyqqh, Dirac1933, Gene Nygaard, Falcorian, Linas, Grammarbot, Brighterorange, FlaBot, Tardis, Srleffler, Chobot, YurikBot, Wavelength, Robertvan1, Johantheghost, Mikeblas, SmackBot, Gilliam, Eug, Complexica, Sbharris, Tsca.bot, Vladislav, OrphanBot, Sadi Carnot, Sunflowr, Cydebot, Msebast~enwiki, Thijs!bot, Headbomb, Pjvpjv, Marek69, Ibison, PhiLiP, JAnDbot, Winndm31, JBdV, Otivaeey, Dirac66, LorenzoB, Conquerist, R'n'B, ARTE, Idioma-bot, VolkovBot, Petergans, Jimmuhk, VVVBot, Jklsc, Abmcdonald, ClueBot, Rotational, ChrisHodgesUK, Interferometrist, SilvonenBot, Addbot, Darko.veberic, Tide rolls, Lightbot, Quantumobserver, JBancroftBrown, AnomieBOT, Jim1138, Royote, Materialscientist, Citation bot, Francescoonwikipedia, LilHelpa, Sheeson, FrescoBot, Craig Pemberton, Phb07jm, R3G1C1D3, Earthandmoon, RjwilmsiBot, EmausBot, Dcirovic, Wikfr, Jfpower, Chrisman62, Boldbdd, Bibcode Bot, Jeancey, Vahram Mekhitarian, ChrisGualtieri, Dexbot, ChuanZheng, Spider55555, Cpt Wise, Tycho Smit, GoldCar and Anonymous: 73

- **Laser** *Source:* https://en.wikipedia.org/wiki/Laser?oldid=728867814 *Contributors:* Bryan Derksen, Zundark, Timo Honkasalo, Koyaanis Qatsi, PierreAbbat, Roadrunner, Ray Van De Walker, Ben-Zin~enwiki, DrBob, Heron, Stevertigo, Mrwojo, Edward, Patrick, Infrogmation, Michael Hardy, Tim Starling, Dante Alighieri, MartinHarper, SGBailey, Wapcaplet, Ixfd64, Eurleif, CesarB, Ahoerstemeier, Baylink, Theresa knott, Jebba, Mark Foskey, Julesd, Lupinoid, Glenn, Whkoh, Rossami, Andres, Palfrey, Brigman, Lee M, Dj ansi, Zimbres, Ec5618, A5, Mbstone, Dysprosia, Tpbradbury, Furrykef, K1Bond007, SEWilco, Fibonacci, Omegatron, Phoebe, Jboyles, Mignon~enwiki, Bloodshedder, FlyByPC, David.Monniaux, Twice25, Mulberry~enwiki, Hankwang, Fredrik, Tomchiukc, Schutz, Sanders muc, Jmabel, Altenmann, Romanm, Merovingian, Yosri, Academic Challenger, Jfire, Jondel, Bkell, Hadal, Calvinchong, Anthony, Angilbas, Xanzzibar, Mattflaschen, Dina, Carnildo, ChanningWalton, Wjbeaty, Ancheta Wis, Giftlite, DocWatson42, Pmaguire, Rudolf 1922, BenFrantzDale, Laurens~enwiki, Everyking, Maha ts, Bensaccount, Quinwound, Eequor, Luigi30, Brockert, SWAdair, DÅ‚ugosz, Bobblewik, Ogxela, K-links, Alanl, Utcursch, Mike R, Jpkoester1, Yath, Antandrus, Beland, Mzajac, Cb6, Kesac, Jokestress, Icairns, GeoGreg, AmarChandra, Gscshoyru, Neutrality, Peter bertok, Ukexpat, Sonett72, Sam nead, Janneok~enwiki, BrianWilloughby, Deglr6328, Adashiel, TheObtuseAngleOfDoom, Udzu, Canterbury Tail, Danh, MToolen, Mike Rosoft, Cypa, Imroy, Sysy, Erc, Discospinster, Rich Farmbrough, Sladen, KittySaturn, FT2, Xezbeth, Alistair1978, Spooky-Mulder, Bender235, Kaisershatner, Loren36, Sgeo, Nabla, RJHall, El C, Joanjoc~enwiki, Hayabusa future, Laurascudder, Sietse Snel, RoyBoy, Femto, Bobo192, Truthflux, NetBot, Fir0002, Blakkandekka, Cmdrjameson, .:Ajvol:., Elipongo, Kappa, TheProject, Snacky, MPerel, Haham hanuka, Hooperbloob, Lysdexia, Espoo, Frodet, Danski14, Alansohn, Gary, TheParanoidOne, Anthony Appleyard, Shawn K. Quinn, Coma28, RPaschotta, Atlant, WakeUp, Ashley Pomeroy, Mr snarf, Equinoxe, Lectonar, Pouya, Water Bottle, Lightdarkness, RoySmith, Spangineer, Snowolf, Wtmitchell, Melaen, Isaac, Fourthords, Wtshymanski, Jason222, Evil Monkey, Ian McEwen, TenOfAllTrades, Cmprince, Gunter, BDD, Versageek, Sleigh, DV8 2XL, Gene Nygaard, Agutie, Dan100, Falcorian, Mhazard9, Infinoid, Gmaxwell, Bacteria, Richard Arthur Norton (1958-), OwenX, Woohookitty, Camw, Guy M, Miaow Miaow, Thorpe, Azov, JonBirge, Cbdorsett, GregorB, SDC, Palica, Mekong Bluesman, Slgrandson, Tslocum, Azkar, Graham87, Magister Mathematicae, Kbdank71, Jclemens, Canderson7, ZeframCochrane, Rjwilmsi, Nightscream, Zbxgscqf, Jake Wartenberg, Vary, Mitchandre, Nneonneo, HappyCamper, SeanMack, Brighterorange, Krash, The wub, Dolphonia, Bhadani, Sango123, Titoxd, FlaBot, Weihao.chiu~enwiki, Duomillia, Arnero, Mathbot, GünniX, Nivix, Fragglet, RexNL, Gurch, Alberrosidus, Goeagles4321, Fresheneesz, Srleffler, SpectrumDT, Zotel, Imnotminkus, Idaltu, Smithbrenon, ...adam..., Chobot, Jaraalbe, Sharkface217, Nagytibi, 334a, Simesa, Gwernol, Fcs, Quicksilvre, Albrozdude, Dermatonet~enwiki, The Rambling Man, Measure, YurikBot, Wavelength, RattusMaximus, Sceptre, Todd Vierling, Hairy Dude, Pmg, Prometheus235, Dimimimon7, Arado, Admiral Roo, CanadianCaesar, Hydrargyrum, Stephenb, Lord Voldemort, Pmurph5, Gaius Cornelius, Shaddack, Ugur Basak, Anomalocaris, David R. Ingham, MosheA, ALoopingIcon, NawlinWiki, Misos, Worldwalker, Adam Martinez, Janke, Grafen, Kvn8907, PaxtonB, Howcheng, MacGyver07, Blueforce4116, Irishguy, JohnFlux, Ergbert, Rudykog, Nick C, Tony1, Dbfirs, Aaron Schulz, Kyle Barbour, Samir, SFC9394, DeadEyeArrow, Slavik81, Nescio, Kkmurray, Stefan Udrea, Searchme, NormanII, Super Rad!, Ali K, Bayerischermann, Nikkimaria, Jwissick, Koshy2000, StealthFox, Hierzuhelfen, GraemeL, Jecowa, JoanneB, Fram, Afn, JLaTondre, Spliffy, Nixer, Tom Duff, Kungfuadam, Paul Erik, Airconswitch, Jknacnud, SkerHawx, Mejor Los Indios, Bibliomaniac15, Luk, Criticality, SmackBot, MattieTK, Kellen, Ashenai, ThreeDee912, Reedy, Slashme, Murray.booth, Tomer yaffe, KnowledgeOfSelf, CelticJobber, David.Mestel, AndyZ, The Photon, Jacek Kendysz, Chairman S., Muchacho Gasolino, Jrockley, Eskimbot, Cla68, BenFranske, Ekschuller, PJM, Gjs238, Kslays, Scott Paeth, Srnec, Ejeffrey, Pzavon, Gilliam, Donama, Andy M. Wang, Chris the speller, Qwasty, Keegan, Iskander32, Agateller, Quinsareth, Persian Poet Gal, Jprg1966, Thumperward, Miquonranger03, Silly rabbit, Papa November, SchfiftyThree, Jerome Charles Potts, Dlohcierekim's sock, Octahedron80, Baa, DHN-bot~enwiki, Sbharris, Colonies Chris, Felipe La Rotta, A. B., Sct72, Dethme0w, Can't sleep, clown will eat me, AntiVan, Scray, OrphanBot, Onorem, TKD, Parent5446, Phaedriel, Photonicsguy, Flyguy649, Fuhghettaboutit, Noumenorian~enwiki, Ianmacm, Makemi, Nakon, TedE, Kntrabssi, EVula, Wcleveland, Dreadstar, Hadmack, M jurrens, HarisM, Ultraexactzz, Salamurai, Ohconfucius, Blahm, Chwech, Nmnogueira, Shadow148, L337p4wn, McDuff, SashatoBot, Rory096, Schnazola, Dbutch20, BorisFromStockdale, Lazertech, Dbtfz, Soap, Kuru, RTejedor, John, Zaphraud, Sosodank, Buchanan-Hermit, Gobonobo, Lazylaces, JoshuaZ, Chodorkovskiy, Mr. Lefty, Beta34, Agathoclea, Slakr, Werdan7, SnoKoneManiac, Joeylawn, Muadd, Dbo789, Cajolingwilhelm, Waggers, AdultSwim, NuncAutNunquam, Amitch, Hu12, Keith-264, OnBeyondZebrax, Jb.ronin, 11K, Pqrstuv, Wizard191, Iridescent, Michaelbusch, Astrobayes, Muhaidib, Lakers, Lasah, Tmangray, Shoeofdeath, Vinsm, R~enwiki, Igoldste, Tony Fox, Hikui87~enwiki, Domitori, Tawkerbot2, MPOxy, Nerfer, Dlohcierekim, Chetvorno, Baieuan, Switchercat, Alexthe5th, Marudand~enwiki, Maziy300, CRGreathouse, Anon user, Tarchon, Rodrigja, Patrick Berry, Mcepjg, Mohammad Sami, FunPika, Hildenja, No ptr, The Font, CBM, Ilikefood, DeLarge, Metaxis, McVities, Vlizzardv, Ballista, WeggeBot, Shizane, Tim1988, Andrew Delong, Handmedown, Myasuda, Oo7565, Grenno, Cydebot, Abeg92, Ryan, Hyperdeath, Gogo Dodo, RomanXNS, Corpx, Ttiotsw, ST47, B4dA1r, Owen Silverstone, Wikipediarules2221, Guitardemon666, Tawkerbot4, Walter Humala, DumbBOT, Chrislk02, Apricotscrub, Waxigloo, Keithf@bel fast.com, Editor at Large, Njan, John Lake, Omicronpersei8, UberScienceNerd, Krylonblue83, Thijs!bot, Epbr123, Daa89563, Mercury~enwiki, Wikid77, IvanStepaniuk, LactoseTI, Hudsonwongzb, Josephseagullstalin, Callmarcus, Gamer007, Headbomb, Sobreira, John254, Tapir Terrific, Wiki fanatic, A3RO, Electron9, TheTruthiness, Ackatsis, Martin Hedegaard, Renamed user 5197261az5af96as6aa, AgentPeppermint, Lunamaria, Noclevername, Thomas Paine1776, Northumbrian, Escarbot, Ileresolu, Gerkleplex, AntiVandalBot, Saimhe, Luna Santin, Opelio, Quintote, Lovibond, Dark-Audit, Sweart1, RapidR, Jj137, TimVickers, Tlabshier, LibLord, Tillman, Random user 8384993, Myanw, Jabeles, Caper13, Res2216firestar, JAnDbot, Happytomato, CombatWombat42, Freepsbane, MER-C, Instinct, Janejellyroll, Rob Mahurin, Seanweir, TheOtherSiguy, Andonic, Hut 8.5, Time3000, East718, Kerotan, Fishclip, LittleOldMe, Mardavich, Meeples, Dog12~enwiki, Magioladitis, VoABot II, Swpb, Think outside the box, CTF83!, Singularity, Fallenangei, Catgut, Indon, Cardamon, Animum, Cyktsui, Pawl Kennedy, BatteryIncluded, First Harmonic, Dirac66, 28421u2232nfenfcenc, Allstarecho, Styrofoam1994, NeonDemon, E104421, DerHexer, Esanchez7587, Joeyman365, Nevit, Kaulike, RiverBed, Charitwo, Robin S, Seba5618, RichMac, Qwerty192837, Gjd001, S3000, Read-write-services, MartinBot, Mythealias,

Mmoneypenny, Mermaid from the Baltic Sea, Fdimer, Dima373, SG Liker, Arjun01, Poeloq, N734LQ, Rettetast, Keith D, CommonsDelinker, Davesf, Nono64, Jackanape Jones, Ctroy36, Kristine anne, J.delanoy, Junggoo, Psychoticfruit, Hill billy wock, JJLESDUDE, Quailman, Bogey97, Vipinmathew, Maurice Carbonaro, Nigholith, LFW, Tazzaler, Quantum ua, Littlebum2002, Dispenser, It Is Me Here, JoeJoeseph, Bot-Schafter, Katalaveno, Peppergrower, Treuzgirl16, Davandron, Gurchzilla, Youngjim, M-le-mot-dit, SJP, Mufka, Chaide, Shoessss, Atropos235, Jackaranga, Juliancolton, Cometstyles, WJBscribe, Neohaven, Cremisis13~enwiki, Freeridr, Gtg204y, Bonadea, Dzenanz, James P Twomey, Banjodog, Withead, Immachargin, Xiahou, Daleious, Wehttam1106, Wiggydo, Dreamweaver9, Cyphern, Gogobera, Deor, VolkovBot, DrDentz, ABF, OHmanEARTHBOUND, Jlaramee, Earendilmm, Soliloquial, QuackGuru, Amh library, Charleca, TXiKiBoT, Killermatt18, FDominec, Narayan82es, Bbltype, A-man262, Ronningt, HIGOCHUICHI, Evanfarrar, Abdullais4u, LeaveSleaves, Figureskatingfan, Turned-Worm, Crazyw00t1, Buffs, Mwilso24, Zain Ebrahim111, Wenli, Andy Dingley, Operating, Valkyryn, Lukus lee, Ecopetition, Amwyll Rwden, Yatoast, SmileToday, Enviroboy, Insanity Incarnate, Raluaces, Brianga, Baggio10~enwiki, Symane, PGWG, Phox63, Gmankiller2, Jellies1324, EverGreg, Chuck Sirloin, FlyingLeopard2014, Gandhi21~enwiki, EmxBot, Aeiouuu, TimProof, John water, Johncarini, Blueking12, Derfugu, Greenlaser, SieBot, Coffee, Tresiden, Imacbelkin, Lightro6, Work permit, BotMultichill, Ellbeecee, Redblitz, PoolsClosed87, Dawn Bard, Caltas, Karaboom, ChevyCha, MarkBolton, RJaguar3, CAIRNSY90, Npeters07, Redhookesb, Vanished User 8a9b4725f8376, Potf13, Calabraxthis, Stonejag, Welsh21, Jonnysimpson, Iamdrevil, RFNEFF, Android Mouse, Flakos17, Tiptoety, JLKrause, Brian R Hunter, Oxymoron83, Antonio Lopez, XxXXMULLIGANXXxx, Nuttycoconut, Orgeat, Lightmouse, Taune, Armedtrader, Iain99, KathrynLybarger, Hobartimus, Macy, Nancy, Anakin101, Smartkid23, Iluvjlpicard, Thinghy, Hamiltondaniel, Abmcdonald, Dust Filter, Allmedia, Nn123645, Joelster, LambOfDog, Chem-awb, LoveMed, Steve, Dhaval akbari, Faithlessthewonderboy, Martarius, Grand2007, Sawt al Hurriyah, Elassint, ClueBot, 7Piguine, NickCT, Avenged Eightfold, Kipkay, Mafuyu~enwiki, Gggrrrnade, Quinxorin, Berryland1jmm, GreenSpigot, Taroaldo, Nebrot, UserDoe, Shinpah1, Doseiai2, Boing! said Zebedee, CounterVandalismBot, Niceguyedc, EconomicsGuy, TypoBoy, DanielleJ, Sandeepjshenoy, Handcloud returns, PaedosAreKewl, Special needs LOL, Pmronchi, Eeekster, Rubin joseph 10, HeroGiant, Farrell0000, Ryan Taylor, MacedonianBoy, MovementLessRestricted, Lunchscale, Drhaddix, Obnoxin, Ember of Light, Ericschulz, Herohtar, Dekisugi, Misskaur, Portobello Prince, Muro Bot, Kakofonous, Thingg, Justdaning, Chovin, Crnorizec, Zaledin, Edebraal, Hummer82793, Nghiaquang2, JOOHD, Sapiens23, XLinkBot, Michaelpkk~enwiki, Womza, Rror, Matthewsasse1, Interferometrist, DrewBear11, AndreNatas, Mikeli88, Chrisbil09, Snowboy9999, Alexius08, Hess88, RyanCross, Emilyisaac, USS Noob Hunter, Wyatt915, Ibakedsomecookies, Crgibson, Sportzplyr9090, Addbot, Xp54321, Willking1979, AVand, DOI bot, Laservet, Meuzzwal, Danielismagic, D367072, Vishnava, CanadianLinuxUser, Leszek Jańczuk, Oursroute, LaaknorBot, Chewychum, Cats are vague, سمرقندي, Sorset, Roux, LinkFA-Bot, Quercus solaris, Naidevinci, 84user, Seeker alpha806, Phill9g7, Numbo3-bot, Apteva, Lion Info, ScienceApe, Matt.T, Luckas-bot, Yobot, Fraggle81, Udayan.choudhury, Gnomerspell, ArchonMagnus, KamikazeBot, IW.HG, Laserslight, Eric-Wester, Sasuke6051, AnomieBOT, KDS4444, A More Perfect Onion, Jim1138, Piano non troppo, AdjustShift, LlywelynII, Yachtsman1, Mrchrisr, Rfwexler, Westerness, Citation bot, GrampaScience, Makele-90, Neurolysis, Lil-Helpa, Xqbot, Konor org, Corrigendas, Capricorn42, TracyMcClark, Gigemag76, Tomdo08, XLostMemoriesx, Inferno, Lord of Penguins, PhysicsR, Scooteemooty, BOOG95, BK107, Noobdestroyah, Omnipaedista, Shirik, RibotBOT, Klknoles, Alexvinny, SCARECROW, Stratocracy, Medicuspetrus, Fixentries, Chaheel Riens, A. di M., FrescoBot, Tobby72, Charles Edwin Shipp, Richbham, Glanhawr~enwiki, Cannolis, Citation bot 1, Pinethicket, LittleWink, Pmokeefe, SiPlus, Σ, Corinne68, Luitgard, TobeBot, Mattyp9999, Lotje, GossamerBliss, MrX, Wikifan798, Earthandmoon, DexDor, Laser-jok, Beyond My Ken, Norlesh, WildBot, Rayman60, EmausBot, Beatnik8983, Dewritech, Jmencisom, BTech United, TuHan-Bot, Dcirovic, Freethron, Kkm010, ZéroBot, Derekleungtzsghei, Fintelia, Dondervogel 2, H3llBot, Unreal7, Fabian Hassler, Hawksfan18, Grandphuba, ChuispastonBot, Philippe BINANT, 28bot, Warharmer, Psydev, ClueBot NG, SusikMkr, Alex-engraver, GlassLadyBug, Ose\fio, Talaga87, Helpful Pixie Bot, Juboroff, Bibcode Bot, Vagobot, ܐܪܡܝܐ, Shawn Worthington Laser Plasma, BattyBot, ܐܪܡܝܐ, ChrisGualtieri, Achyut55, BOT, Layzeeboi, TylerDurden8823, Davisonkirby, Kelseyknecht, Dexbot, ColinCullis, LzrPundit, CuriousMind01, Lugia2453, Cheerioswithmilk, TheRealJoeWiki, Aladdin Ali Baba, Tmlmr34, LMANSH, ImTrollingYouDude, BeachComber1972, HFEO, Rkaup, Iztwoz, BenStein69, Gatitbat, DaPhil, VALID REALITY, OmniArticleEditor, ܐܪܡܝܐ, Shipandreceive, AnotherNewAccount, Mfb, Dx-Man12, Kindly Proofreader, Monkbot, SkateTier, Plesantdreams, Edzavala, IamCool117, Sourgosling, Rhlius, UareNumber6, Pangkakit, Izkala, Themidget17, KasparBot, Limitless undying love, Thorkall, ScientistMr, Academicguru22, GoldCar, EAWH, Target360YT and Anonymous: 1159

- **Photon structure function** *Source:* https://en.wikipedia.org/wiki/Photon_structure_function?oldid=718119683 *Contributors:* Vegaswikian, Bgwhite, JHCaufield, Lambiam, Leyo, Carriearchdale, Yobot, AnomieBOT, HolgerFiedler, John of Reading, Timetraveler3.14, Snotbot, BG19bot, Mogism, Profb39 and Anonymous: 6

- **Ballistic photon** *Source:* https://en.wikipedia.org/wiki/Ballistic_photon?oldid=440035574 *Contributors:* Arcadian, Srleffler, A876, Alaibot, JL-Bot, Tom.vettenburg, Srguy, Locobot and Anonymous: 4

- **Photonic molecule** *Source:* https://en.wikipedia.org/wiki/Photonic_molecule?oldid=732822223 *Contributors:* Velella, Bhny, Headbomb, Lfstevens, I JethroBT, Paradoctor, Yobot, AnomieBOT, Citation bot, Omnipaedista, Quondum, Bibcode Bot, Dexbot, Reatlas and Anonymous: 16

- **Two-photon physics** *Source:* https://en.wikipedia.org/wiki/Two-photon_physics?oldid=734475936 *Contributors:* Xerxes314, Karol Langner, Wtmitchell, Woohookitty, Edison, Krishnavedala, Phmer, SCZenz, Whobot, Crystallina, SmackBot, Jpvinall, Bluebot, Sergio.ballestrero, Only, Lambiam, Paul Fisher, Headbomb, EagleFan, Potatoswatter, Cuzkatzimhut, Brews ohare, SchreiberBike, Yobot, AnomieBOT, Are you ready for IPv6?, Kotika98, Marshallsumter, Davdde, Rainald62, Puzl bustr, RjwilmsiBot, Quondum, One.Ouch.Zero, Helpful Pixie Bot, Bibcode Bot, Zedshort, Kryomaxim, Profb39, Zerberos, The Bible in Metre, MARCOS BUIRA PARDO, Shankarsivarajan, Jordan-daniel 20161227, SpencerKlein and Anonymous: 16

19.7.2 Images

- **File:060815_polaroid.svg** *Source:* https://upload.wikimedia.org/wikipedia/commons/b/b5/060815_polaroid.svg *License:* CC-BY-SA-3.0 *Contributors:* Transferred from en.wikipedia to Commons by Johnman239 using CommonsHelper. *Original artist:* SyntaxError55 at English Wikipedia

- **File:28-06-2015_Problem_7.59.jpg** *Source:* https://upload.wikimedia.org/wikipedia/commons/9/90/28-06-2015_Problem_7.59.jpg *License:* CC BY-SA 4.0 *Contributors:* Own work *Original artist:* Rajettan

- **File:A_city_illuminated_by_colorful_artificial_lighting_at_night.jpg** *Source:* https://upload.wikimedia.org/wikipedia/commons/2/2c/A_city_illuminated_by_colorful_artificial_lighting_at_night.jpg *License:* CC BY-SA 4.0 *Contributors:* Own work *Original artist:* Rhlius
- **File:Aleksandr_Prokhorov.jpg** *Source:* https://upload.wikimedia.org/wikipedia/commons/4/4c/Aleksandr_Prokhorov.jpg *License:* Public domain *Contributors:* http://nobelprize.org/nobel_prizes/physics/laureates/1964/prokhorov-bio.html *Original artist:* Nobel foundation
- **File:Aphrodita_aculeata_(Sea_mouse).jpg** *Source:* https://upload.wikimedia.org/wikipedia/commons/c/c2/Aphrodita_aculeata_%28Sea_mouse%29.jpg *License:* CC BY-SA 3.0 *Contributors:* Own work *Original artist:* MichaelMaggs
- **File:Atmospheric_electromagnetic_opacity.svg** *Source:* https://upload.wikimedia.org/wikipedia/commons/3/34/Atmospheric_electromagnetic_opacity.svg *License:* Public domain *Contributors:* Vectorized by User:Mysid in Inkscape, original NASA image from File:Atmospheric electromagnetic transmittance or opacity.jpg. *Original artist:* NASA (original); SVG by Mysid.
- **File:Bohr-atom-PAR.svg** *Source:* https://upload.wikimedia.org/wikipedia/commons/5/55/Bohr-atom-PAR.svg *License:* CC-BY-SA-3.0 *Contributors:* Transferred from en.wikipedia to Commons. *Original artist:* Original uplo:JabberWok]] at en.wikipedia
- **File:Bose_gas_quantities.png** *Source:* https://upload.wikimedia.org/wikipedia/commons/b/b1/Bose_gas_quantities.png *License:* Public domain *Contributors:* Own work *Original artist:* PAR at wikipedia
- **File:Calcite-HUGE.jpg** *Source:* https://upload.wikimedia.org/wikipedia/commons/3/37/Calcite-HUGE.jpg *License:* Public domain *Contributors:* http://en.wikipedia.org/wiki/File:Calcite-HUGE.jpg *Original artist:* Alkivar
- **File:Calcite.jpg** *Source:* https://upload.wikimedia.org/wikipedia/commons/7/7a/Calcite.jpg *License:* Public domain *Contributors:* ? *Original artist:* ?
- **File:Cauchy_schwarz_inequality.svg** *Source:* https://upload.wikimedia.org/wikipedia/commons/8/81/Cauchy_schwarz_inequality.svg *License:* Public domain *Contributors:* self-made, based on http://en.wikipedia.org/wiki/Image:060819_cauchy_schwarz_inequality.png (PD) *Original artist:* dnet
- **File:Christiaan_Huygens,_by_Caspar_Netscher.jpg** *Source:* https://upload.wikimedia.org/wikipedia/commons/c/c3/Christiaan_Huygens%2C_by_Caspar_Netscher.jpg *License:* Public domain *Contributors:* Picture made by uploader *Original artist:* Caspar Netscher (circa 1639–1684)
- **File:Cloud_in_the_sunlight.jpg** *Source:* https://upload.wikimedia.org/wikipedia/commons/8/81/Cloud_in_the_sunlight.jpg *License:* CC BY 2.0 *Contributors:* hey son, get out of the clouds *Original artist:* Ibrahim Iujaz from Rep. Of Maldives
- **File:Coherent_899_dye_laser.jpg** *Source:* https://upload.wikimedia.org/wikipedia/commons/b/b7/Coherent_899_dye_laser.jpg *License:* CC BY 2.5 *Contributors:* ? *Original artist:* ?
- **File:Commercial_laser_lines.svg** *Source:* https://upload.wikimedia.org/wikipedia/commons/4/48/Commercial_laser_lines.svg *License:* CC BY-SA 3.0 *Contributors:*

  The data and its references can be found in the spreadsheet Commercial laser lines.xls (unfortunately Wikipedia does not allow uploading spreadsheets). Currently most of the data is taken from Weber's book *Handbook of laser wavelengths* [#cite_note-1 [1]], with newer data in particular for semiconductor lasers. For quasi-cw lasers (e.g. metal vapor lasers) the length of the full line gives the mean power. Uses File:Linear visible spectrum.svg

  Original artist: Danh
- **File:Commons-logo.svg** *Source:* https://upload.wikimedia.org/wikipedia/en/4/4a/Commons-logo.svg *License:* CC-BY-SA-3.0 *Contributors:* ? *Original artist:* ?
- **File:DIN_4844-2_Warnung_vor_Laserstrahl_D-W010.svg** *Source:* https://upload.wikimedia.org/wikipedia/commons/1/16/DIN_4844-2_Warnung_vor_Laserstrahl_D-W010.svg *License:* Public domain *Contributors:* Own work *Original artist:* Torsten Henning
- **File:Diode_laser.jpg** *Source:* https://upload.wikimedia.org/wikipedia/commons/d/d9/Diode_laser.jpg *License:* Public domain *Contributors:* Jet Propulsion Laboratory website: http://technology.jpl.nasa.gov/gallery/index.cfm?page=imageDetail&ItemID=120&catId=8 (archive) *Original artist:* ?
- **File:Direct.jpg** *Source:* https://upload.wikimedia.org/wikipedia/commons/f/fa/Direct.jpg *License:* Public domain *Contributors:* Own work by the original uploader *Original artist:* Phmer at English Wikipedia
- **File:Doppelspalt.svg** *Source:* https://upload.wikimedia.org/wikipedia/commons/0/02/Doppelspalt.svg *License:* Public domain *Contributors:*
- Doppelspalt.jpg *Original artist:* Doppelspalt.jpg: Opasson
- **File:EM_spectrum.svg** *Source:* https://upload.wikimedia.org/wikipedia/commons/f/f1/EM_spectrum.svg *License:* CC-BY-SA-3.0 *Contributors:* ? *Original artist:* ?
- **File:EM_spectrumrevised.png** *Source:* https://upload.wikimedia.org/wikipedia/commons/3/30/EM_spectrumrevised.png *License:* CC BY-SA 3.0 *Contributors:* File:EM spectrum.svg and File:Linear visible spectrum.svg *Original artist:* Philip Ronan, Gringer
- **File:Edit-clear.svg** *Source:* https://upload.wikimedia.org/wikipedia/en/f/f2/Edit-clear.svg *License:* Public domain *Contributors:* The *Tango! Desktop Project*. *Original artist:*

  The people from the Tango! project. And according to the meta-data in the file, specifically: "Andreas Nilsson, and Jakub Steiner (although minimally)."
- **File:Einstein_patentoffice.jpg** *Source:* https://upload.wikimedia.org/wikipedia/commons/a/a0/Einstein_patentoffice.jpg *License:* Public domain *Contributors:* Cropped from original at the Historical Museum of Berne. *Original artist:* Lucien Chavan [#cite_note-author-1 [1]] (1868 - 1942), a friend of Einstein's when he was living in Berne.
- **File:Electromagneticwave3D.gif** *Source:* https://upload.wikimedia.org/wikipedia/commons/4/4c/Electromagneticwave3D.gif *License:* CC BY-SA 3.0 *Contributors:* Own work *Original artist:* Lookang many thanks to Fu-Kwun Hwang and author of Easy Java Simulation = Francisco Esquembre

19.7. TEXT AND IMAGE SOURCES, CONTRIBUTORS, AND LICENSES

- **File:Electromagneticwave3Dfromside.gif** *Source:* https://upload.wikimedia.org/wikipedia/commons/a/ad/Electromagneticwave3Dfromside.gif *License:* CC BY-SA 3.0 *Contributors:* Own work *Original artist:* Lookang many thanks to Fu-Kwun Hwang and author of Easy Java Simulation = Francisco Esquembre
- **File:ExperimentCouder-Young.png** *Source:* https://upload.wikimedia.org/wikipedia/commons/9/90/ExperimentCouder-Young.png *License:* CC BY-SA 3.0 *Contributors:* Own work *Original artist:* Krauss
- **File:F2versusQ2.jpg** *Source:* https://upload.wikimedia.org/wikipedia/commons/6/69/F2versusQ2.jpg *License:* CC BY-SA 4.0 *Contributors:* Own work *Original artist:* Profb39
- **File:F2versusxtwoQ2.jpg** *Source:* https://upload.wikimedia.org/wikipedia/commons/5/56/F2versusxtwoQ2.jpg *License:* CC BY-SA 4.0 *Contributors:* Own work *Original artist:* Profb39
- **File:FELIX.jpg** *Source:* https://upload.wikimedia.org/wikipedia/commons/1/14/FELIX.jpg *License:* CC BY-SA 3.0 *Contributors:* Own work *Original artist:* China Crisis
- **File:FarNearFields-USP-4998112-1.svg** *Source:* https://upload.wikimedia.org/wikipedia/commons/5/5d/FarNearFields-USP-4998112-1.svg *License:* Public domain *Contributors:* US Patent 6657596 *Original artist:* Goran M Djuknic
- **File:Fgamfeyndia.png** *Source:* https://upload.wikimedia.org/wikipedia/commons/a/a9/Fgamfeyndia.png *License:* CC BY-SA 3.0 *Contributors:* Own work *Original artist:* Profb39
- **File:Fluctuation.jpg** *Source:* https://upload.wikimedia.org/wikipedia/commons/d/d7/Fluctuation.jpg *License:* Public domain *Contributors:* Own work *Original artist:* Philipum at English Wikipedia
- **File:Folder_Hexagonal_Icon.svg** *Source:* https://upload.wikimedia.org/wikipedia/en/4/48/Folder_Hexagonal_Icon.svg *License:* Cc-by-sa-3.0 *Contributors:* ? *Original artist:* ?
- **File:Four_Lasers_Over_Paranal.jpg** *Source:* https://upload.wikimedia.org/wikipedia/commons/5/5f/Four_Lasers_Over_Paranal.jpg *License:* CC BY 4.0 *Contributors:* European Southern Observatory *Original artist:* ESO/F. Kamphues
- **File:Free-to-read_lock_75.svg** *Source:* https://upload.wikimedia.org/wikipedia/commons/8/80/Free-to-read_lock_75.svg *License:* CC0 *Contributors:*
  Adapted from <a href='//en.wikipedia.org/wiki/File:Open_Access_logo_PLoS_white_green.svg' class='image' title='Open_Access_logo_PLoS_white_green.svg'><img alt='Open_Access_logo_PLoS_white_green.svg' src='//upload.wikimedia.org/wikipedia/commons/thumb/9/90/Open_Access_logo_PLoS_white_green.svg/9px-Open_Access_logo_PLoS_white_green.svg.png' width='9' height='14' srcset='//upload.wikimedia.org/wikipedia/commons/thumb/9/90/Open_Access_logo_PLoS_white_green.svg/14px-Open_Access_logo_PLoS_white_green.svg.png 1.5x, //upload.wikimedia.org/wikipedia/commons/thumb/9/90/Open_Access_logo_PLoS_white_green.svg/18px-Open_Access_logo_PLoS_white_green.svg.png 2x' data-file-width='640' data-file-height='1000' /></a>
  Original artist:
  This version:Trappist_the_monk (talk) (Uploads)
- **File:Gamma-ray-microscope.svg** *Source:* https://upload.wikimedia.org/wikipedia/commons/6/63/Gamma-ray-microscope.svg *License:* CC-BY-SA-3.0 *Contributors:* Own work *Original artist:* Radeksonic
- **File:Gold_leaf_electroscope_diagram.svg** *Source:* https://upload.wikimedia.org/wikipedia/commons/8/8a/Gold_leaf_electroscope_diagram.svg *License:* Public domain *Contributors:*
- Gold_leaf_electroscope_diagram.jpg *Original artist:* Gold_leaf_electroscope_diagram.jpg: Luke FM (talk)
- **File:Gould_notebook_001.jpg** *Source:* https://upload.wikimedia.org/wikipedia/en/d/dd/Gould_notebook_001.jpg *License:* Fair use *Contributors:* ? *Original artist:* ?
- **File:Guassian_Dispersion.gif** *Source:* https://upload.wikimedia.org/wikipedia/commons/5/56/Guassian_Dispersion.gif *License:* CC0 *Contributors:* This mathematical image was created with Mathematica *Original artist:* Teply
- **File:Heinrich_Rudolf_Hertz.jpg** *Source:* https://upload.wikimedia.org/wikipedia/commons/5/50/Heinrich_Rudolf_Hertz.jpg *License:* Public domain *Contributors:* http://wellcomeimages.org/indexplus/imageM0014750.html *Original artist:* Robert Krewaldt
- **File:Heisenberg_gamma_ray_microscope.svg** *Source:* https://upload.wikimedia.org/wikipedia/commons/b/bc/Heisenberg_gamma_ray_microscope.svg *License:* CC-BY-SA-3.0 *Contributors:* Wikimedia commons *Original artist:* parri
- **File:Heisenbergbohr.jpg** *Source:* https://upload.wikimedia.org/wikipedia/commons/1/1a/Heisenbergbohr.jpg *License:* Public domain *Contributors:* http://www.fnal.gov/pub/inquiring/timeline/images/heisenbergbohr.jpg shown on http://www.fnal.gov/pub/inquiring/timeline/05.html *Original artist:* Fermilab, U.S. Department of Energy
- **File:Helium_neon_laser_spectrum.svg** *Source:* https://upload.wikimedia.org/wikipedia/commons/8/85/Helium_neon_laser_spectrum.svg *License:* CC-BY-SA-3.0 *Contributors:*
- Helium_neon_laser_spectrum.png *Original artist:*
- derivative work: Papa November (<a href='//commons.wikimedia.org/wiki/User_talk:Papa_November' title='User talk:Papa November'>talk</a>)
- **File:History_of_laser_intensity.svg** *Source:* https://upload.wikimedia.org/wikipedia/commons/6/6c/History_of_laser_intensity.svg *License:* Public domain *Contributors:* No machine-readable source provided. Own work assumed (based on copyright claims). *Original artist:* No machine-readable author provided. Slashme assumed (based on copyright claims).
- **File:Husimi_distribution_squeezed_state.jpg** *Source:* https://upload.wikimedia.org/wikipedia/commons/2/28/Husimi_distribution_squeezed_state.jpg *License:* CC BY-SA 3.0 *Contributors:* Own work *Original artist:* Sijothankam
- **File:Interference_of_a_quantum_particle_with_itself.gif** *Source:* https://upload.wikimedia.org/wikipedia/commons/7/7d/Interference_of_a_quantum_particle_with_itself.gif *License:* CC0 *Contributors:* Own work *Original artist:* Thierry Dugnolle

- **File:LASER.jpg** *Source:* https://upload.wikimedia.org/wikipedia/commons/b/b9/LASER.jpg *License:* CC BY 2.5 *Contributors:* Own work *Original artist:* ???
- **File:Laser,_quantum_principle.ogv** *Source:* https://upload.wikimedia.org/wikipedia/commons/5/54/Laser%2C_quantum_principle.ogv *License:* CC BY-SA 3.0 *Contributors:* Own work *Original artist:* Jubobroff
- **File:Laser.svg** *Source:* https://upload.wikimedia.org/wikipedia/commons/1/1f/Laser.svg *License:* CC-BY-SA-3.0 *Contributors:* Own work *Original artist:* User:Tatoute
- **File:Laser_DSC09088.JPG** *Source:* https://upload.wikimedia.org/wikipedia/commons/4/4b/Laser_DSC09088.JPG *License:* CC-BY-SA-3.0 *Contributors:* en:Kastler-Brossel Laboratory at en:Paris VI: Pierre et Marie Curie *Original artist:* Copyright © 2004 David Monniaux
- **File:Laser_label_2.jpg** *Source:* https://upload.wikimedia.org/wikipedia/commons/4/4e/Laser_label_2.jpg *License:* CC BY-SA 3.0 *Contributors:* File:Laser label.jpg *Original artist:* User:BOY
- **File:Laser_play.jpg** *Source:* https://upload.wikimedia.org/wikipedia/commons/0/0e/Laser_play.jpg *License:* CC BY 2.0 *Contributors:* Beams in Fog + Car Windshield *Original artist:* Jeff Keyzer from San Francisco, CA, USA
- **File:Laser_sizes.jpg** *Source:* https://upload.wikimedia.org/wikipedia/commons/6/6b/Laser_sizes.jpg *License:* Public domain *Contributors:* ? *Original artist:* ?
- **File:Laserlink_hss46.jpg** *Source:* https://upload.wikimedia.org/wikipedia/commons/b/bd/Laserlink_hss46.jpg *License:* CC-BY-SA-3.0 *Contributors:* Self-photographed *Original artist:* Sanngetall
- **File:Light-wave.svg** *Source:* https://upload.wikimedia.org/wikipedia/commons/a/a1/Light-wave.svg *License:* CC-BY-SA-3.0 *Contributors:* No machine-readable source provided. Own work assumed (based on copyright claims). *Original artist:* No machine-readable author provided. Gpvos assumed (based on copyright claims).
- **File:Light_cone_colour.svg** *Source:* https://upload.wikimedia.org/wikipedia/commons/5/56/Light_cone_colour.svg *License:* Public domain *Contributors:* Own work *Original artist:* Incnis Mrsi 10:15, 4 June 2008 (UTC)
- **File:Light_dispersion_conceptual_waves350px.gif** *Source:* https://upload.wikimedia.org/wikipedia/commons/7/73/Light_dispersion_conceptual_waves350px.gif *License:* Public domain *Contributors:* Light dispersion conceptual waves.gif *Original artist:* LucasVB
- **File:Light_spectrum.svg** *Source:* https://upload.wikimedia.org/wikipedia/commons/e/eb/Light_spectrum.svg *License:* CC-BY-SA-3.0 *Contributors:*
- Light_spectrum.png *Original artist:* Light_spectrum.png: Original uploader was Denelson83 at en.wikipedia
- **File:Linear_visible_spectrum.svg** *Source:* https://upload.wikimedia.org/wikipedia/commons/d/d9/Linear_visible_spectrum.svg *License:* Public domain *Contributors:* Own work *Original artist:* Gringer
- **File:Lying_down_on_the_VLT_platform.jpg** *Source:* https://upload.wikimedia.org/wikipedia/commons/d/d0/Lying_down_on_the_VLT_platform.jpg *License:* CC BY 4.0 *Contributors:* http://www.eso.org/public/images/gerd_huedepohl_4/ *Original artist:* G. Hüdepohl/ESO
- **File:MESSENGER_-_MLA.jpg** *Source:* https://upload.wikimedia.org/wikipedia/commons/8/8e/MESSENGER_-_MLA.jpg *License:* Public domain *Contributors:* http://nssdc.gsfc.nasa.gov/database/MasterCatalog?sc=2004-030A&ex=5 *Original artist:* ?
- **File:Mach-Zehnder_photons_animation.gif** *Source:* https://upload.wikimedia.org/wikipedia/commons/a/a0/Mach-Zehnder_photons_animation.gif *License:* CC BY 3.0 *Contributors:* Own work *Original artist:* user:Geek3
- **File:Merge-arrow.svg** *Source:* https://upload.wikimedia.org/wikipedia/commons/a/aa/Merge-arrow.svg *License:* Public domain *Contributors:* ? *Original artist:* ?
- **File:Mergefrom.svg** *Source:* https://upload.wikimedia.org/wikipedia/commons/0/0f/Mergefrom.svg *License:* Public domain *Contributors:* ? *Original artist:* ?
- **File:Military_laser_experiment.jpg** *Source:* https://upload.wikimedia.org/wikipedia/commons/a/a0/Military_laser_experiment.jpg *License:* Public domain *Contributors:*
  This Image was released by the United States Air Force with the ID 090809-F-5527s-0001 <a class='external text' href='//commons.wikimedia.org/w/index.php?title=Category:Files_created_by_the_United_States_Air_Force_with_known_IDs,,&,,filefrom=090809-F-5527s-0001#mw-category-media'>(next)</a>.
  This tag does not indicate the copyright status of the attached work. A normal copyright tag is still required. See Commons:Licensing for more information.
  Original artist: US Air Force
- **File:Millikan.jpg** *Source:* https://upload.wikimedia.org/wikipedia/commons/2/2f/Millikan.jpg *License:* Public domain *Contributors:* http://nobelprize.org/nobel_prizes/physics/laureates/1923/millikan-bio.html *Original artist:* Nobel foundation
- **File:Moon_clementine_lidar.jpg** *Source:* https://upload.wikimedia.org/wikipedia/commons/8/85/Moon_clementine_lidar.jpg *License:* Public domain *Contributors:* plotted using GMT and gltm2bpr.tab from PDS superimposed on map from http://solarviews.com/cap/moon/moonmap.htm *Original artist:* Martin Pauer (Power)
- **File:Mudflats-polariser.jpg** *Source:* https://upload.wikimedia.org/wikipedia/commons/a/a3/Mudflats-polariser.jpg *License:* CC-BY-SA-3.0 *Contributors:* ? *Original artist:* ?
- **File:Noise_squeezed_states.jpg** *Source:* https://upload.wikimedia.org/wikipedia/commons/d/da/Noise_squeezed_states.jpg *License:* CC-BY-SA-3.0 *Contributors:* dissertation *Original artist:* Gerd Breitenbach
- **File:Nuvola_apps_kalzium.svg** *Source:* https://upload.wikimedia.org/wikipedia/commons/8/8b/Nuvola_apps_kalzium.svg *License:* LGPL *Contributors:* Own work *Original artist:* David Vignoni, SVG version by Bobarino
- **File:Onde_electromagnetique.svg** *Source:* https://upload.wikimedia.org/wikipedia/commons/3/35/Onde_electromagnetique.svg *License:* CC-BY-SA-3.0 *Contributors:* Self, based on Image:Onde electromagnetique.png *Original artist:* SuperManu

19.7. TEXT AND IMAGE SOURCES, CONTRIBUTORS, AND LICENSES 173

- **File:Phase_change_-_en.svg** *Source:* https://upload.wikimedia.org/wikipedia/commons/0/0b/Phase_change_-_en.svg *License:* Public domain *Contributors:* Own work *Original artist:* F l a n k e r, penubag
- **File:Phase_distribution_squeezed_coherent_states_subpoisson.jpg** *Source:* https://upload.wikimedia.org/wikipedia/commons/c/c2/Phase_distribution_squeezed_coherent_states_subpoisson.jpg *License:* CC-BY-SA-3.0 *Contributors:* dissertation *Original artist:* Gerd Breitenbach
- **File:Phillipp_Lenard_in_1900.jpg** *Source:* https://upload.wikimedia.org/wikipedia/commons/1/1d/Phillipp_Lenard_in_1900.jpg *License:* Public domain *Contributors:* Encyclopaedia Britannica. Original source AIP Emilio Segrè Visual Archives, American Institute of Physics. *Original artist:* Not mentioned in any source
- **File:Photoelectric_effect.svg** *Source:* https://upload.wikimedia.org/wikipedia/commons/f/f5/Photoelectric_effect.svg *License:* CC-BY-SA-3.0 *Contributors:* en:Inkscape *Original artist:* Wolfmankurd
- **File:Photoelectric_effect_diagram.svg** *Source:* https://upload.wikimedia.org/wikipedia/commons/d/db/Photoelectric_effect_diagram.svg *License:* CC BY 3.0 *Contributors:* This vector image was created with Inkscape. *Original artist:* Klaus-Dieter Keller
- **File:Photon-photon_scattering.svg** *Source:* https://upload.wikimedia.org/wikipedia/commons/5/5a/Photon-photon_scattering.svg *License:* Public domain *Contributors:* Kurt Gottfried; Victor Frederick Weisskopf (1986) *Concepts of particle physics*, 2, Oxford University Press, p. 266 ISBN: 0195033930. *Original artist:* Krishnavedala
- **File:Photon_numbers_squeezed_coherent_states_subpoisson.jpg** *Source:* https://upload.wikimedia.org/wikipedia/commons/d/d3/Photon_numbers_squeezed_coherent_states_subpoisson.jpg *License:* CC-BY-SA-3.0 *Contributors:* dissertation *Original artist:* Gerd Breitenbach
- **File:Photon_numbers_squeezed_vacuum.jpg** *Source:* https://upload.wikimedia.org/wikipedia/commons/2/27/Photon_numbers_squeezed_vacuum.jpg *License:* CC-BY-SA-3.0 *Contributors:* dissertation *Original artist:* Gerd Breitenbach
- **File:PierreGassendi.jpg** *Source:* https://upload.wikimedia.org/wikipedia/commons/c/c8/PierreGassendi.jpg *License:* Public domain *Contributors:* Œuvres complètes de Voltaire – le siècle de Louis XIV. *Original artist:* Louis-Édouard Rioult
- **File:Portal-puzzle.svg** *Source:* https://upload.wikimedia.org/wikipedia/en/f/fd/Portal-puzzle.svg *License:* Public domain *Contributors:* ? *Original artist:* ?
- **File:Position_and_momentum_of_a_Gaussian_initial_state_for_a_QHO,_balanced.gif** *Source:* https://upload.wikimedia.org/wikipedia/en/f/fe/Position_and_momentum_of_a_Gaussian_initial_state_for_a_QHO%2C_balanced.gif *License:* CC0 *Contributors:* made with Mathematica

  Original artist:

  Teply
- **File:Position_and_momentum_of_a_Gaussian_initial_state_for_a_QHO,_narrow.gif** *Source:* https://upload.wikimedia.org/wikipedia/en/5/5c/Position_and_momentum_of_a_Gaussian_initial_state_for_a_QHO%2C_narrow.gif *License:* CC0 *Contributors:* made with Mathematica

  Original artist:

  Teply
- **File:Position_and_momentum_of_a_Gaussian_initial_state_for_a_QHO,_wide.gif** *Source:* https://upload.wikimedia.org/wikipedia/en/6/68/Position_and_momentum_of_a_Gaussian_initial_state_for_a_QHO%2C_wide.gif *License:* CC0 *Contributors:* made with Mathematica

  Original artist:

  Teply
- **File:Prism_flat_rainbow.jpg** *Source:* https://upload.wikimedia.org/wikipedia/commons/2/25/Prism_flat_rainbow.jpg *License:* CC0 *Contributors:* Own work *Original artist:* Kelvinsong
- **File:Propagation_of_a_de_broglie_plane_wave.svg** *Source:* https://upload.wikimedia.org/wikipedia/commons/f/ff/Propagation_of_a_de_broglie_plane_wave.svg *License:* Public domain *Contributors:* Own work *Original artist:* Maschen
- **File:Propagation_of_a_de_broglie_wave.svg** *Source:* https://upload.wikimedia.org/wikipedia/commons/2/21/Propagation_of_a_de_broglie_wave.svg *License:* Public domain *Contributors:* Own work *Original artist:* Maschen
- **File:Propagation_of_a_de_broglie_wavepacket.svg** *Source:* https://upload.wikimedia.org/wikipedia/commons/b/b8/Propagation_of_a_de_broglie_wavepacket.svg *License:* CC BY-SA 3.0 *Contributors:* Own work *Original artist:* Maschen
- **File:Prototype_photon_counting_system,_c_1980s._(9660571969).jpg** *Source:* https://upload.wikimedia.org/wikipedia/commons/4/41/Prototype_photon_counting_system%2C_c_1980s._%289660571969%29.jpg *License:* CC BY-SA 2.0 *Contributors:* Prototype photon counting system, c 1980s. *Original artist:* Science Museum London / Science and Society Picture Library
- **File:QHO-coherent3-amplitudesqueezed2dB-animation-color.gif** *Source:* https://upload.wikimedia.org/wikipedia/commons/c/c2/QHO-coherent3-amplitudesqueezed2dB-animation-color.gif *License:* CC BY 3.0 *Contributors:* This graphic was created with matplotlib. *Original artist:* Geek3
- **File:Quantum_mechanics_travelling_wavefunctions_wavelength.svg** *Source:* https://upload.wikimedia.org/wikipedia/commons/f/f1/Quantum_mechanics_travelling_wavefunctions_wavelength.svg *License:* CC0 *Contributors:* Own work *Original artist:* Maschen
- **File:Question_book-new.svg** *Source:* https://upload.wikimedia.org/wikipedia/en/9/99/Question_book-new.svg *License:* Cc-by-sa-3.0 *Contributors:*

  Created from scratch in Adobe Illustrator. Based on Image:Question book.png created by User:Equazcion *Original artist:*

  Tkgd2007

- **File:Refraction-with-soda-straw.jpg** *Source:* https://upload.wikimedia.org/wikipedia/commons/b/b9/Refraction-with-soda-straw.jpg *License:* CC-BY-SA-3.0 *Contributors:* Bcrowell *Original artist:* User Bcrowell on en.wikipedia
- **File:Science.jpg** *Source:* https://upload.wikimedia.org/wikipedia/commons/5/54/Science.jpg *License:* Public domain *Contributors:* ? *Original artist:* ?
- **File:Sequential_superposition_of_plane_waves.gif** *Source:* https://upload.wikimedia.org/wikipedia/commons/d/db/Sequential_superposition_of_plane_waves.gif *License:* CC0 *Contributors:* made with Mathematica *Original artist:* Teply
- **File:Starfield_Optical_Range_-_sodium_laser.jpg** *Source:* https://upload.wikimedia.org/wikipedia/commons/5/5d/Starfire_Optical_Range_-_sodium_laser.jpg *License:* Public domain *Contributors:* ? *Original artist:* ?
- **File:Stimulated_Emission.svg** *Source:* https://upload.wikimedia.org/wikipedia/commons/0/09/Stimulated_Emission.svg *License:* GFDL *Contributors:* Own work *Original artist:* V1adis1av
- **File:Stimulatedemission.png** *Source:* https://upload.wikimedia.org/wikipedia/commons/8/8a/Stimulatedemission.png *License:* CC-BY-SA-3.0 *Contributors:* en:Image:Stimulatedemission.png *Original artist:* User:(Automated conversion),User:DrBob
- **File:Stylised_Lithium_Atom.svg** *Source:* https://upload.wikimedia.org/wikipedia/commons/e/e1/Stylised_Lithium_Atom.svg *License:* CC-BY-SA-3.0 *Contributors:* based off of Image:Stylised Lithium Atom.png by Halfdan. *Original artist:* SVG by Indolences. Recoloring and ironing out some glitches done by Rainer Klute.
- **File:THEL-ACTD.jpg** *Source:* https://upload.wikimedia.org/wikipedia/commons/0/06/THEL-ACTD.jpg *License:* Public domain *Contributors:* The original image was uploaded on de.wikipedia as de:Bild:THEL shoot2.jpg, from US Army Space & Missile Defense Command *Original artist:* ?
- **File:Uncertainty_principle.gif** *Source:* https://upload.wikimedia.org/wikipedia/commons/4/47/Uncertainty_principle.gif *License:* CC0 *Contributors:* Own work *Original artist:* Thierry Dugnolle
- **File:Vertex_correction.svg** *Source:* https://upload.wikimedia.org/wikipedia/commons/8/87/Vertex_correction.svg *License:* Public domain *Contributors:* ? *Original artist:* User:Harmaa
- **File:VisibleEmrWavelengths.svg** *Source:* https://upload.wikimedia.org/wikipedia/commons/e/e2/VisibleEmrWavelengths.svg *License:* Public domain *Contributors:* created by me *Original artist:* maxhurtz
- **File:Wave-particle_duality.gif** *Source:* https://upload.wikimedia.org/wikipedia/commons/7/7d/Wave-particle_duality.gif *License:* CC0 *Contributors:* Own work *Original artist:* Thierry Dugnolle
- **File:Wave-particle_duality.ogv** *Source:* https://upload.wikimedia.org/wikipedia/commons/e/e4/Wave-particle_duality.ogv *License:* CC BY-SA 3.0 *Contributors:* Own work *Original artist:* Jubobroff
- **File:Wave_packet_squeezed_states.jpg** *Source:* https://upload.wikimedia.org/wikipedia/commons/f/fc/Wave_packet_squeezed_states.jpg *License:* CC-BY-SA-3.0 *Contributors:* dissertation *Original artist:* Gerd Breitenbach
- **File:Wavelet.gif** *Source:* https://upload.wikimedia.org/wikipedia/commons/9/92/Wavelet.gif *License:* CC BY-SA 3.0 *Contributors:* Own work *Original artist:* Thierry Dugnolle
- **File:Wigner_function_squeezed_states.jpg** *Source:* https://upload.wikimedia.org/wikipedia/commons/7/78/Wigner_function_squeezed_states.jpg *License:* CC-BY-SA-3.0 *Contributors:* dissertation *Original artist:* Gerd Breitenbach
- **File:Wikisource-logo.svg** *Source:* https://upload.wikimedia.org/wikipedia/commons/4/4c/Wikisource-logo.svg *License:* CC BY-SA 3.0 *Contributors:* Rei-artur *Original artist:* Nicholas Moreau
- **File:Wikiversity-logo.svg** *Source:* https://upload.wikimedia.org/wikipedia/commons/9/91/Wikiversity-logo.svg *License:* CC BY-SA 3.0 *Contributors:* Snorky (optimized and cleaned up by verdy_p) *Original artist:* Snorky (optimized and cleaned up by verdy_p)
- **File:Wind-turbine-icon.svg** *Source:* https://upload.wikimedia.org/wikipedia/commons/a/ad/Wind-turbine-icon.svg *License:* CC BY-SA 3.0 *Contributors:* Own work *Original artist:* Lukipuk
- **File:YAL-1A_Airborne_Laser_unstowed_crop.jpg** *Source:* https://upload.wikimedia.org/wikipedia/commons/c/c8/YAL-1A_Airborne_Laser_unstowed_crop.jpg *License:* Public domain *Contributors:* Selected ALTB Photos. *Airborne Laser Test Bed.* MDA. Retrieved on 29 June 2013. *Original artist:* US Missile Defense Agency
- **File:Young_Diffraction.png** *Source:* https://upload.wikimedia.org/wikipedia/commons/8/8a/Young_Diffraction.png *License:* Public domain *Contributors:* ? *Original artist:* ?

19.7.3 Content license

- Creative Commons Attribution-Share Alike 3.0

Made in the USA
Coppell, TX
02 November 2023